BUILDING SERVICES HANDBOOK

BUILDING SERVICES HANDBOOK

Second edition

Fred Hall
and
Roger Greeno

BUTTERWORTH
HEINEMANN

OXFORD AMSTERDAM BOSTON LONDON NEW YORK PARIS
SAN DIEGO SAN FRANCISCO SINGAPORE SYDNEY TOKYO

Butterworth-Heinemann
An imprint of Elsevier Science
Linacre House, Jordan Hill, Oxford OX2 8DP
200 Wheeler Road, Burlington, MA 01803

First published 2001
Reprinted 2001, 2002
Second edition 2003

British Library Cataloguing in Publication Data
A catalogue record for this title is available from the British Library

ISBN 0 7506 6143 7

For information on all Butterworth-Heinemann publications
visit our website at www.bh.com

Typeset by Keyword Typesetting Services Ltd, Wallington, Surrey
Printed and bound in Great Britain by Biddles Ltd. *www.biddles.co.uk*

CONTENTS

Part Nine   Gas Installation, Components and Controls 287

Part Ten   Electrical Supply and Installations 331

PREFACE

The capital and installation costs of building services in modern buildings can take up 50% of the total construction budget. For highly serviced buildings such as sports centres, this figure can easily exceed 75%. Services can also take up 15% of a building's volume. Therefore building services cannot be ignored. Architects have learnt to accept and accommodate the increased need for pipes, ducts and cabling encroaching on to their designs. Some with reluctance, not least Louis Kahn when writing in *World Architecture* in 1964: 'I do not like ducts, I do not like pipes. I hate them so thoroughly, I feel that they have to be given their place. If I just hated them and took no care, I think they would invade the building and completely destroy it.' Not all architects have chosen to compete with the ducting and mechanical plant. Some have followed the examples of Renzo Piano and Richard Rogers by integrating it with the construction and making it a feature of the building, viz. the Pompidou Centre in Paris and the Lloyds Building in London.

Building services are the dynamics in a static structure, providing movement, communications, facilities and comfort. As they are unavoidable, it is imperative that architects, surveyors, builders, structural engineers, planners, estate managers and all those concerned with the construction of buildings have a knowledge and appreciation of the subject.

This book incorporates a wide range of building services. It provides a convenient reference for all construction industry personnel. It is an essential reference for the craftsman, technician, construction site manager, facilities manager and building designer. For students of building crafts, national certificates and diplomas, undergraduates and professional examinations, this book will substantiate study notes and be an important supplement to lectures.

The services included in this book are cold and hot water supplies, heating, ventilation, air conditioning, drainage, sanitation, refuse and sewage disposal, gas, electricity, oil installation, fire services, transportation, accommodation for services, energy recovery and alternative energy. The emphasis throughout is economic use of text with a high proportion of illustrations to show the principles of installation in a comprehensive manner. Where appropriate, subjects are supplemented with references for further reading into legislative and national standards. Most topics have design applications with charts and formulae to calculate plant and equipment ratings or sizes.

This book has been developed from the second edition of *Essential Building Services and Equipment* by Frederick E. Hall. Fred endorsed this with thanks to his '. . . late wife for her patience and understanding during the preparation of the first edition.' I would like to add my sincere thanks to my former colleague, Fred, for allowing me to use his material as the basis for this new presentation. It is intended as a complementary volume to the *Building Construction Handbook* by Roy Chudley and Roger Greeno, also published by Butterworth-Heinemann.

Roger Greeno, Guildford, 2000

PREFACE TO SECOND EDITION

The success of the first edition as a reader for building and services further and higher education courses, and as a general practice reference, has permitted further research and updating of material in this new publication.

This new edition retains the existing pages as established reference, updates as necessary and develops additional material in response to evolving technology with regard to the introduction of new British Standards, European Standards, Building Regulations, Water Regulations and good practice guidance. Where appropriate, references are provided to these documents for further specific reading.

Roger Greeno, Guildford, 2003

1 COLD WATER AND SUPPLY SYSTEMS

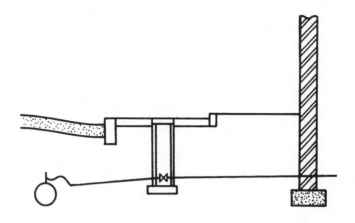

RAIN CYCLE – SOURCES OF WATER SUPPLY

FILTRATION OF WATER

STERILISATION AND SOFTENING

STORAGE AND DISTRIBUTION OF WATER

VALVES AND TAPS

JOINTS ON WATER PIPES

WATER MAINS

DIRECT SYSTEM OF COLD WATER SUPPLY

INDIRECT SYSTEM OF COLD WATER SUPPLY

BACKFLOW PROTECTION

SECONDARY BACKFLOW PROTECTION

COLD WATER STORAGE CISTERNS

COLD WATER STORAGE CALCULATIONS

BOOSTED COLD WATER SYSTEMS

DELAYED ACTION FLOAT VALVE

PIPE SIZING BY FORMULA

PIPE SIZES AND RESISTANCES

HYDRAULICS AND FLUID FLOW

Surface sources – Lakes, streams, rivers, reservoirs, run off from roofs and paved areas.

Underground sources – Shallow wells, deep wells, artesian wells, artesian springs, land springs.

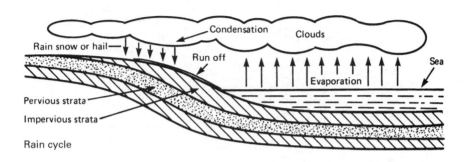

Rain cycle

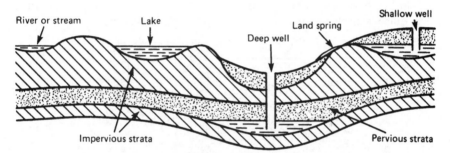

Surface and normal underground supplies

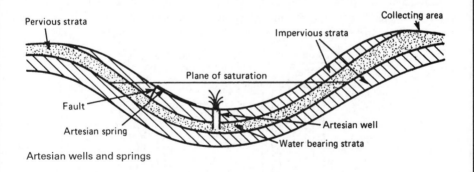

Artesian wells and springs

Filtration of Water

Pressure filter – rate of filtration 4 to 12 m^3 per m^2 per hour. To backwash, valve A is closed and valves B and C opened. Compressed air clears the sand of dirt. Diameter = 2·4 m.

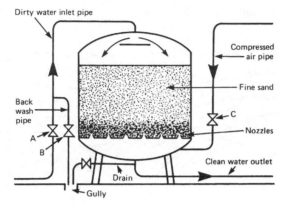

Slow sand filter bed – rate of filtration 0·2 to 1·15 m^3 per m^2 per hour. Filter beds can occupy large areas and the top layer of sand will require removal and cleaning at periodic intervals.

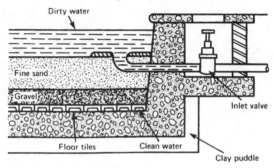

Small domestic filter – the unglazed porcelain cylinder will arrest very fine particles of dirt and even micro-organisms. The cylinder can be removed and sterilised in boiling water for 10 minutes.

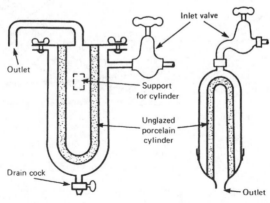

Sterilisation by chlorine injection – water used for drinking must be sterilised. Chlorine is generally used for this purpose to destroy organic matter. Minute quantities (0·1 to 0·3 p.p.m.) are normally added after the filtration process.

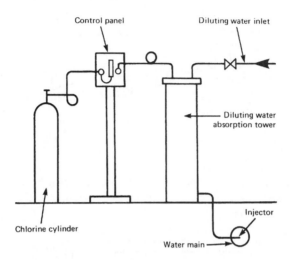

Softening of hard water by base exchange process – sodium zeolites exchange their sodium base for calcium (chalk) or magnesium bases in the water. Sodium zeolite plus calcium carbonate or sulphate becomes calcium zeolite plus sodium carbonate or sulphate. To regenerate, salt is added; calcium zeolite plus sodium chloride (salt) becomes sodium zeolite plus calcium chloride which is flushed away.

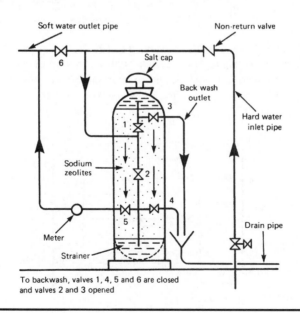

5

Gravitational distribution – the water from upland gathering grounds is impounded in a reservoir. From this point the water is filtered and chlorinated before serving an inhabited area at lower level. There are no pumping costs.

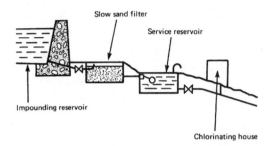

Pumped distribution – water extracted from a river is pumped into a settlement tank, subsequently filtered and chlorinated. Pump maintenance and running costs make this process more expensive than gravity systems.

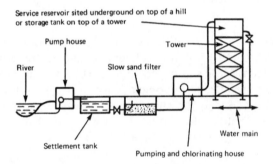

Ring main distribution – water mains supplying a town or village may be in the form of a grid. This is preferable to radial distribution as sections can be isolated with minimal disruption to the remaining system and there is no more opportunity for water to maintain a flow.

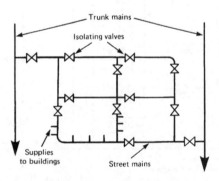

The globe-type stop valve is used to control the flow of water at high pressure. To close the flow of water the crutch head handle is rotated slowly in a clockwise direction gradually reducing the flow, thus preventing sudden impact and the possibility of vibration and water hammer.

The gate or sluice valve is used to control the flow of water on low pressure installations. The wheel head is rotated clockwise to control the flow of water, but this valve will offer far less resistance to flow than a globe valve. With use the metallic gate will wear and on high pressure installations would vibrate.

The drain valve has several applications and is found at the lowest point in pipe systems, boilers and storage vessels.

For temperatures up to 100°C valves are usually made from brass. For higher temperatures gun metal is used. Brass contains 50% zinc and 50% copper. Gun metal contains 85% copper, 5% zinc and 5% tin.

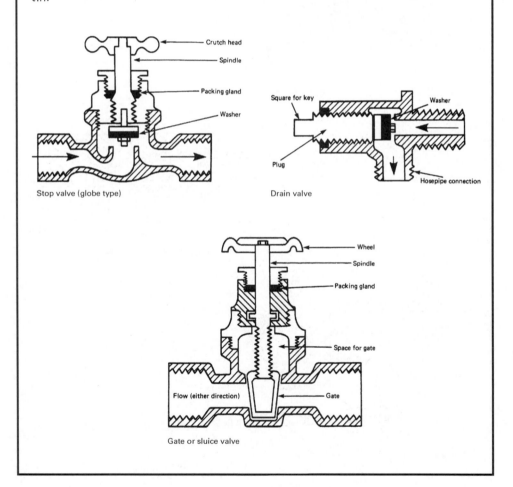

Stop valve (globe type)

Drain valve

Gate or sluice valve

Float valves are automatic flow control devices fitted to cisterns to maintain an appropriate volume of water. Various types are in use. The diaphragm type is the least noisy as there is less friction between moving parts. The Portsmouth and Croydon-type valves have a piston moving horizontally or vertically respectively, although the latter is obsolete and only likely to be found in very old installations. Water outlets must be well above the highest water level (type `B' air gap) to prevent back siphonage of cistern water into the main supply. Nozzle diameters reduce as the pressure increases. High, medium and low pressure valves must be capable of closing against pressures of 1380, 690 and 275 kPa respectively.

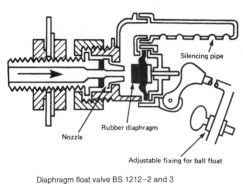

Diaphragm float valve BS 1212–2 and 3

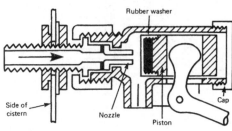

Portsmouth/piston float valve BS 1212–1

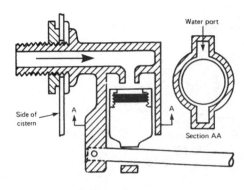

Croydon float valve

The pillar tap is used to supply water to basins, baths, bidets and sinks. Combined hot and cold pillar taps are available with fixed or swivel outlet. The outlet of these taps must be bi-flow, i.e. separate waterways for hot and cold water to prevent crossflow of water within the pipework.

The bib tap is for wall fixing, normally about 150 mm above a sanitary appliance. The 'Supatap' bib tap permits a change of washer without shutting off the water supply. It is also available in pillar format. They are very easy to turn on or off and therefore suitable for the disabled.

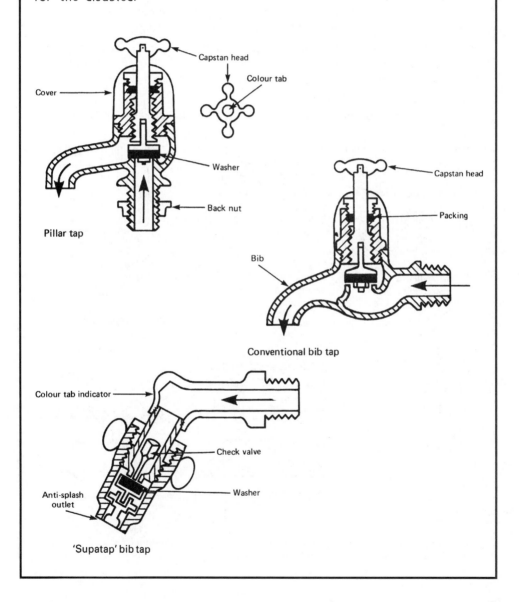

Pillar tap

Conventional bib tap

'Supatap' bib tap

Joints on Water Pipes

Copper pipes may be jointed by bronze welding. Non-manipulative compression joints are used on pipework above ground and manipulative compression joints are used on underground pipework. The latter are specifically designed to prevent pipes pulling out of the joint. Push-fit joints are made from polybutylene. These provide simplicity of use and savings in time. Capillary joints have an integral ring of soft solder. After cleaning the pipe and fitting with wire wool and fluxing, heat application enables the solder to flow and form a joint. Solder alloy for drinking water supplies must be lead free, i.e. copper and tin.

The Talbot joint is a push-fit joint for polythene pipes. A brass ferrule or support sleeve in the end of the pipe retains the pipe shape.

Threaded joints on steel pipes are sealed by non-toxic jointing paste and hemp or polytetrafluorethylene (PTFE) tape. A taper thread on the pipe will help to ensure a water-tight joint. Union joints permit slight deflection without leakage.

Lead pipes are no longer acceptable due to the risk of poisoning.

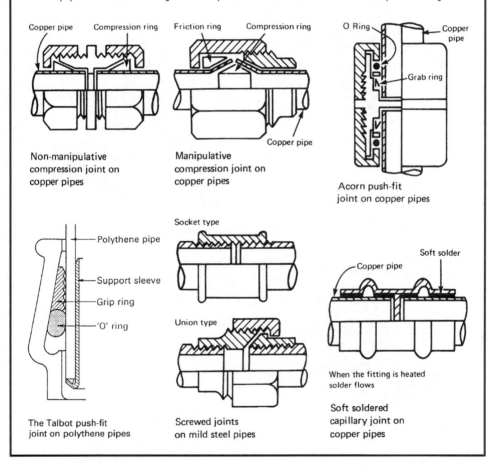

Copper pipe Compression ring

Non-manipulative
compression joint on
copper pipes

Friction ring Compression ring

Copper pipe

Manipulative
compression joint on
copper pipes

O Ring Copper pipe

Grab ring

Acorn push-fit
joint on copper pipes

Polythene pipe

Support sleeve

Grip ring

'O' ring

The Talbot push-fit
joint on polythene pipes

Socket type

Union type

Screwed joints
on mild steel pipes

Soft solder

Copper pipe

When the fitting is heated
solder flows

Soft soldered
capillary joint on
copper pipes

Water mains have been manufactured from a variety of materials. The material selected must be compatible with the water constituents, otherwise corrosion and decomposition of the pipes may occur. Contemporary materials which suit most waters are ductile cast iron to BS EN 545 and uPVC to BS PAS (Publicly Available Specification) No. 27. The water undertaking or authority must be consulted prior to laying mains to determine suitable materials, laying techniques and pipe diameter. Firefighting and hydrant requirements will prioritise the criteria with a minimum pressure of 30 m head (300 kPa) from a 75 mm diameter pipe supplied from both ends, or 100 mm diameter from one end only. Bedding of mains is usually a surround of shingle to accommodate any movement. uPVC pipes are pigmented blue for easy identification in future excavations and cast iron has a blue plastic tape attached for the same reason.

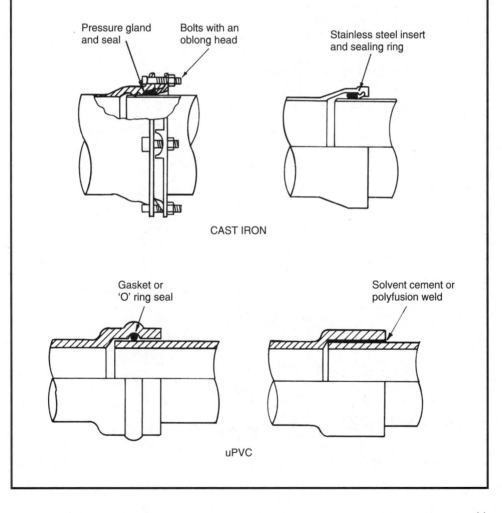

Pressure gland and seal

Bolts with an oblong head

Stainless steel insert and sealing ring

CAST IRON

Gasket or 'O' ring seal

Solvent cement or polyfusion weld

uPVC

Connection to Water Main

The water authority requires at least 7 days' written notice for connection to their supply main. The main is drilled and tapped live with special equipment, which leaves a plug valve ready for connection to the communication pipe. A goose neck or sweeping bend is formed at the connection to relieve stresses on the pipe and valve. At or close to the property boundary, a stop valve is located with an access compartment and cover at ground level. A meter may also be located at this point. The communication and supply pipe should be snaked to allow for settlement in the ground. During warm weather, plastic pipes in particular should be snaked to accommodate contraction after backfilling.

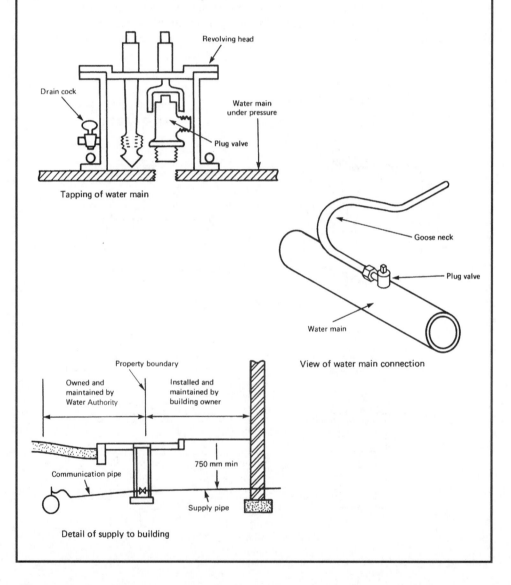

Revolving head

Drain cock

Water main under pressure

Plug valve

Tapping of water main

Goose neck

Plug valve

Water main

View of water main connection

Property boundary

Owned and maintained by Water Authority

Installed and maintained by building owner

750 mm min

Communication pipe

Supply pipe

Detail of supply to building

Water meters are installed at the discretion of the local water authority. Most require meters on all new build and conversion properties, plus existing buildings which have been substantially altered. In time, in common with other utilities, all buildings will have metered water supply. Meters are either installed in the communication pipe, or by direct annular connection to the stopvalve. If underground location is impractical, the water authority may agree internal attachment to the rising main.

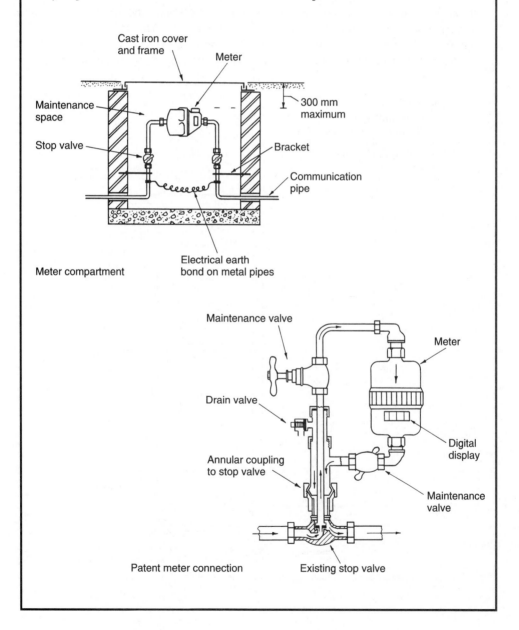

Cast iron cover and frame

Meter

Maintenance space

300 mm maximum

Stop valve

Bracket

Communication pipe

Meter compartment

Electrical earth bond on metal pipes

Maintenance valve

Meter

Drain valve

Annular coupling to stop valve

Digital display

Maintenance valve

Patent meter connection

Existing stop valve

Direct System of Cold Water Supply

For efficient operation, a high pressure water supply is essential particularly at periods of peak demand. Pipework is minimal and the storage cistern supplying the hot water cylinder need only have 115 litres capacity. The cistern may be located within the airing cupboard or be combined with the hot water cylinder. Drinking water is available at every draw-off point and maintenance valves should be fitted to isolate each section of pipework. With every outlet supplied from the main, the possibility of back siphonage must be considered.

Back siphonage can occur when there is a high demand on the main. Negative pressure can then draw water back into the main from a submerged inlet, e.g. a rubber tube attached to a tap or a shower fitting without a check valve facility left lying in dirty bath water.

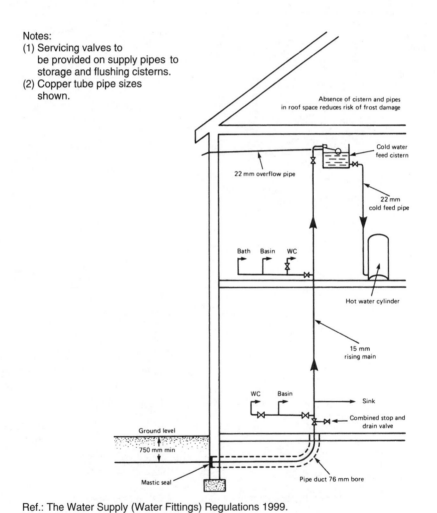

Notes:
(1) Servicing valves to be provided on supply pipes to storage and flushing cisterns.
(2) Copper tube pipe sizes shown.

Absence of cistern and pipes in roof space reduces risk of frost damage

22 mm overflow pipe

Cold water feed cistern

22 mm cold feed pipe

Bath Basin WC

Hot water cylinder

15 mm rising main

WC Basin Sink

Combined stop and drain valve

Ground level

750 mm min

Mastic seal

Pipe duct 76 mm bore

Ref.: The Water Supply (Water Fittings) Regulations 1999.

The indirect system of cold water supply has only one drinking water outlet, at the sink. The cold water storage cistern has a minimum capacity of 230 litres, for location in the roof space. In addition to its normal supply function, it provides an adequate emergency storage in the event of water main failure. The system requires more pipework than the direct system and is therefore more expensive to install, but uniform pressure occurs at all cistern-supplied outlets. The water authorities prefer this system as it imposes less demand on the main. Also, with fewer fittings attached to the main, there is less chance of back siphonage. Other advantages of lower pressure include less noise and wear on fittings, and the opportunity to install a balanced pressure shower from the cistern.

Notes:
(1) Servicing valves to be provided on supply pipes to storage and flushing cisterns.
(2) Copper tube pipe sizes shown.

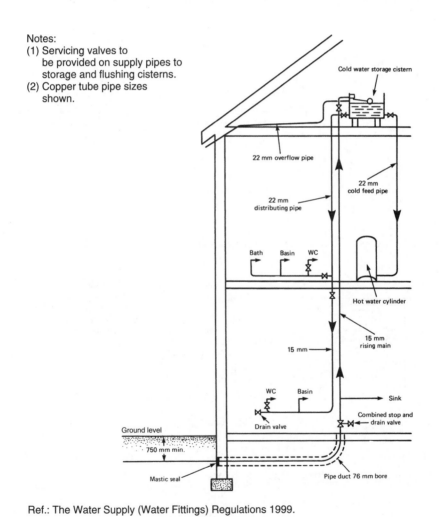

Ref.: The Water Supply (Water Fittings) Regulations 1999.

Backflow Protection

All pipework systems and associated appliances must be protected against backflow or back siphonage. The Water Byelaws require that float valve controlled inlets to cisterns and tanks have sufficient air gap to prevent stored water in a cistern contacting the water supply. Likewise, draw-offs at taps must have sufficient air gap above water being used in a sanitary appliance.

A type `A´ air gap is required where the appliance water is contaminated, e.g. in a bath or basin. A type `B´ air gap is required where stored water may be contaminated, e.g. in a cistern.

Air gap dimensions vary depending on the type of appliance installed and the diameter of the supply fitting. For specific applications see BS 6281-1 and BS 6281-2 for type `A´ and `B´ respectively.

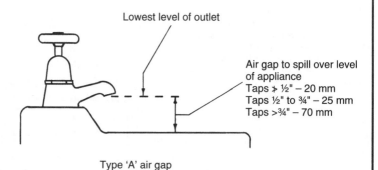

Lowest level of outlet

Air gap to spill over level of appliance
Taps ≯ ½" – 20 mm
Taps ½" to ¾" – 25 mm
Taps >¾" – 70 mm

Type 'A' air gap

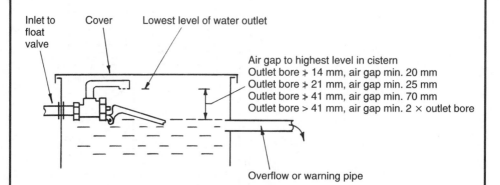

Inlet to float valve   Cover   Lowest level of water outlet

Air gap to highest level in cistern
Outlet bore ≯ 14 mm, air gap min. 20 mm
Outlet bore ≯ 21 mm, air gap min. 25 mm
Outlet bore ≯ 41 mm, air gap min. 70 mm
Outlet bore > 41 mm, air gap min. 2 × outlet bore

Overflow or warning pipe

Type 'B' air gap

Ref: BS 6281: Devices without moving parts for the prevention of contamination of water by backflow.

Secondary backflow or back siphonage protection is an alternative or supplement to the provision of air gaps. It is achieved by using mechanical devices such as double check valves or a vacuum breaker in the pipeline. Special arrangements of pipework with branches located above the spill level of appliances are also acceptable.

Ref: BS 6282, Devices with moving parts for the prevention of contamination of water by backflow.

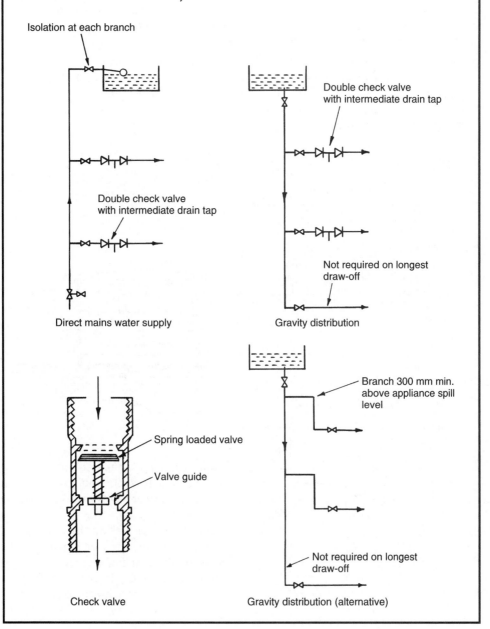

Isolation at each branch

Double check valve with intermediate drain tap

Double check valve with intermediate drain tap

Not required on longest draw-off

Direct mains water supply

Gravity distribution

Branch 300 mm min. above appliance spill level

Spring loaded valve

Valve guide

Not required on longest draw-off

Check valve

Gravity distribution (alternative)

Cold Water Storage Cisterns

Cisterns can be manufactured from galvanised mild steel (large non-domestic capacities), polypropylene or glass reinforced plastics. They must be well insulated and supported on adequate bearers to spread the concentrated load. Plastic cisterns will require uniform support on boarding over bearers. A dustproof cover is essential to prevent contamination.

For large buildings, cisterns are accommodated in a purpose-made plant room at roof level or within the roof structure. This room must be well insulated and ventilated, and be provided with thermostatic control of a heating facility.

Where storage demand exceeds 4500 litres, cisterns must be duplicated and interconnected. In the interests of load distribution this should be provided at much lower capacities. For maintenance and repairs each cistern must be capable of isolation and independent operation.

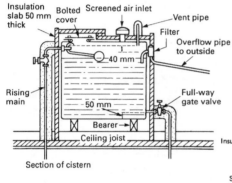

Section of cistern

Ref. BS 7181: Specification for storage cisterns up to 500 l actual capacity for water supply for domestic purposes.

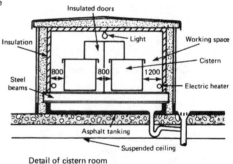

Detail of cistern room

Duplicated cisterns

Refs. BS 417: Specification for galvanised low carbon steel cisterns, cistern lids, tanks and cylinders.
BS 4213: Specification for cold water storage and combined feed and expansion cisterns (polyolefin or olefin copolymer) up to 500 l capacity used for domestic purposes.

Cold water storage data is provided to allow for up to 24 hour interruption of mains water supply.

Building purpose	Storage/person/24 hrs	
Boarding school	90 litres	
Day school	30	
Department store with canteen	45	(3)
Department store without canteen	40	(3)
Dwellings	90	(1)
Factory with canteen	45	
Factory without canteen	40	
Hostel	90	
Hotel	135	(2) (3)
Medical accommodation	115	
Office with canteen	45	
Office without canteen	40	
Public toilets	15	
Restaurant	7 per meal	

Notes: (1) 115 or 230 litres min. see pages 14 and 15

(2) Variable depending on classification.

(3) Allow for additional storage for public toilets and restaurants.

At the design stage the occupancy of a building may be unknown. Therefore the following can be used as a guide:

Building purpose	Occupancy
Dept. store	1 person per 30 m^2 net floor area
Factory	30 persons per WC
Office	1 person per 10 m^2 net floor area
School	40 persons per classroom
Shop	1 person per 10 m^2 net floor area

E.g. A 1000 m^2 (net floor area) office occupied only during the day therefore allow 10 hours' emergency supply.

$1000/10 = 100$ persons $\times 40$ litres $= 4000$ litres (24 hrs)

$\qquad\qquad\qquad\qquad\qquad\qquad\quad = 1667$ litres (10 hrs)

For medium and high rise buildings, there is often insufficient mains pressure to supply water directly to the upper floors. Boosting by pump from a break tank is therefore usually necessary and several more of these tanks may be required as the building rises, depending on the pump capacity. A break pressure cistern is also required on the down service to limit the head or pressure on the lower fittings to a maximum of 30 m (approx. 300 kPa). The drinking water header pipe or storage vessel supplies drinking water to the upper floors. As this empties and the water reaches a predetermined low level, the pipeline switch engages the duty pump. A float switch in the break tank protects the pumps from dry running if there is an interruption to mains supply. The various pipe sections are fitted with isolating valves to facilitate maintenance and repairs.

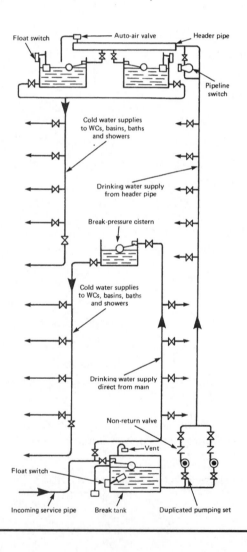

As an alternative to the drinking water header pipe, an auto-pneumatic cylinder may be used. Compressed air in the cylinder forces water up to the float valves and drinking water outlets on the upper floors. As the cylinder empties a low pressure switch engages the duty pump. When the pump has replenished the cylinder, a high pressure switch disengages the pump. In time, some air is absorbed by the water. As this occurs, a float switch detects the high water level in the cylinder and activates an air compressor to regulate the correct volume of air. Break pressure cisterns may be supplied either from the storage cisterns at roof level or from the rising main. A pressure reducing valve is sometimes used instead of a break pressure cistern.

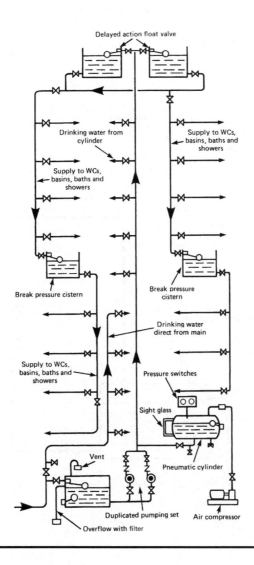

In modest rise buildings of several storeys where water is in fairly constant demand, water can be boosted from a break tank by a continuously running pump. The installation is much simpler and less costly than the previous two systems as there is less need for specialised items of equipment. Sizing of the pump and its delivery rating are critical, otherwise it could persistently overrun, or at the other extreme be inadequate. Modern pumps have variable settings allowing considerable scope around the design criteria. The pump is normally scheduled to run on a timed programme, e.g. in an office block it may commence an hour before normal occupancy and run on for a couple of hours after. Water delivery should be just enough to meet demand. When demand is low a pressure regulated motorised bleed valve opens to recirculate water back to the break tank.

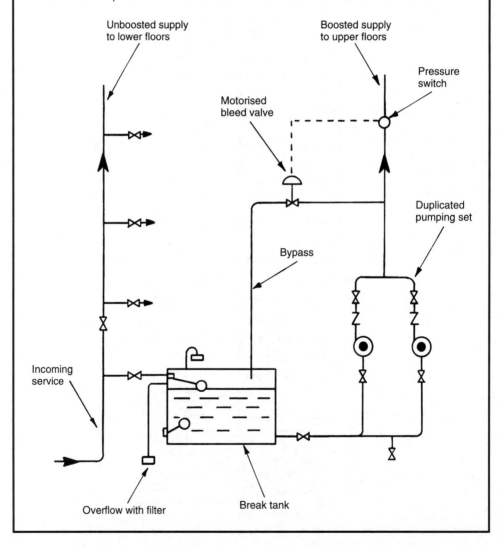

If normal float valves are used to regulate cistern water supply from an auto-pneumatic cylinder (page 21), then cylinder and pump activity will be frequent and uneconomic. Therefore to regulate activity and deliveries to the cistern, a delayed action float valve mechanism is fitted to the storage cistern.

Stage 1. Water filling the cistern lifts hemi-spherical float and closes the canister valve.

Stage 2. Water overflows into the canister, raises the ball float to close off water supply.

Stage 3. As the cistern empties, the ball float remains closed until low water level releases the hemi-spherical float. As this float valve drops, water is released from the canister to open the ball float valve to replenish the cistern from the pneumatic supply.

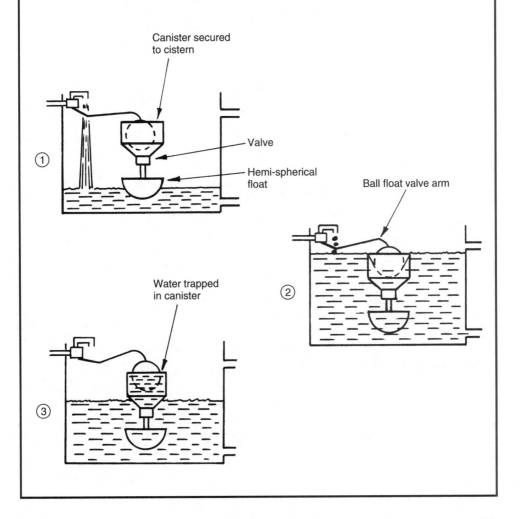

Canister secured
to cistern

Valve

Hemi-spherical
float

Ball float valve arm

Water trapped
in canister

①

②

③

23

Pipe Sizing by Formula

Thomas Box formula:

$$d = \sqrt[5]{\frac{q^2 \times 25 \times L \times 10^5}{H}}$$

where: d = diameter (bore) of pipe (mm)
     q = flow rate (l/s)
     H = head or pressure (m)
     L = length (effective) of pipe (m)
        (actual length + allowance for bends, tees, etc.)

e.g.

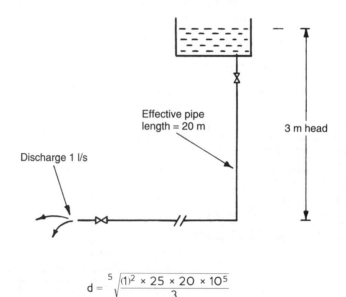

Effective pipe length = 20 m

3 m head

Discharge 1 l/s

$$d = \sqrt[5]{\frac{(1)^2 \times 25 \times 20 \times 10^5}{3}}$$

$$d = \sqrt[5]{16\,666\,667} = 27 \cdot 83 \text{ mm}$$

The nearest commercial size above this is 32 mm bore steel or 35 mm outside diameter copper.

Note: Head in metres can be converted to pressure in kPa by multiplying by gravity, e.g. 3 m × 9.81 = 29.43 kPa (approx. 30 kPa).

| Steel pipe (inside dia.) | | Copper tube (mm) | | Polythene (mm) | |
Imperial (")	Metric (mm)	Outside dia.	Bore	Outside dia.	Bore
$\frac{1}{2}$	15	15	13·5	20	15
$\frac{3}{4}$	20	22	20	27	22
1	25	28	26	34	28
$1\frac{1}{4}$	32	35	32	42	35
$1\frac{1}{2}$	40	42	40		
2	50	54	51·5		
$2\frac{1}{2}$	65	67	64·5		
3	80	76	73·5		

Approximate equivalent pipe lengths of some fittings (m).

Pipe bore (mm)	Elbow	Tee	Stop valve	High pressure float valve
15	0·6	0·7	4·5	75
20	0·8	1·0	7	50
25	1·0	1·5	10	40
32	1·4	2·0	13	35
40	1·7	2·5	16	21
50	2·3	3·5	22	20

Notes: Figure given for a tee is the change of direction; straight through has no significant effect. These figures are only intended as a guide, they will vary between materials and design of fittings.

Recommended flow rates for various sanitary appliances (litres/sec)

WC cistern	0·11
Hand basin	0·15
Hand basin (spray tap)	0·03
Bath (19 mm tap)	0·30
Bath (25 mm tap)	0·60
Shower	0·11
Sink (13 mm tap)	0·19
Sink (19 mm tap)	0·30
Sink (25 mm tap)	0·40

Loading units are factors which can be applied to a variety of appliances. They have been established by considering the frequency of use of individual appliances and the desired water flow rate.

Appliance	Loading units
Hand basin	1·5 to 3 (depends on application)
WC cistern	2
Washing machine	3
Dishwasher	3
Shower	3
Sink (13 mm tap)	3
Sink (19 mm tap)	5
Bath (19 mm tap)	10
Bath (25 mm tap)	22

By determining the number of appliances on a pipework system and summating the loading units, an equivalent flow in litres per second can be established from the following conversion graph:

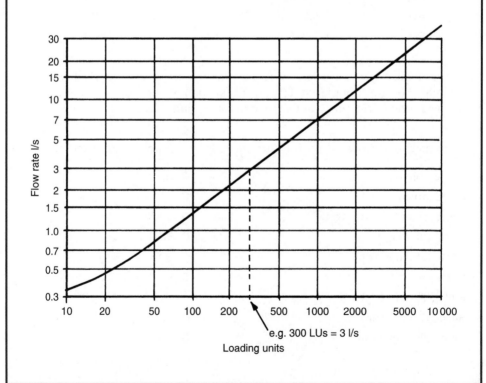

e.g. 300 LUs = 3 l/s

Pressure or head loss in pipework systems can be expressed as the relationship between available pressure (kPa) or head (m) and the effective length (m) of pipework. The formula calculation on page 24 can serve as an example:

Head = 3 m. Effective pipe length = 20 m. So, 3/20 = 0·15 m/m

By establishing the flow rate from loading units or predetermined criteria (1 l/s in our example), a nomogram may be used to obtain the pipe diameter. The chart below is for illustration and general use. For greater accuracy, pipe manufacturers' design data should be consulted for different materials and variations in water temperatures.

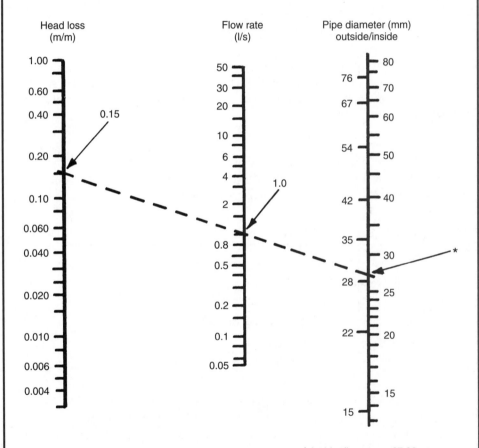

* Inside diameter = 27.83 mm
(see page 24)

Ref. BS 6700: Specification for design, installation, testing and maintenance of services supplying water for domestic use within buildings and their curtilage.

Hydraulics

Hydraulics is the experimental science concerning the study of energy in fluid flow. That is, the force of pressure required to overcome the resistance to fluid flowing through pipes, caused by the friction between the pipe and liquid movement.

The total energy of the liquid flowing in a pipe declines as the pipe length increases, mainly due to friction between the fluid and the pipe wall. The amount of energy or pressure loss will depend on:

- Smoothness/roughness of the internal pipe wall.
- Diameter of pipe or circumference of internal pipe wall.
- Length of pipe.
- Velocity of fluid flow.
- Amount of turbulence in the flow.
- Viscosity and temperature of fluid.

Theories relating to pressure loss by fluids flowing in pipes are diverse, but an established relationship is that the pressure losses (h) caused by friction are proportional to the square of the velocity of flow (v):

$$h \propto v^2$$

From this, for a pipe of constant size it can be seen that by developing the proportional relationship, a doubling (or more) of pressure will increase the velocity accordingly:

h (m)	v (m/s)
4	1·5
8	2·12 ($1.5 \times \sqrt{2}$)
12	2·60 ($1.5 \times \sqrt{3}$)
16	3·00 ($1.5 \times \sqrt{4}$) or ($2.12 \times \sqrt{2}$)
24	3·66 ($1.5 \times \sqrt{6}$) or ($2.60 \times \sqrt{2}$)
32	4·24 ($1.5 \times \sqrt{8}$) or ($3.00 \times \sqrt{2}$) etc., etc.

Also, it can be shown that if the condition (temperature and viscosity) of a fluid in a pipe remains constant, the discharge through that pipe is directly proportional to the square root of the fifth power of its diameter:

$$\sqrt{d^5}$$

This relationship can be identified in the Thomas Box pipe sizing formula shown on page 24.

Reynolds number – a coefficient of friction based on the criteria for similarity of motion for all fluids. Relevant factors are related by formula:

$$\frac{\text{density} \times \text{velocity} \times \text{linear parameter (diameter)}}{\text{viscosity}}$$

This is more conveniently expressed as:

$$R = \frac{\rho v d}{\mu}$$

Where: R = Reynolds number
ρ = fluid density (kg/m^3)
v = velocity (m/s)
d = diameter of pipe (m)
μ = viscosity of the fluid (Pa s) or (Ns/m^2)

Whatever the fluid type or temperature, an R value of less than 2000 is considered streamline or laminar. A value greater than 2000 indicates that the fluid movement is turbulent.

E.g. 1. A 12 mm diameter pipe conveying fluid of density 1000 kg/m^3 and viscosity of 0·013 Pa s at 2 m/s flow velocity has a Reynolds number of:

$$\frac{1000 \times 2 \times 0·012}{0·013} = 1846 \text{ (streamline flow)}$$

D'Arcy formula – used for calculating the pressure head loss of a fluid flowing full bore in a pipe, due to friction between fluid and pipe surface.

$$h = \frac{4 \, fL \, v^2}{2 \, g \, d}$$

Where: h = head loss due to friction (m)
f = coefficient of friction
L = length of pipe (m)
v = average velocity of flow (m/s)
g = gravitational acceleration (9.81 m/s^2)
d = internal diameter of pipe (m)

Note: 'f', the D'Arcy coefficient, ranges from about 0·005 (smooth pipe surfaces and streamline flow) to 0·010 (rough pipe surfaces and turbulent flow). Tables can be consulted, although a mid value of 0·0075 is appropriate for most problem solving.

E.g. 2. A 12 mm diameter pipe, 10 m long, conveying a fluid at a velocity of flow of 2 m/s

$$\text{Head loss} = \frac{4 \times 0·0075 \times 10 \times 2^2}{2 \times 9·81 \times 0·012} = 5·09m$$

Depending on the data available, it is possible to transpose the D'Arcy formula for other purposes. For example, it may be used to calculate pipe diameter in this format:

$$d = \frac{4\,f\,L\,v^2}{2\,g\,h}$$

Flow rate (Q) – the discharge rate or flow rate of a fluid in a pipe is expressed as the volume in cubic metres (V) flowing per second (s). Q (m^3/s) is dependent on the pipe cross sectional area dimensions (m^2) and the velocity of fluid flow (m/s). Q may also be expressed in litres per second, where 1 m^3/s = 1000 l/s.

A liquid flowing at an average velocity (v) in a pipe of constant area (A) discharging a length (L) of liquid every second (s), has the following relationship:

$$Q = V \div s \qquad \text{where } V = L \times A \qquad \text{and} \qquad v = L \div s$$

So, $\quad Q = L \times A \div s \qquad$ where $v = L \div s$, $\qquad \therefore \qquad Q = v \times A$

Q = flow rate (m^3/s), v = velocity of flow (m/s) and
A = cross sectional area of pipe (m^2)

E.g. 1. The quantity of water flowing through a 12 mm diameter pipe at 2 m/s will be:

$Q = v \times A$, where $A = \pi r^2$
$Q = 2 \times 0.000113 = 0.000226$ m^3/s or 0.226 l/s

Relative discharge of pipes – this formula may be used to estimate the number of smaller branch pipes that can be successfully supplied by one main pipe:

$$N = \sqrt{(D \div d)^5}$$

where N = number of short branch pipes
D = diameter of main pipe (mm)
d = diameter of short branch pipes (mm)

E.g. 2. The number of 32 mm short branch pipes that can be served from one 150 mm main will be:

$$N = \sqrt{(150 \div 32)^5} = 47$$

E.g. 3. The size of water main required to supply 15, 20 mm short branch pipes will be by formula transposition:

$D = d \times \sqrt[5]{N^2}$
$D = 20 \times \sqrt[5]{15^2} = 59$ (65 mm nearest standard)

2 HOT WATER SUPPLY SYSTEMS

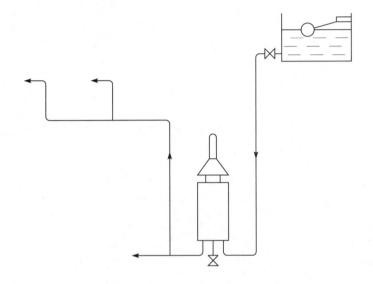

DIRECT SYSTEM OF HOT WATER SUPPLY

INDIRECT SYSTEM OF HOT WATER SUPPLY

UNVENTED HOT WATER STORAGE SYSTEM

EXPANSION AND TEMPERATURE RELIEF VALVES

HOT WATER STORAGE CYLINDERS

PRIMATIC HOT WATER STORAGE CYLINDER

MEDIUM AND HIGH RISE BUILDING SUPPLY SYSTEMS

TYPES OF BOILER

SECONDARY CIRCULATION

DUPLICATION OF PLANT

ELECTRIC AND GAS WATER HEATERS

SOLAR HEATING OF WATER

HOT WATER STORAGE CAPACITY

BOILER RATING

PIPE SIZING

PRESSURISED SYSTEMS

CIRCULATION PUMP RATING

LEGIONNAIRES' DISEASE IN HOT WATER SYSTEMS

SEDBUK

The hot water from the boiler mixes directly with the water in the cylinder. If used in a 'soft' water area the boiler must be rust-proofed. This system is not suited to 'hard' waters, typical of those extracted from boreholes into chalk or limestone strata. When heated the calcium precipitates to line the boiler and primary pipework, eventually 'furring up' the system to render it ineffective and dangerous. The storage cylinder and associated pipework should be well insulated to reduce energy losses. If a towel rail is fitted, this may be supplied from the primary flow and return pipes.

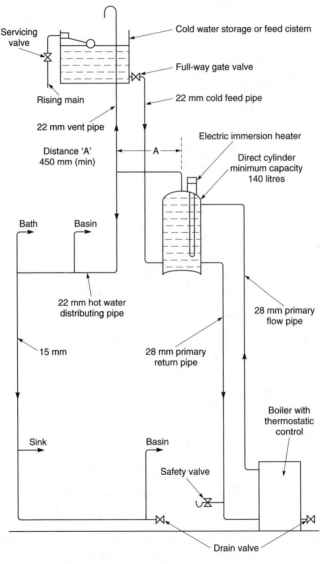

Note: All pipe sizes shown are for copper outside diameter.

Indirect System of Hot Water Supply

This system is used in 'hard' water areas to prevent scaling or 'furring' of the boiler and primary pipework. Unlike the direct system, water in the boiler and primary circuit is not drawn off through the taps. The same water circulates continuously throughout the boiler, primary circuit and heat exchange coil inside the storage cylinder. Fresh water cannot gain access to the higher temperature areas where precipitation of calcium would occur. The system is also used in combination with central heating, with flow and return pipes to radiators connected to the boiler. Boiler water temperature may be set by thermostat at about 80°C.

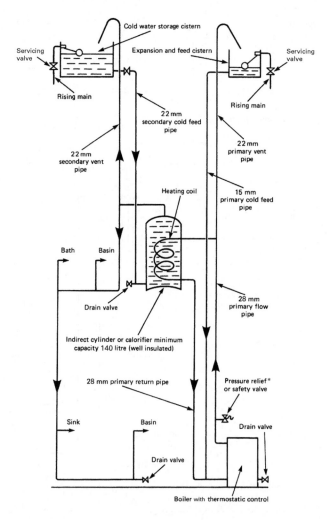

*A safety valve is not normally required on indirect open vent systems, as in the unlikely occurrence of the primary flow and vent becoming obstructed, water expansion would be accommodated up the cold feed pipe.

The Building Regulations, Approved Document J, permit the installation of packaged unit unvented hot water storage systems which have been accredited by the British Board of Agrément (BBA) or other European Organisation for Technical Approvals (EOTA) member bodies. Components should satisfy BS 7206: Specification for unvented hot water storage units and packages. A system of individual approved components is also acceptable. Safety features must include:

1. Flow temperature control between 60 and 65°C.
2. 95°C limit thermostat control of the boiler to close off the fuel supply if the working thermostat fails.
3. Expansion and temperature relief valves to operate at 95°C.
4. Check valves on water main connections.

The system is less space consuming than conventional systems and saves installation costs as there are no cold water storage and expansion cisterns. In addition to satisfying the Building Regulations, the local water authority should be consulted for approval and to ensure that there is adequate mains pressure.

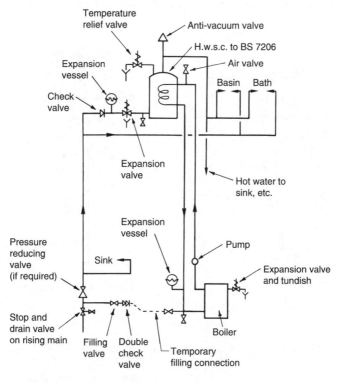

Unvented system with hot water storage capacity in excess of 15 litres, with a sealed primary circuit

Expansion Valve and Temperature Relief Valve

Expansion devices in hot water systems are designed as a safe means for discharging water when system operating parameters are exceeded, i.e. in conditions of excess pressure and/or temperature.

Expansion valve – Care should be taken when selecting expansion or pressure relief valves. They should be capable of withstanding 1.5 times the maximum pressure to which they are subjected, with due regard for water mains pressure increasing overnight as demand decreases.

Temperature relief valve – These should be fitted to all unvented hot water storage vessels exceeding 15 litres capacity. They are normally manufactured as a combined temperature and pressure relief valve. In addition to the facility for excess pressure to unseat the valve, a temperature sensing element is immersed in the water to respond at a pre set temperature of 95°C.

Discharge from these devices should be safely controlled and visible, preferably over a tundish as shown on page 75.

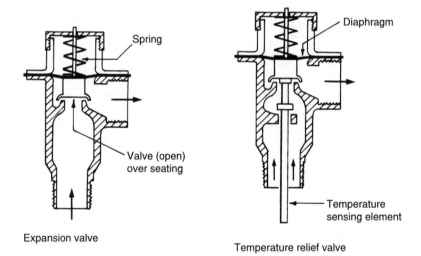

Expansion valve

Temperature relief valve

Ref. BS 6283: Safety and control devices for use in hot water systems.

BS 1566: Copper indirect cylinders for domestic purposes, BS 1566-1: Double feed indirect and BS 1566-2: Single feed indirect (primatic).

BS 699: Copper direct cylinders for domestic purposes.

BS 417-2: Specification for galvanised low carbon steel cisterns, cistern lids, tanks and cylinders.

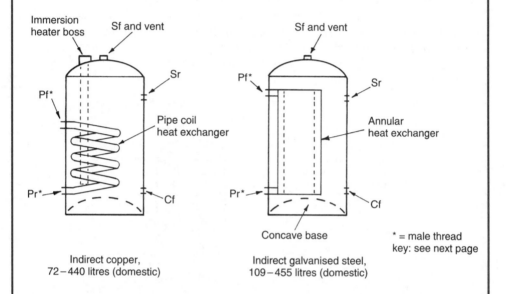

Indirect copper,
72–440 litres (domestic)

Indirect galvanised steel,
109–455 litres (domestic)

Concave base

* = male thread
key: see next page

Direct cylinders have no coil or annular heat exchangers. They can be identified with female pipe threads for the primary flow and return connections. For domestic use: copper – 74 to 450 litres capacity, galvanised steel – 73 to 441 litres capacity. Direct and indirect cylinders for industrial and commercial applications are manufactured in copper and galvanised steel in capacities up to 4500 litres.

Notes:

(1) Copper and galvanised (zinc plated) steel pipes and components should not be used in the same installation. In addition to electrolytic action between the dissimilar metals, pitting corrosion caused by tiny particles of dissolved copper settling on the galvanising will produce local cells which dissolve the zinc and expose the steel to rusting.

(2) Copper and galvanised steel cylinders normally incorporate an aluminium and a magnesium sacrificial anode, respectively. These are designed to deteriorate over sufficient time to allow a protective coating of lime scale to build up on the exposed surfaces.

Primatic Hot Water Storage Cylinder

BS 1566-2: Specification for single feed indirect cylinders.

An indirect hot water system may be installed using a 'primatic' or single feed indirect cylinder. Conventional expansion and feed cistern, primary cold feed and primary vent pipes are not required, therefore by comparison, installation costs are much reduced. Only one feed cistern is required to supply water to be heated indirectly, by water circulating in an integral primary heater. Feed water to the primary circuit and boiler is obtained from within the cylinder, through the primary heater. The heat exchanger inside the cylinder has three air locks which prevent mixing of the primary and secondary waters. No corrosion inhibitors or system additives should be used where these cylinders are installed.

Key:
Sf = Secondary flow pipe
Pf = Primary flow pipe
Pr = Primary return pipe
He = Heat exchanger
Cf = Cold feed pipe

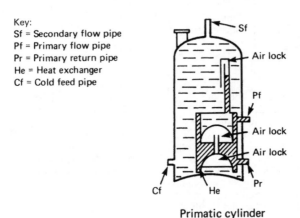

Primatic cylinder

Installation of primatic cylinder

For larger buildings a secondary circuit will be required to reduce 'dead-legs' and to maintain an effective supply of hot water at all outlets. Convection or thermo-siphonage may provide circulation, but for a more efficient service a circulatory pump will be necessary. In buildings which are occupied for only part of the day, e.g. schools, offices, etc., a time control or programmer can be used to regulate use of the pump. Also, one of the valves near the pump should be motorised and automatically shut off with the pump and boiler when hot water is not required. All secondary circuits should be well insulated to reduce heat losses through the pipework. A heating installation can operate in conjunction with this system, but may require duplication of boilers or separate boilers for each function.

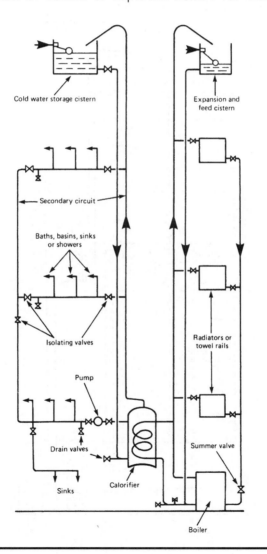

Sealed Indirect Hot Water System for a High Rise Building

For convenience and to reduce wear on fittings, the maximum head of water above taps and other outlets is 30 m. This is achieved by using intermediate or break pressure cisterns for each sub-circuit. Head tanks are provided to ensure sufficient volume of stored hot water and adequate delivery to the upper floors. Compared with conventional installations a considerable amount of pipework and fitting time can be saved by using an expansion vessel to absorb expansion of water in the primary circuit. However, the boiler and calorifiers must be specified to a high quality standard to withstand the water pressure. All pipework and equipment must be well insulated.

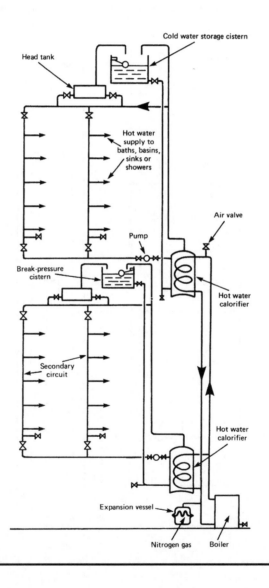

Cast iron sectional – made up of a series of hollow sections, joined together with left- and right-hand threaded nipples to provide the heat capacity required. When installed, the hollow sections contain water which is heated by energy transfer through the cast iron from the combusted fuel. Applications: domestic to large industrial boilers.

Steel shell, fire or flame tube – hot combusted fuel and gases discharge through multiple steel tubes to the extract flue. Heat energy from the burnt fuel transfers through the tube walls into cylindrical waterways. Tubes may be of annular construction with water surrounding a fire tube core. Uses: commercial and industrial buildings.

Copper or steel water tube – these reverse the principle of fire tubes. Water circulates in a series of finned tubes whilst the combusted fuel effects an external heat transfer. These are typical of the heat exchangers in domestic boilers.

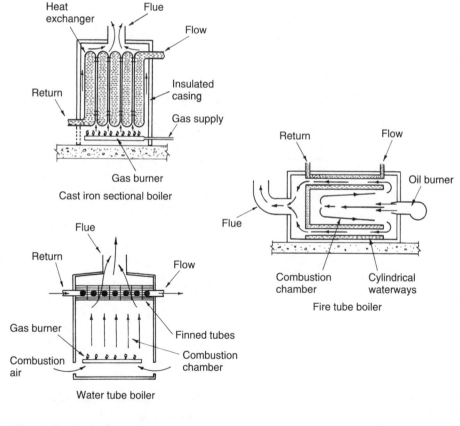

Cast iron sectional boiler

Fire tube boiler

Water tube boiler

All of these boiler types may be fired by solid fuel, gas or oil.

Condensing Gas Boilers

Condensing boilers have a greater area of heat transfer surface than conventional boilers. In addition to direct transfer of heat energy from the burning fuel, heat from the flue gases is used as secondary heating to the water jacket. Instead of the high temperature (200–250°C) flue gases and water vapour discharging to atmosphere, they are recirculated around the water jacket by a fan. This fan must be fitted with a sensor to prevent the boiler firing in the event of failure. Condensation of vapour in the flue gases is drained to a suitable outlet. The overall efficiency is about 90%, which compares well with the 75% expected of conventional boilers. However, purchase costs are higher, but fuel savings should justify this within a few years.

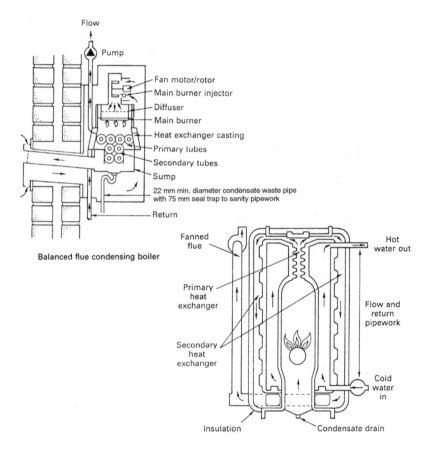

Balanced flue condensing boiler

Conventional flue condensing boiler

Refs: BS 6798: Specification for installation of gas-fired boilers of rated input not exceeding 70 kWnet. Building Regulations, Approved Document H1: Foul Water Drainage, Section 1 – Sanitary pipework.

This system saves considerably in installation time and space, as there is no need for cisterns in the roof space, no hot water storage cylinder and associated pipework. The 'combi' gas boiler functions as an instantaneous water heater only heating water as required, thereby effecting fuel savings by not maintaining water at a controlled temperature in a cylinder. Water supply is from the mains, providing a balanced pressure at both hot and cold water outlets. This is ideal for shower installations. Boiler location may be in the airing cupboard, leaving more space in the kitchen. The system is sealed and has an expansion vessel which is normally included in the manufacturer's pre-plumbed, pre-wired package for simple installation. Further control details are shown on page 95.

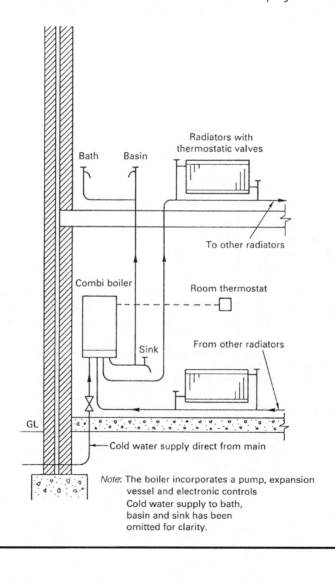

Bath   Basin

Radiators with thermostatic valves

To other radiators

Combi boiler

Room thermostat

Sink

From other radiators

GL

Cold water supply direct from main

Note: The boiler incorporates a pump, expansion vessel and electronic controls
Cold water supply to bath, basin and sink has been omitted for clarity.

Secondary Circulation

To prevent user inconvenience waiting for the cold water 'dead-leg' to run off and to prevent water wastage, long lengths of hot water distribution pipework must be avoided. Where cylinder to tap distances are excessive, a pumped secondary flow and return circuit may be installed with minimal 'dead-legs' branching to each tap. The pipework must be fully insulated and the circulation pump timed to run throughout the working day, e.g. an office system could be programmed with the boiler controls, typically 8.00 am to 6.00 pm, 5 days a week. A non-return valve prevents reverse circulation when the pump is not in use.

Pipe dia. (inside)	Max. length of secondary flow without a return (m)
<19 mm	12·0
19 to 25 mm	7·5
>25 mm	3·0

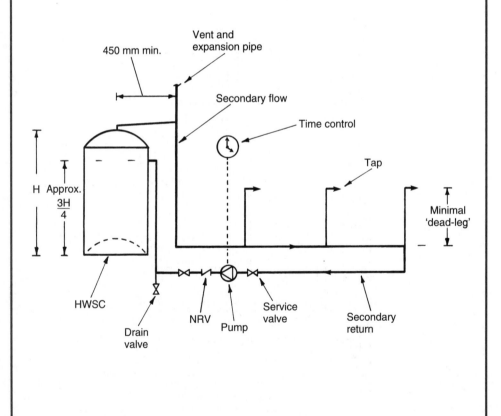

Dual installations or duplication of plant and equipment is required in buildings where operating efficiency is of paramount concern. With this provision, the supply of hot water in hotels, commercial buildings, offices, etc. is ensured at all times, as it is most unlikely that all items of plant will malfunction simultaneously. It may also be necessary to divide the design capacity of plant to reduce the concentration of structural loads. Each boiler and calorifier may be isolated for repair or renewal without disturbing the function of the others. Therefore when designing the system it is usual to oversize plant by up to one-third, to ensure the remaining plant has reasonable capacity to cope with demand. There is also the facility to economise by purposely isolating one boiler and calorifier during periods when a building is only part occupied.

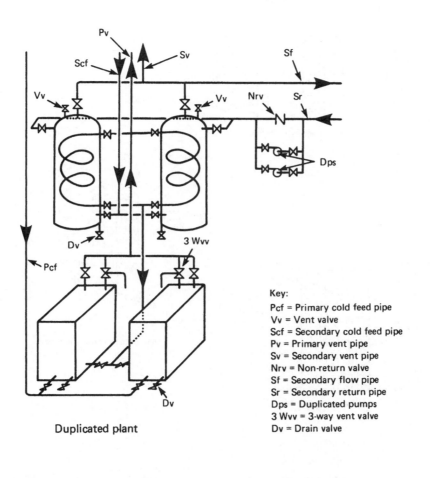

Duplicated plant

Key:
Pcf = Primary cold feed pipe
Vv = Vent valve
Scf = Secondary cold feed pipe
Pv = Primary vent pipe
Sv = Secondary vent pipe
Nrv = Non-return valve
Sf = Secondary flow pipe
Sr = Secondary return pipe
Dps = Duplicated pumps
3 Wvv = 3-way vent valve
Dv = Drain valve

An electric immersion heater may be used within a conventional hot water storage cylinder. Alternatively, individual or self-contained open outlet heaters may be located over basins, baths or sinks. Combined cistern-type heaters can be used to supply hot water to several sanitary appliances. Energy conservation is achieved with an integral thermostat set between 60 and 65°C. This temperature is also sufficient to kill any bacteria. The immersion heater must be electrically earth bonded and the cable supplying the heating element must be adequate for the power load. A cable specification of 2·5 mm^2 is normally adequate with a 20 amp double pole control switch supplied direct from the consumer's unit or fuse box. Overload protection at the consumers unit is a 16 amp fuse or circuit breaker for a 3 kW element and 20 amp for a 4 kW element.

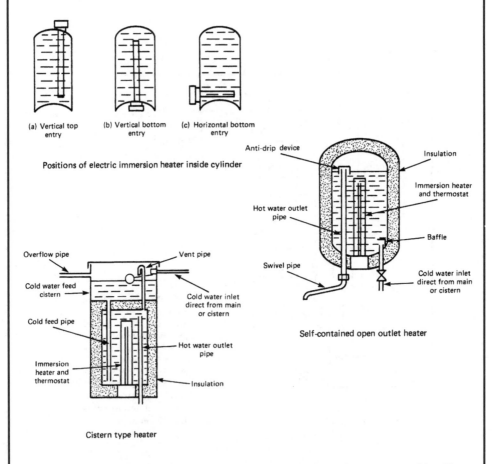

(a) Vertical top entry

(b) Vertical bottom entry

(c) Horizontal bottom entry

Positions of electric immersion heater inside cylinder

Anti-drip device

Insulation

Hot water outlet pipe

Immersion heater and thermostat

Baffle

Swivel pipe

Cold water inlet direct from main or cistern

Self-contained open outlet heater

Overflow pipe

Vent pipe

Cold water feed cistern

Cold water inlet direct from main or cistern

Cold feed pipe

Immersion heater and thermostat

Hot water outlet pipe

Insulation

Cistern type heater

Ref. BS 3198: Specification for copper hot water storage combination units for domestic purposes.

The cistern-type heater should be located with the water level at least 1·5 m above the water draw-off taps. If there is insufficient space to accommodate this combination unit, a smaller pressure-type water heater may be fitted. These are small enough to locate under the sink or elsewhere in the kitchen. They have two immersion heaters, the upper element of 500 watts rating is for general use supplying hot water to the basin, sink and other small appliances. The lower element of 2500 watts may be on a timed control to provide sufficient hot water for baths. The pressure heater is supplied with cold water from a high level cistern.

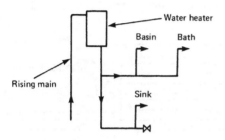

Installation of electric cistern type heater

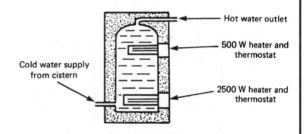

Pressure-type electric water heater

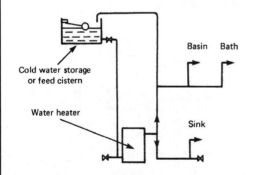

Installation of pressure-type electric water heater

Instantaneous water heaters are relatively compact non-storage units suitable for use with individual sinks, basins and showers. For user safety they are fitted with a pressure switch to disconnect the electricity if the water supply is interrupted and a thermal cut-out to prevent the water overheating. Mains pressure to these units should be maintained below 400 kPa (4 bar). In some high pressure supply areas this will require a pressure reducing valve to be installed on the service pipe. Some expansion of hot water will occur whilst the unit is in use. This can be contained if there is a least 3 metres of pipework before the unit and the closest cold water draw-off. If this is impractical, an expansion vessel may be used. For more details of electric shower installations see pages 253 and 254.

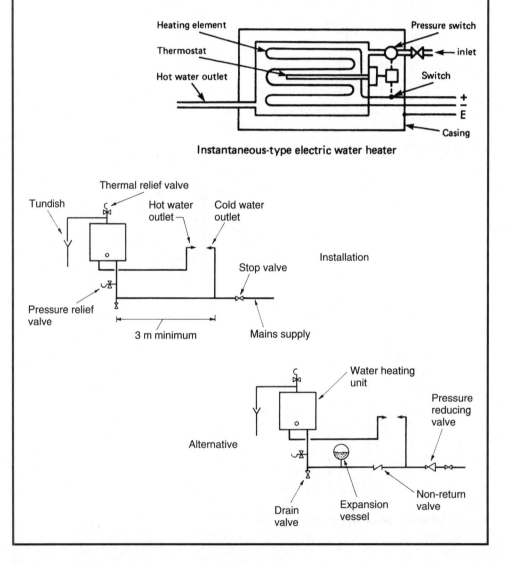

Instantaneous-type electric water heater

Industrial, commercial and domestic demand for electricity is considerably reduced overnight. Therefore during this time, the electricity supply companies can market their spare capacity as off-peak electricity by selling it at a reduced rate – approximately half the cost of standard day time tariff. Supplies are adapted to operate through a programmer or time control which diverts the electricity to a special off-peak or white meter, usually from midnight to 7 a.m. In order to maximise the benefit, slightly larger than standard capacity hot water storage cylinders of 162 or 190 litres are recommended. To conserve energy, these cylinders must be thoroughly insulated and the immersion heaters fitted with integral thermostatic control. If supplementary hot water is required during the day, this can be provided by a secondary immersion heater at standard supply tariff.

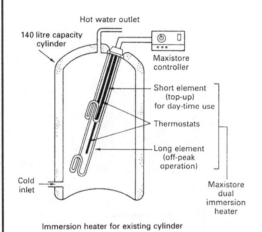

Immersion heater for existing cylinder

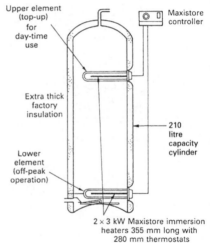

Special package unit

The secondary immersion heater or boost heater is close to the top of the cylinder to ensure that only a limited quantity of water is heated at standard tariff. To maximise economy, the off-peak thermostat is set at 65°C and the boost thermostat at 60°C.

When the hot water outlet is opened, cold water flows through a venturi fitting. The venturi contains a diaphragm which responds to the flow differential pressure and this opens the gas valve. A pilot flame ignites gas flowing through the burner which heats the water as it passes through the heat exchanger. Installation can be direct from the water main or from a cold water storage cistern. A multipoint system has the hot water outlet suppling several appliances.

A gas circulator can be used to heat water in a storage cylinder. They are usually fitted with an economy or three-way valve. This gives optional use of water circulation through a high or low return pipe for variable hot water storage volume. Domestic installations may be in the kitchen, with vertical flow and return pipes to a storage cylinder in the airing cupboard.

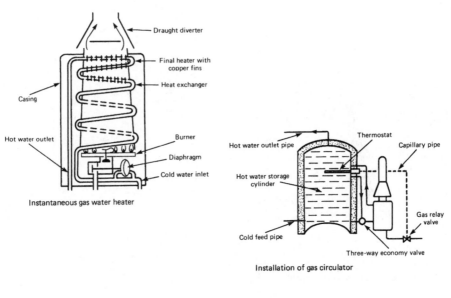

Instantaneous gas water heater

Installation of gas circulator

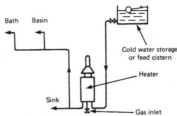

Installation of instantaneous gas water heater

Ref: BS EN 26: Gas fired instantaneous water heaters for the production of domestic hot water, fitted with atmospheric burners.

The storage type of gas water heater is a self-contained unit and is therefore simpler and quicker to install than a gas circulator. Capacities range from 75 to 285 litres. The smaller units are single-point heaters for supplying hot water to an individual sink or basin. Larger, higher rated storage heaters can be used to supply hot water to a bath, basin, sink and shower. These are called multi-point heaters. They may also be installed in flats up to three storeys, with cold water supplied from one cistern. A vent pipe on the cold feed will prevent siphonage. To prevent hot water from the heaters on the upper floors flowing down to the heater on the ground floor, the branch connection on the cold feed pipe must be above the heaters.

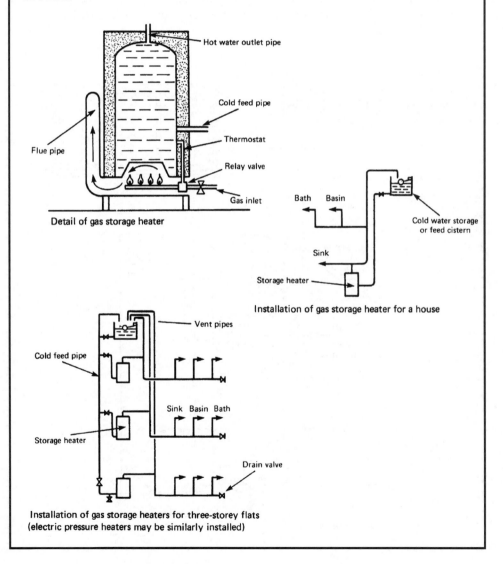

Detail of gas storage heater

Installation of gas storage heater for a house

Installation of gas storage heaters for three-storey flats
(electric pressure heaters may be similarly installed)

Solar Heating of Water

Solar energy can contribute significantly to hot water requirements. In some countries it is the sole source of energy for hot water. In the UK its efficiency varies with the fickle nature of the weather, but fuel savings of about 40% are possible. For domestic application, the collector should be 4 to 6 m^2 in area, secured at an angle of 40° to the horizontal and facing south. The solar cylinder capacity of about 200 litres is heated to 60°C. The cylinder and associated pipework must be very well insulated and the solar part of the system should contain a blend of water and non-toxic anti-freeze. The pump is switched on when the temperature of water at point X exceeds that at point Y by 2 to 3°C. The solar cylinder and the conventional cylinder may be fitted on the same level, or to save space a combined solar/conventional cylinder can be obtained from specialist suppliers.

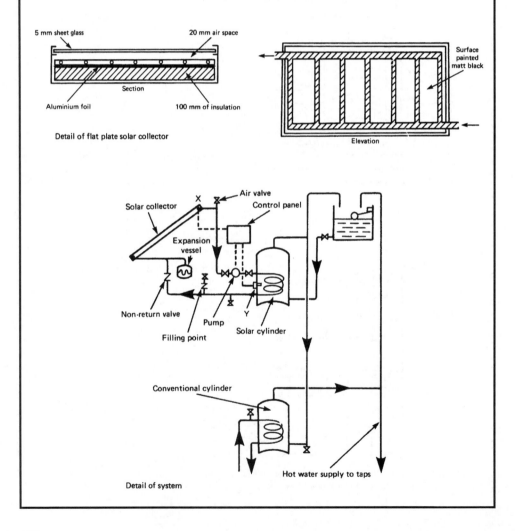

5 mm sheet glass 20 mm air space

Section

Aluminium foil   100 mm of insulation

Detail of flat plate solar collector

Surface painted matt black

Elevation

X Air valve
Solar collector   Control panel
Expansion vessel
Non-return valve
Pump   Y
Filling point   Solar cylinder

Conventional cylinder

Hot water supply to taps

Detail of system

The capacity of hot water storage vessels must be adequate for the building purpose. Exact requirements are difficult to determine, but reasonable estimates are possible. These should include provision for rate of energy consumption (see table below) and the time taken to reheat the water to the required storage temperature (see boiler rating calculation – next page). Many buildings have variable use and inconsistent demands. This often creates an overdesign situation, unless care is taken to establish peak use periods and the system calculations adjusted accordingly. With these building types, non-storage instantaneous fittings may be preferred.

For most buildings the following table can be used as guidance:

Building purpose	Storage capacity (litres/person)	Energy consumption (kW/person)
Dwellings:		
single bath	30	0·75
multi-bath	45	1·00
Factory/Office	5	0·10
Hotels	35*	1·00
Hostels	30	0·70
Hospitals	35*	1·00
Schools/Colleges:		
day	5	0·10
boarding	25	0·70
Sports pavilions	35	1·00

* Average figures

E.g. A student hall of residence (hostel) to accommodate 50 persons.

Capacity: 50 × 30 = 1500 litres
Energy consumption: 50 × 0·70 = 35 kW

The nearest capacity storage vessel can be found from manufacturers' catalogues or by reference to BS 1566. For convenience, two or three cylinders of equivalent capacity may be selected.

Boiler Rating

Boilers are rated in kilowatts, where 1 watt equates to 1 joule of energy per second, i.e. W = J/s. Many manufacturers still use the imperial measure of British thermal units per hour for their boilers. For comparison purposes 1 kW equates to 3412 Btu/h.

Rating can be expressed in terms of gross or net heat input into the appliance. Values can be calculated by multiplying the fuel flow rate (m^3/s) by its calorific value (kJ/m^3 or kJ/kg). Input may be gross if the latent heat due to condensation of water is included in the heat transfer from the fuel. Where both values are provided in the appliance manufacturer's information, an approximate figure for boiler operating efficiency can be obtained, e.g. if a gas boiler has gross and net input values of 30 and 24 kW respectively, the efficiency is 24/30 × 100/1 = 80%.

Oil and solid fuel appliances are normally rated by the maximum declared energy output (kW), whereas gas appliances are rated by net heat input rate (kW[net]).

Calculation of boiler power:

$$kW = \frac{kg\ of\ water \times S.h.c. \times Temp.\ rise}{Time\ in\ seconds}$$

where: 1 litre of water weighs 1 kg
 S.h.c. = specific heat capacity of water, 4·2 kJ/kgK
 K = degrees Kelvin temperature interval
 Temp. rise = rise in temperature that the boiler will need to increase the existing mixed water temperature (say 30°C) to the required storage temperature (say 60°C).
 Time in seconds = time the boiler takes to achieve the temperature rise. 1 to 2 hours is typical, use 1·5 hours in this example.

From the example on the previous page, storage capacity is 1500 litres, i.e. 1500 kg of water. Therefore:

$$Boiler\ power = \frac{1500 \times 4 \cdot 2 \times (60 - 30)}{1 \cdot 5 \times 3600} = 35\ kW\ net$$

Given the boiler has an efficiency of 80%, it will be gross input rated:

$$35 \times 100/80 = 43 \cdot 75\ kW$$

Note: The boiler operating efficiency is the relationship between a unit of fuel energy consumed to produce a unit of heat energy in the appliance hot water. It is not to be compared with the seasonal efficiency of a boiler (SEDBUK), see page 59.

The water in primary flow and return pipework may circulate by convection. This produces a relatively slow rate of movement of about 0·2 m/s, depending on pipe length and location of boiler and cylinder. Modern systems are more efficient, incorporating a circulation pump to create a water velocity of between 0·75 and 3·0 m/s. This permits smaller pipe sizes and will provide a faster thermal response.

Inside diameter of pipe	Velocity min.	Velocity max. (copper)	Velocity max. (steel)
<50 mm	0·75 m/s	1·0 m/s	1·5 m/s
>50 mm	1·25 m/s	1·5 m/s	3·0 m/s

Exceeding these recommendations may lead to excessive system noise and possible pipe erosion.

E.g. using the Copper Development Association design chart shown on the next page, with the boiler rating from the previous example of 43.75 kW gross heat input and 35 kW net heat input.

$$\text{Mass flow rate (kg/s)} = \frac{\text{Boiler net heat input}}{\text{S.h.c.} \times \text{Temp. diff. (pf–pr)}}$$

Temperature difference between primary flow (pf) and primary return (pr) in pumped water circuits is usually about 10 K, i.e. 80°C–70°C. With convected circulation the return temperature will be about 60°C.

$$\text{Mass flow rate} = \frac{35}{4·2 \times 10} = 0·83 \text{ kg/s}$$

On the design chart, co-ordinating 0·83 kg/s with a pumped flow rate of 1 m/s indicates a 42 mm inside diameter copper tube. (35 mm is just too small.)

By comparison, using convected circulation of, say, 0·15 m/s and a mass flow rate with a 20 K temperature difference of 0·42 kg/s, the pipe size would be 76 mm.

Water Flow Resistance Through Copper Tube

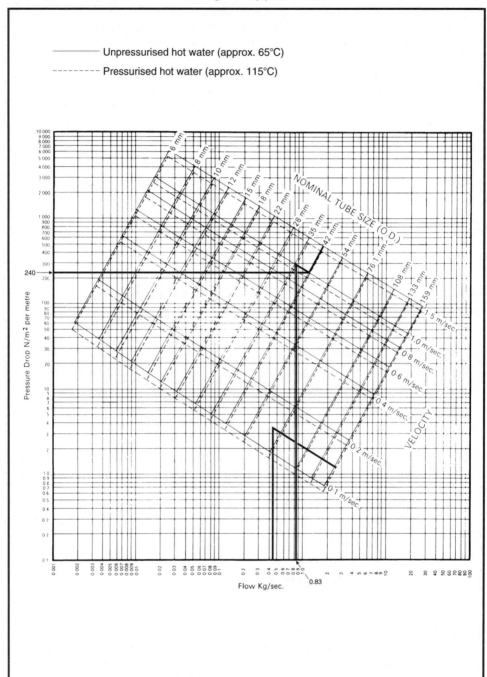

Reproduced with the kind permission of the Copper Development Association.

Circulatory pumps produce minimal pressure in the primary flow and return, but the flow rate is considerably enhanced. The pressure can be ascertained from design charts as a pressure drop in N/m^2 per metre or pascals per metre. 1 N/m^2 equates to 1 pascal (Pa).

From the design chart, circulation in a 42 mm copper tube at 1 m/s produces a pressure drop of 240 Pa per metre. An estimate of the primary flow and return effective pipe length (see page 25) is required to establish the total resistance that the pump must overcome. For example, if the effective pipe length is 20 m:

240 × 20 = 4800 Pa or 4·8 kPa.

Therefore the pump specification would be 0·83 kg/s at 4·8 kPa.

Manufacturers' catalogues can be consulted to select a suitable pump. To provide for flexibility in installation, a degree of variable performance is incorporated into each model of pump. This range of characteristics can be applied by several different control settings as shown in the following graphic.

Pump performance chart:

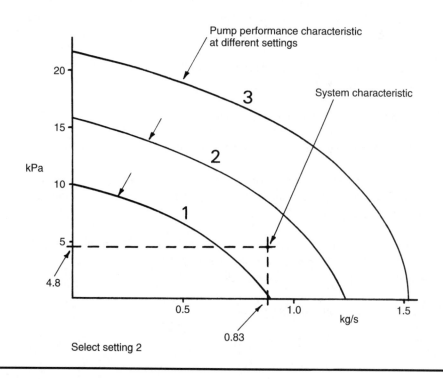

Legionnaires' Disease in Hot Water Systems

Bacterial growths which cause Legionnaires' disease develop in warm, moist, natural conditions such as swamps. They have adapted to living in the built environment in the artificial atmosphere of air conditioning and hot water systems. A large number of outbreaks of the disease have occurred, with some people suffering a prolonged illness similar to pneumonia. The elderly are particularly vulnerable and many have died, hence the name of the illness which was attributed to a group of retired legionnaires who were infected whilst attending a reunion in Philadelphia, USA, in 1976. Numerous other outbreaks and subsequent deaths have led to strict maintenance and installation controls of services installations. This has been effected by the Health and Safety Executive under the Health and Safety at Work, etc. Act and the Workplace (Health, Safety and Welfare) Regulations. The following measures are recommended for use with hot water systems:

1. Stored hot water temperature 60 to 65°C throughout the storage vessel.

2. Routine maintenance involving heating the water to 70°C as a precaution.

3. Changing the design of cylinders and calorifiers with concave bases. These are suspect, as the lower recesses could provide areas of reduced water temperature with little or no movement.

4. Connections to storage vessels should encourage through movement of water.

5. Pipework 'dead-legs' to be minimal.

6. All pipework to be insulated to reduce water temperature losses.

7. Where secondary circulation is required, supplementary trace element heating tape should be applied to maintain a minimum water temperature of 50°C.

8. Showers with recessed/concave outlet roses to be avoided. Other designs to have a self-draining facility to avoid inhalation of contaminated moisture droplets.

9. Spray taps – similar provision to 8.

SEDBUK is the acronym for *Seasonal Efficiency of Domestic Boilers in the United Kingdom*. It has developed under the Government's Energy Efficiency Best Practice Programme to provide a manufacturers' data base which represents the efficiency of gas- and oil-fired domestic boilers sold in the UK. See website: www.boilers.org.uk, or www.sedbuk.com. This voluntary site is updated monthly and it contains over 75% of new and existing products.

SEDBUK must not be confused with the operating efficiencies which are sometimes quoted in manufacturers' literature. These compare gross and net heat input values – see page 54. SEDBUK is the average annual in-use efficiency achieved in typical domestic conditions. The principal parameters included in the SEDBUK calculation are:
• type of boiler
• fuel ignition system
• internal store size
• type/grade of fuel.

Also included are the operating influences:
• typical patterns of usage – daily, weekly, etc.
• climatic variations.

Quoted SEDBUK figures are based on standard laboratory tests from manufacturers, certified by an independent Notified Body which is accredited for boiler testing to European Standards.

Efficiency bands:

Band	SEDBUK range (%)
A	> 90
B	86–90
C	82–86
D	78–82
E	74–78
F	70–74
G	< 70

See next page for the minimum acceptable band values for different fuel and installation types.

SEDBUK and SAP

Building Regulations, Approved Document L1: Conservation of fuel and power in dwellings, requires reasonable boiler efficiency for installations in new dwellings and for replacement equipment in existing dwellings. The following values apply:

Fuel system and boiler type	Min. SEDBUK value (%)
Natural gas	78
Natural gas back boiler	75
Liquid petroleum gas (LPG)	80
LPG back boiler	77
Oil	85
Oil combination boiler	82
Solid fuel	See HETAS certification

The SEDBUK database is an essential reference when calculating part of the Government's Standard Assessment Procedure for Energy Rating of Dwellings (SAP rating). Additional factors to be considered are: ventilation, heat losses through the fabric (U values) and solar gains. To comply with the Building Regulations, builders are required to submit energy rating calculations to the local building control authority. This data is also available for prospective house buyers and tenants for comparison purposes when assessing anticipated annual fuel costs for hot water and heating. SAP values vary from 1 to 120, with 80 considered the minimum expectation of new dwellings. SAP worksheets are available in Appendix M to Approved Document L1 of the Building Regulations.

Recognised organisations for accrediting 'competent persons' as installers of domestic hot water and central heating systems:

Gas – Council for Registered Gas Installers (CORGI).

Oil – Oil Firing Technical Association for the Petroleum Industry (OFTEC).

Solid fuel – Heating Equipment Testing and Approval Scheme (HETAS).

Refs:

Building Regulations, Approved document L1 – Conservation of fuel and power in dwellings, 2002.

The Government's Standard Assessment Procedure for Energy Rating of Dwellings, 2001.

(Both published by The Stationery Office.)

3 HEATING SYSTEMS

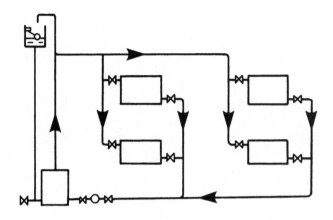

Radiators and convectors are the principal means of heat emission in most buildings. Less popular alternatives include exposed pipes and radiant panels for use in warehousing, workshops and factories, where appearance is not important. Embedded panels of pipework in the floor screed can also be used to create 'invisible' heating, but these have a slow thermal response as heat energy is absorbed by the floor structure.

Despite the name radiator, no more than 40% of the heat transferred is by radiation. The remainder is convected, with a small amount conducted through the radiator brackets into the wall. Originally, radiators were made from cast iron in three forms: hospital, column and panel. Hospital radiators were so called because of their smooth, easy to clean surface, an important specification in a hygienic environment. Column radiators vary in the number of columns. The greater the number, the greater the heat emitting surface. Cast iron radiators are still produced to special order, but replicas in cast aluminium can be obtained. Cast iron panels have been superseded by pressed profiled steel welded panels. These are much slimmer and easier to accommodate than cast iron in the modern house. In addition to the corrugated profile, finned backing will also increase the heating surface and contribute to a higher convected output. Pressed steel radiators are made in single, double and triple panels.

Convectors have a steel casing containing a finned heat exchanger. About 90% of the heat emission is convected and this may be enhanced if a thermostatically controlled fan is also located in the casing. They are more effective than radiators for heating large rooms, and in this situation their extra bulk can be accommodated.

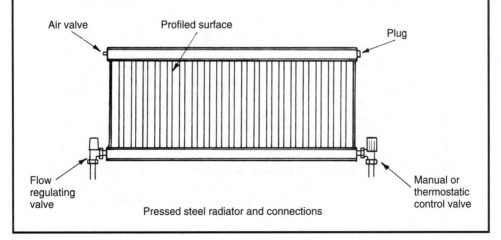

Pressed steel radiator and connections

In temperate and cold climates where there is insufficient warmth from the sun during parts of the year, heat losses from the human body must be balanced. These amount to the following approximate proportions: radiation 45%, convection 30% and evaporation 25%. Internal heat gains from machinery, lighting and people can contribute significantly, but heat emitters will provide the main contribution in most buildings.

Enhancement of radiator performance can be achieved by placing a sheet of reflective foil on the wall between the fixing brackets. Emitter location is traditionally below window openings, as in older buildings the draughts were warmed as they infiltrated the ill-fitting sashes. With quality double glazed units this is no longer so important and in the absence of a window, locating a shelf above the radiator will prevent pattern staining of the wall due to convective currents. Radiant panels and strips suspend from the ceiling in industrial premises and other situations where wall space is unavailable.

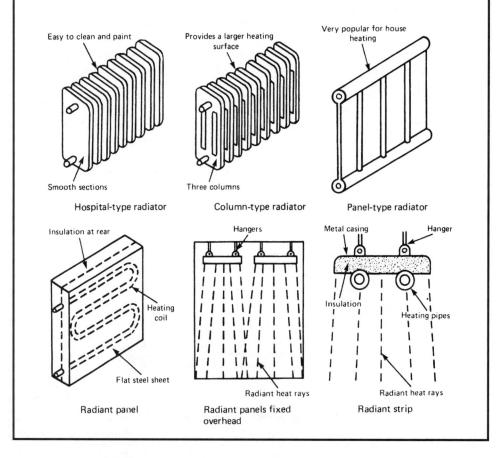

Hospital-type radiator

Column-type radiator

Panel-type radiator

Radiant panel

Radiant panels fixed overhead

Radiant strip

Radiant and convector skirting heaters are unobtrusive at skirting level and provide uniform heat distribution throughout a room. Natural convectors have a heating element at a low level within the casing. This ensures that a contained column of warm air gains velocity before discharging to displace the cooler air in the room. Fan convectors may have the heater at high level with a variable speed fan located below. In summer, the fan may also be used to create air circulation. Overhead unit heaters are used in workshops to free the wall space for benches, machinery, etc. A variation may be used as a warm air curtain across doorways and shop entrances. Individual unit heaters may have a thermostatically controlled inlet valve or a bank of several units may be controlled with zoning and diverter valves to regulate output in variable occupancy situations.

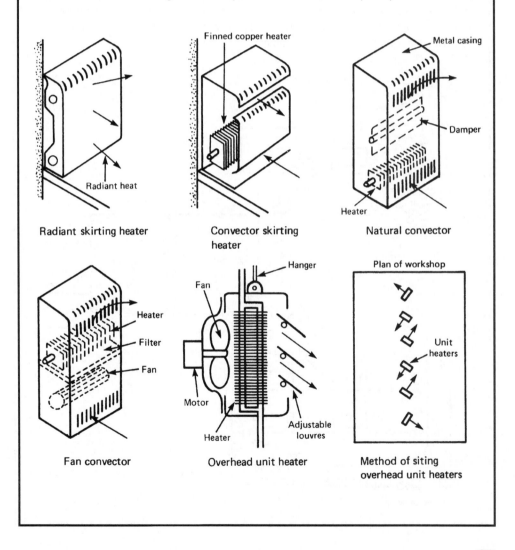

Radiant skirting heater

Convector skirting heater

Natural convector

Fan convector

Overhead unit heater

Method of siting overhead unit heaters

In low temperature, hot water heating systems the boiler water temperature is thermostatically controlled to about 80°C. Systems may be 'open' with a small feed and expansion cistern or mains fed 'sealed' with an expansion vessel.

The type of system and pipe layout will depend on the building purpose and space available for pipework. A ring or loop circuit is used for single storey buildings. Drop and ladder systems are used for buildings of several storeys. The drop system has the advantage of being self-venting and the radiators will not become air locked. Traditional solid fuelled systems operate by convection or gravity circulation (otherwise known as thermo-siphonage). Contemporary practice is to install a pump for faster circulation and a more rapid and effective thermal response. This will also complement modern fuel controls on the boiler and allow for smaller pipe sizes. The additional running costs are minimal.

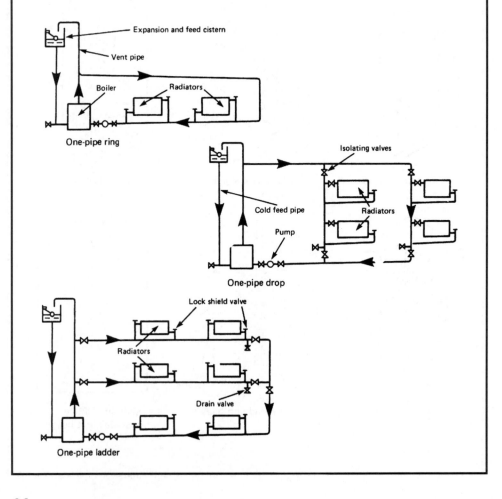

The one- and two-pipe parallel systems are useful where pipework can be accommodated within a floor structure, a raised floor or a suspended ceiling. The disadvantage with all one-pipe systems is the difficulty of supplying hot water to the radiators furthest from the boiler. As the heat is emitted from each radiator, cooling water returns to mix with the hot water supplying subsequent radiators, gradually lowering the temperature around the circuit. Eventually the last or 'index' radiator receives lukewarm water at best, necessitating a very large radiator to provide any effect. Pumped circulation may help, but it will require a relatively large diameter pipe to retain sufficient hot water to reach the 'index' radiators. Two-pipe systems are less affected, as the cool water from each radiator returns directly to the boiler for reheating. However, radiators will need flow balancing or regulating to ensure an even distribution of hot water. The reverse-return or equal travel system requires the least regulating, as the length of pipework to and from each radiator at each floor level is equal. In all systems the circulating pump is normally fitted as close to the boiler as possible, either on the heating flow or return. Most pump manufacturers recommend location on the higher temperature flow.

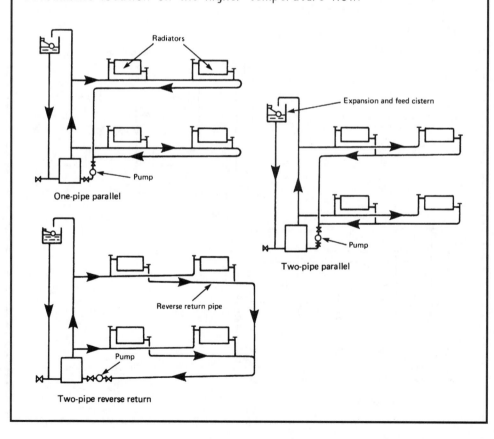

Radiators

Pump

One-pipe parallel

Expansion and feed cistern

Pump

Two-pipe parallel

Reverse return pipe

Pump

Two-pipe reverse return

The two-pipe upfeed system is used when it is impractical to locate pipes horizontally at high level. The main heating distribution pipes can be placed in a floor duct or within a raised floor. The two-pipe drop is used where a high level horizontal flow pipe can be positioned in a roof space or in a suspended ceiling, and a low level return within a ground floor or basement ceiling. This system has the advantage of self-venting. The two-pipe high level return system is particularly appropriate for installation in refurbishments to existing buildings with solid ground floors. In this situation it is usually too time consuming, impractical and possibly structurally damaging to cut a trough or duct in the concrete.

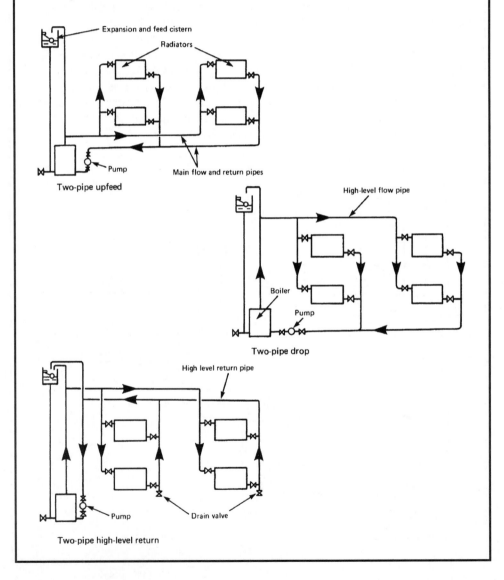

Two-pipe upfeed

Two-pipe drop

Two-pipe high-level return

Pumped small bore heating systems have 28 or 22 mm outside diameter copper tube for the main heating flow and return pipework, with 15 mm o.d. branches to each radiator. This compares favourably with the old gravity/convection circulation systems which sometimes required pipes of over 50 mm diameter to effect circulation. If cylinder and boiler are separated vertically by floor levels, there will be sufficient pressure for hot water to circulate by convection through the primary flow and return pipes. However, most modern systems combine a pumped primary and heating flow with circulation regulated by thermostats and motorised valves. Variations in one and two pipe systems are shown on pages 66–68. Two pipe systems are always preferred for more effective hot water distribution.

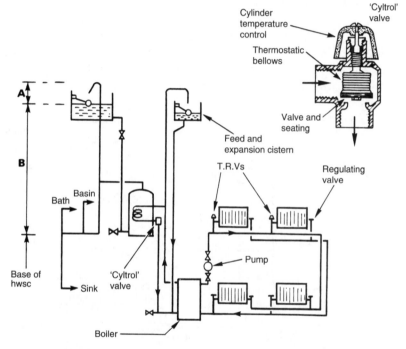

Small bore heating system

Notes:

1. 'Cyltrol' valve to be as close as possible to hwsc, to sense hot water return temperature and maintain stored water at about 55°C. Where used with a solid fuel boiler, an unvalved radiator or towel rail is connected across the primary pipes to dissipate excess heat when the 'cyltrol' closes.

2. Min. height of expansion pipe above cistern water level (A) = (B) in metres × 40 mm + 150 mm. E.g. if (B), cistern water level to base of hwsc is 2·5 m, then (A) is 2·5 × 40 mm + 150 mm = 250 mm.

Low Temperature Microbore Hot Water Heating System

The microbore system also has pumped circulation through 28 or 22 mm o.d. copper tube main flow and return pipes to radiators. The diameter depending on the number and rating of emitters connected. The difference between this system and conventional small bore is the application of a centrally located manifold between boiler and emitters. Manifolds are produced with standard tube connections for the flow and return and several branches of 6, 8, 10 or 12 mm outside diameter. A combined manifold is also available. This is more compact, having a blank in the middle to separate flow from return. Manifolds are generally allocated at one per floor. Systems may be open vented or fitted with an expansion vessel. The advantage of microbore is ease and speed of installation, as long lengths of small diameter soft copper tubing are produced in coils. It is also unobtrusive where exposed, very easily concealed and is less damaging to the structure when holes are required. Water circulation noise may be noticeable as velocity is greater than in small bore systems. Pumped circulation is essential due to the high resistance to water flow in the small diameter pipes.

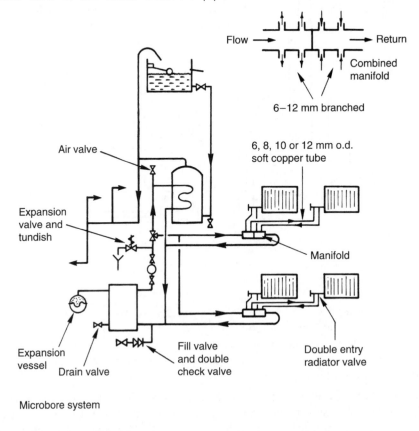

Flow → ← Return

Combined manifold

6–12 mm branched

6, 8, 10 or 12 mm o.d. soft copper tube

Air valve

Expansion valve and tundish

Manifold

Expansion vessel

Drain valve

Fill valve and double check valve

Double entry radiator valve

Microbore system

This is an alternative method for distributing hot water. It can be effected by using two separate pumps from the boiler flow: one to supply the hot water storage cylinder and the other the heating circuit. Grundfos Pumps Ltd. have developed a purpose-made dual pump for this purpose, which is integrated into one body. This system conveniently replaces the conventional single pump and associated two or three port motorised distribution valves. Each pump is dedicated to hot water or heating and individually controlled by cylinder or room thermostat. The correct flow and pressure can be regulated to the characteristics of the specific circuit.

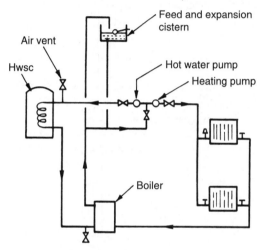

Conventional open vent system

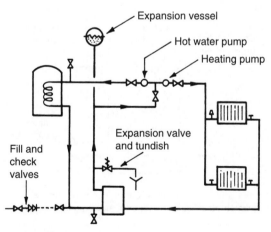

Typical sealed system

Air Elimination in Hot Water and Heating Systems

In conventional low pressure systems, air and other gases produced by heating water should escape through the vent and expansion pipe. Air must be removed to prevent the possibility of air locks, corrosion and noise. To assist air removal, a purpose made device resembling a small canister may be used to concentrate the gases. This simple fitting is located on the boiler flow and vent pipe to contain the water velocity and ensure efficient concentration and release of air into the vent.

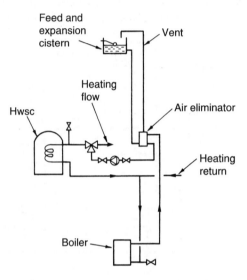

Application of air eliminator

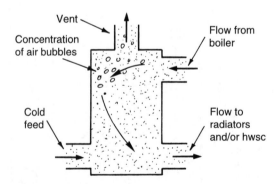

Air eliminator (approx. 100 mm high × 75 mm dia. with standard 22 mm o.d. copper tube connections)

The system consists of 15 mm or 22 mm o.d. annealed copper pipes embedded in the floor, ceiling or walls. This has the benefit of avoiding unsightly pipes and radiators. Heat distribution is uniform, providing a high standard of thermal comfort as heat is emitted from the building fabric. However, thermal response is slow as the fabric takes time to heat up and to lose its heat. Thermostatic control is used to maintain the following surface temperatures:

Floors – 27°C
Ceilings – 49°C
Walls – 43°C

Joints on copper pipes must be made by capillary soldered fittings or by bronze welding. Unjointed purpose made plastic pipes can also be used. Before embedding the pipes they should be hydraulically tested to a pressure of 1400 kPa for 24 hours.

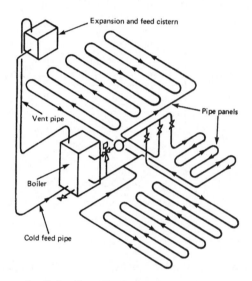

Installation of panel heating system

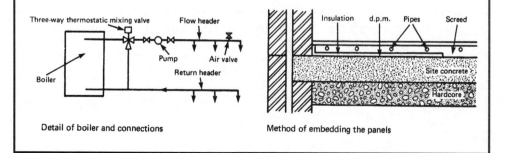

Detail of boiler and connections

Method of embedding the panels

Current practice is to use jointless polyethylene tube embedded in a 70 mm cement and sand screed (50 mm minimum cover to tube). In suspended timber floors the tube may be elevated by clipping tracks or brackets with metallic reflective support trays, prior to fixing the chipboard decking.

Boiler flow temperature for underfloor heating is about 50°C, whilst that for hot water storage and radiators is about 80°C. Therefore, where the same boiler supplies both hot water storage cylinder and/or radiators and underfloor heating, a motorised thermostatic mixing valve is required to blend the boiler flow and underfloor heating return water to obtain the optimum flow temperature.

Extract from performance tables for a design room temperature of 21°C with a blended flow temperature of 50°C:

Solid floor – Pipe dia. (mm)	Pipe spacing (mm)	Output (W/m^2)
15	100	82
15	200	67
18	300	55
Suspended floor –		
15	300*	47

*Assumes two pipe runs between floor joists spaced at 600 mm centres.

For a room with a solid floor area of 13·5 m^2 requiring a heating input of 758 watts (see page 101), the output required from the underfloor piping is:

$$758 \div 13{\cdot}5 = 56{\cdot}15 \text{ watts/m}^2$$

Therefore, 15 mm diameter pipe at 200 mm spacing (67 W/m^2) is more than adequate, whilst 18 mm diameter pipe at 300 mm spacing (55 W/m^2) is just below.

In any water heating system, provision must be made for the expansion of water. A combined expansion and feed cistern is the traditional means. This will have normal expansion space under usual boiler firing conditions of about 4% of the total volume of water in the system, plus a further third as additional expansion space for high boiler firing. Although the expansion can be accommodated up to the overflow level, there should be at least 25 mm between overflow and the fully expanded water level.

Contemporary sealed systems have an expansion vessel connected close to the boiler. It contains a diaphragm and a volume of air or nitrogen to absorb the expansion. To conserve wear on the diaphragm, location is preferred on the cooler return pipe and on the negative side of the pump. System installation is simpler and quicker than with an expansion cistern. The air or nitrogen is pressurised to produce a minimum water pressure at the highest point on the heating system of 10 kPa (approx. 1 m head of water). This is necessary, otherwise when filling the system, water would fill the vessel leaving no space for expansion.

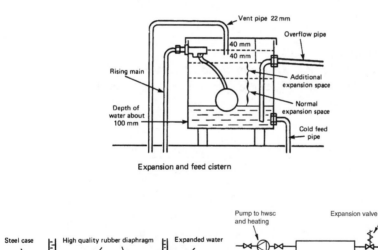

Expansion and feed cistern

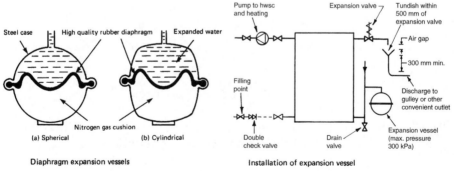

Diaphragm expansion vessels

Installation of expansion vessel

Expansion Vessels

Expansion vessels are produced to BS 6144. They must be correctly sized to accommodate the expansion of heated water without the system safety/pressure relief valve operating. The capacity of an expansion vessel will depend on the static pressure (metres head from the top of the system to the expansion vessel), the system maximum working pressure (same setting as p.r.v.) obtained from manufacturer's details and the volume of water in the system (approx. 15 litres per kW of boiler power).

Capacity can be calculated from the following formula:

$$V = \frac{e \times C}{1 - P_i/P_f}$$

where: V = vessel size (litres)

e = expansion factor (see table)

C = capacity of system (litres)

P_i = static pressure (absolute)*

P_f = max. working pressure (absolute)*

* absolute pressure is 1 atmosphere (atm) of approx. 100 kPa, plus system pressure.

E.g. C = 100 litres

P_i = 1·5 atm or 150 kPa (5m head static pressure)

P_f = 1·9 atm or 190 kPa (9m head static pressure)

Water temp. = 80°C

Temp. °C	Exp. factor
50	0·0121
60	0·0171
70	0·0227
80	0·0290
90	0·0359

$$V = \frac{0 \cdot 029 \times 100}{1 - 150/190} = 13 \cdot 80 \text{ litres}$$

Ref: BS 6144, Specification for expansion vessels using an internal diaphragm, for unvented hot water supply systems.

Solar space heating must be complemented with a very high standard of thermal insulation to the building fabric. The solar panel shown on page 52 for hot water provision will need a much larger area, typically 40 m² for a 3 to 4 bedroom detached estate house. A solar tank heat exchanger of about 40 m³ water capacity is located in the ground. It is fitted with a preset safety type valve which opens to discharge water to waste if it should overheat. The solar panel and associated pipework are mains filled and supplemented with a glycol or anti-freeze additive.

With diminishing fossil fuel resources and inevitable rising fuel prices, solar heating is encouraged as a supplement or even an alternative to conventionally fuelled systems. For use as the sole energy for a heating system there is still considerable scope for research and development. Technological developments are improving, particularly with the 'heat bank' or storage facility shown. In time it may become viable even with the UK's limited solar energy in winter months.

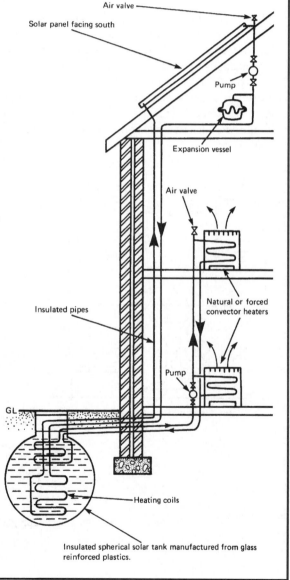

Air valve

Solar panel facing south

Pump

Expansion vessel

Air valve

Natural or forced convector heaters

Insulated pipes

Pump

GL

Heating coils

Insulated spherical solar tank manufactured from glass reinforced plastics.

High Temperature, Hot Water Heating Systems

Pressurisation allows water to be heated up to 200°C without the water changing state and converting to steam. This permits the use of relatively small diameter pipes and heat emitters, but for safety reasons these systems are only suitable in commercial and industrial situations. Even then, convectors are the preferred emitter as there is less direct contact with the heating surface. Alternatively, radiators must be encased or provision made for overhead unit heaters and suspended radiant panels. All pipes and emitters must be specified to the highest standard.

Water can be pressurised by steam or nitrogen. Pressurised steam is contained in the upper part of the boiler. To prevent the possibility of the pressurised water 'flashing' into steam, a mixing pipe is required between the heating flow and return. Nitrogen gas is contained in a pressure vessel separate from the boiler. It is more popular than steam as a pressurising medium, being easier to control, clean, less corrosive and less compatible with water. Air could be an alternative, but this is more corrosive than nitrogen and water soluble.

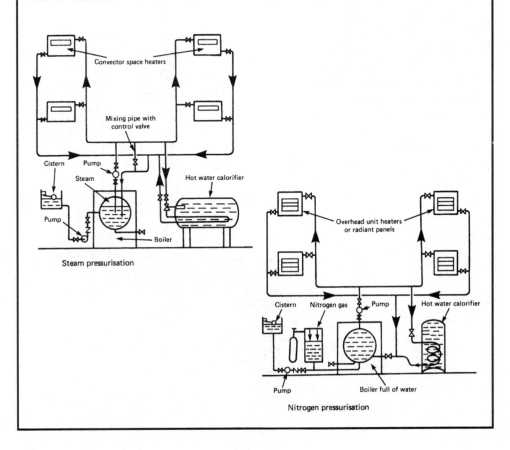

Steam pressurisation

Nitrogen pressurisation

When pressurising with nitrogen it is important that the pressure increases in line with temperature. If it is allowed to deviate the water may 'flash', i.e. convert to steam, causing system malfunction and possible damage to equipment.

To commission the system:

1. Water is pumped from the feed and spill cistern.
2. Air is bled from high levels and emitters.
3. Air is bled from the pressure vessel until the water level is at one-third capacity.
4. Nitrogen is charged into the pressure vessel at half design working pressure.
5. Boiler fired and expansion of hot water causes the water volume and nitrogen pressure in the vessel to double.

Note: Pressure vessel must be carefully designed to accommodate expanded water – approximately 4% of its original volume.

Safety features include a pressure control relay. This opens a motorised valve which lets excess water spill into the feed cistern if the boiler malfunctions and overheats. It also detects low pressure, possibly from system leakage and engages the feed pump to replenish the water and pressure.

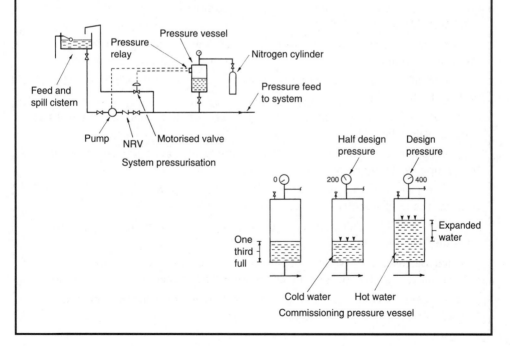

System pressurisation

Commissioning pressure vessel

Steam was the energy source of the Victorian era. At this time electricity and associated equipment that we now take for granted were in the early stages of development. Steam was generated in solid fuel boilers to power engines, drive machines and for a variety of other applications, not least as a medium for heat emitters. In this latter capacity it functioned well, travelling over long distances at high velocity (24–36 m/s) without the need for a pump.

By contemporary standards it is uneconomic to produce steam solely for heating purposes. However, it can be used for heating where steam is available from other processes. These include laundering, sterilising, kitchen work, manufacturing and electricity generation. Most of these applications require very high pressure, therefore pressure reducing valves will be installed to regulate supply to heating circuits.

Steam systems maximise the latent heat properties of water when evaporating. This is approximately 2260 kJ/kg at boiling point, considerably more than the sensible heat property of water at this temperature of approximately 420 kJ/kg. Because of this high heat property, the size of heat emitters and associated pipework can be considerably less than that used for hot water systems.

Steam terminology:

Absolute pressure – gauge pressure + atmospheric pressure (101·325 kN/m^2 or kPa).

Latent heat – heat which produces a change of state without a change in temperature, i.e. heat which converts water to steam.

Sensible heat – heat which increases the temperature of a substance without changing its state.

Enthalpy – total heat of steam expressed as the sum of latent heat and sensible heat.

Dry steam – steam which has been completely evaporated, contains no droplets of liquid water.

Wet steam – steam with water droplets in suspension, present in the steam space, typically in pipes and emitters.

Flash steam – condensate re-evaporating into steam after passing through steam traps.

Saturated steam – steam associated with or in contact with the water in the boiler or steam drum over the boiler.

Superheated steam – steam which is reheated or has further heat added after it leaves the boiler.

Classification – low pressure, 35 kPa–170 kPa (108–130°C).
medium pressure, 170 kPa–550 kPa (130–160°C).
high pressure, over 550 kPa (160°C and above).
Note: Gauge pressures shown.

Systems can be categorised as gravity or mechanical. In both, the
steam flows naturally from boiler to emitters without the need for a
pump. In the mechanical system a positive displacement pump is used
to lift condensed steam (condensate) into the boiler. Steam pressure
should be as low as possible as this will increase the latent heat
capacity. A steam trap prevents energy loss at each emitter. These
are fitted with a strainer or filter to contain debris and will require
regular cleaning. A sight glass after each trap gives visual indication
that the trap is functioning correctly, i.e. only condensate is passing.
On long pipe runs a 'drip relay' containing steam valve, strainer,
trap, sight glass and a gate valve will be required to control
condensing steam. This is represented by the strainer and trap in the
mechanical system shown below. Expansion loops or bellows will also
be required on long pipe runs to absorb thermal movement. All
pipework and accessories must be insulated to a very high standard.

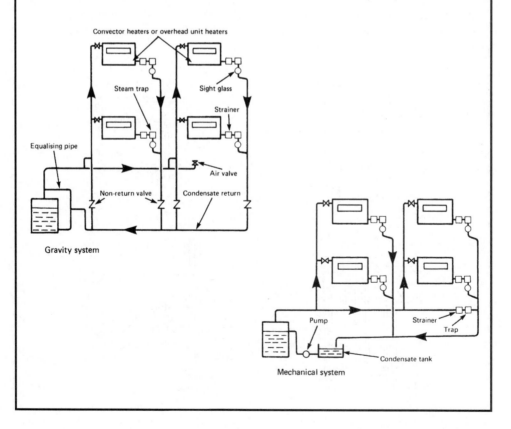

Gravity system

Mechanical system

Steam Traps

The purpose of a steam trap is to separate steam from condensate, retaining the energy efficient steam in distribution pipework and emitters. Traps are produced in various forms and sizes to suit all situations, some of which are shown below. The thermostatic and bi-metallic types are for relatively small applications such as radiators and unit heaters. The bucket and ball-float types are more suited to separating larger volumes of condensate and steam at the end of long pipe runs and in calorifiers.

Thermostatic – bellows expand or contract in response to steam or condensate repectively. Lower temperature condensate passes through.

Bi-metallic – condensate flows through the trap until higher temperature steam bends the strip to close the valve.

Bucket – condensate sinks the bucket. This opens the valve allowing steam pressure to force water out until the valve closes.

Ball-float – the copper ball rises in the presence of condensate opening the valve to discharge water until steam pressure closes the valve.

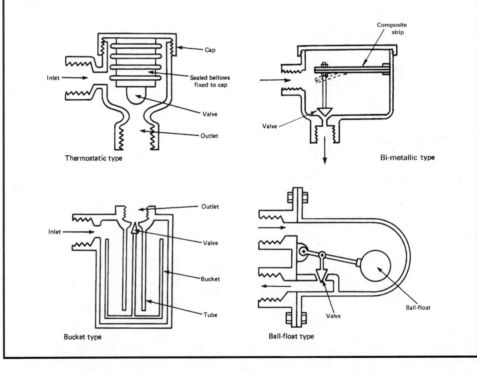

Non-storage type – used for providing instantaneous hot water for space heating. The steam tube bundle or battery occupies a relatively large area compared to the surrounding amount of water. To avoid temperature override and to control the steam flow, a thermostat and modulating valve must be fitted.

Storage type – these are used to store hot water for manufacturing processes and/or washing demands. Unlike non-storage calorifiers, these have a low steam to water ratio, i.e. a relatively small battery of steam pipes surrounded by a large volume of water.

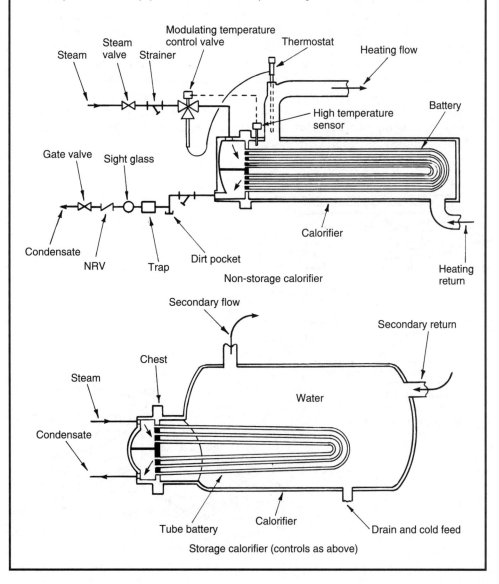

Non-storage calorifier

Storage calorifier (controls as above)

A district heating system is in principle an enlarged system of heating one building, extended to heat several buildings. It can be sufficiently large enough to heat a whole community or even a small town from one centralised boiler plant. Centralising plant and controls saves space in individual buildings. An effective plant management service will ensure the equipment is functioning to peak efficiency. Each building owner is required to pay a standing charge for the maintenance of plant and to subscribe for heat consumed through an energy metered supply, similar to other utilities. An energy meter differs from a capacity or volume meter by monitoring the heat energy in the water flow, as this will vary in temperature depending on the location of buildings. The boiler and associated plant should be located in close proximity to buildings requiring a high heat load, e.g. an industrial estate. Long runs of heating pipes are required and these must be well insulated. They are normally located below ground but may be elevated around factories. Systems can incorporate industrial waste incinerators operating in parallel with conventional boilers and may also use surplus hot water from turbine cooling processes in power stations or electricity generators. This is known as Combined Heat and Power.

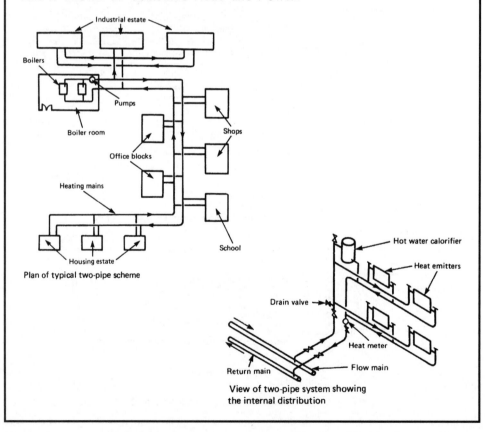

Plan of typical two-pipe scheme

View of two-pipe system showing the internal distribution

The three-pipe system is similar to the two-pipe system except for an additional small diameter flow pipe connected to the boilers. This is laid alongside the larger diameter flow pipe and has a separate circulation pump. This smaller flow pipe is used during the summer months when space heating is not required, although in the intermediate seasons it could supply both with limited application to heating. It should have enough capacity to supply the heating coils in the hot water storage cylinders plus a small reserve. It can be seen as an economy measure to reduce hot water heating volume, energy loss from the larger diameter pipe and pump running costs. A common large diameter return pipe can be used.

Pipes must be at least 450 mm below the surface as protection from vehicle loads. They must also be well insulated against heat loss and frost damage if water is not circulating. Insulation must be waterproof and the pipes protected from corrosion. Inevitably there will be some heat losses from the mains pipework. This will approximate to 15% of the system heating load.

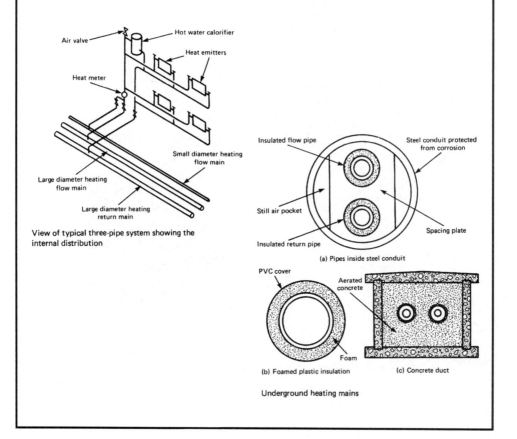

View of typical three-pipe system showing the internal distribution

(a) Pipes inside steel conduit

(b) Foamed plastic insulation

(c) Concrete duct

Underground heating mains

The four-pipe system supplies both hot water and space heating as two separate systems. Individual hot water storage cylinders are not required, as large capacity calorifiers are located in the boiler plant room and possibly at strategic locations around the district being served. This considerably simplifies the plumbing in each building as cold water storage cisterns are also unnecessary, provided all cold water outlets can be supplied direct from the main. However, the boiler plant room will be considerably larger to accommodate the additional components and controls. Excavation and installation costs will also be relatively expensive, but system flexibility and closure of the heating mains and associated boilers during the summer months should provide economies in use.

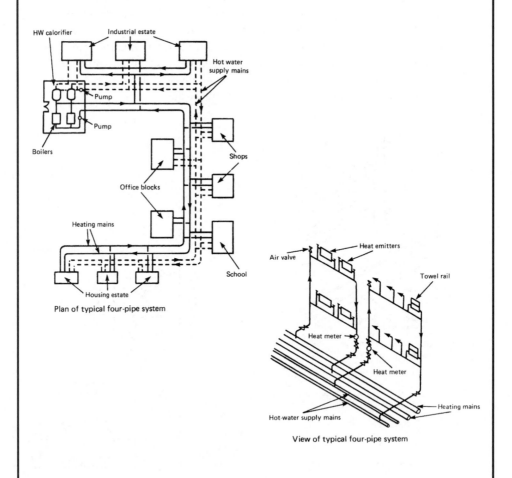

Plan of typical four-pipe system

View of typical four-pipe system

Potential for more economic use of electricity generating plant can be appreciated by observing the energy waste in the large plumes of condensing water above power station cooling towers. Most power stations are only about 50% efficient, leaving a considerable margin for reprocessing the surplus hot water.

Combining electricity generation with a supply of hot water has become viable since the deregulation and privatisation of electricity supply. Prior to this, examples were limited to large factory complexes and remote buildings, e.g. prisons, which were independent of national power generation by special licence. Until the technology improves, it is still only practical for large buildings or expansive collections of buildings such as university campuses and hospitals.

Surplus energy from oil- or gas-fired engine driven alternators occurs in hot water from the engine cooling system and the hot exhaust gases. In a CHP system the rate of heat energy produced is directly related to the amount of electricity generated. There will be times when available hot water is insufficient. Therefore a supplementary energy source from a conventional boiler will be required.

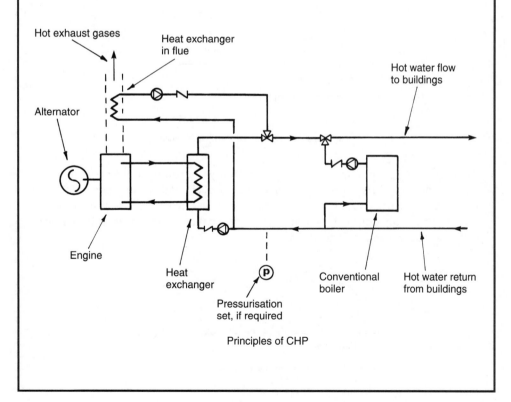

Principles of CHP

All pipe materials expand and contract when subject to temperature change. This linear change must be accommodated to prevent fatigue in the pipework, movement noise, dislocation of supports and damage to the adjacent structure.

Expansion devices:
• Natural changes in direction.
• Axial expansion bellows.
• Expansion loops.

Bellows and loops are not normally associated with domestic installations.

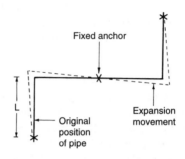

Natural changes in direction or offsets

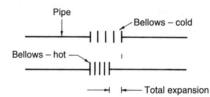

Axial expansion bellows responding to hot water

Bellows are factory-made fittings normally installed 'cold-drawn' to the total calculated expansion for hot water and steam services. The bellows can then absorb all anticipated movement by contraction. Where the pipe content is cold or refrigerated fluids, the bellows are compressed during installation.

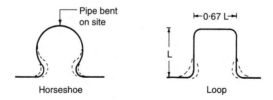

Site made loops or horseshoe

Coefficients of linear expansion for common pipework materials:

Material	Coeff. of expansion (m/mK × 10^{-6})
Cast iron	10·22
Copper	16·92
Mild steel	11·34
PVC (normal impact)	55·10
PVC (high impact)	75·10
Polyethylene (low density)	225·00
Polyethylene (high density)	140·20
ABS (acrylonitrile butadiene styrene)	110·20

E.g. An 80 mm diameter steel pipe of 20 m fixed length is subject to a temperature increase from 20°C to 80°C (60 K).

Formula:

Expansion = Original length × coeff. of expansion × Temp. diff.

$$= 20 \times 11{\cdot}34 \times 10^{-6} \times 60$$

$$= 0{\cdot}0136 \text{ m or } 13{\cdot}6 \text{ mm}$$

Single offset:

$$L = 100 \sqrt{zd}$$

L = see previous page

z = expansion (m)

d = pipe diameter (m)

$$L = 100 \sqrt{0{\cdot}0136} \times 0{\cdot}080 = 3{\cdot}30 \text{ m minimum.}$$

Loops:

$$L = 50 \sqrt{zd}$$

$$L = 50 \sqrt{0{\cdot}0136} \times 0{\cdot}080 = 1{\cdot}65 \text{ m minimum.}$$

Top of loop = 0·67 × L = 1·10 m minimum.

Notes:
- Provide access troughs or ducts for pipes in screeds (Chapter 14).
- Sleeve pipework through holes in walls, floors and ceilings (see pages 283 and 468 for fire sealing).
- Pipework support between fixed anchors to permit movement, i.e. loose fit brackets and rollers.
- Place felt or similar pads between pipework and notched joists.
- Branches to fixtures to be sufficient length and unconstrained to prevent dislocation of connections.
- Allow adequate space between pipework and structure.

Thermostatic Control of Heating Systems

Thermostatic control of heating and hot water systems reduces consumers' fuel bills, regulates the thermal comfort of building occupants and improves the efficiency of heat producing appliances. Approved Document L to the Building Regulations effects these provisions. This has the additional objective of limiting noxious fuel gases in the atmosphere and conserving finite natural fuel resources.

A room thermostat should be sited away from draughts, direct sunlight and heat emitters, at between 1·2 and 1·5 m above floor level. Thermostatic radiator valves may also be fitted to each emitter to provide independent control in each room. A less expensive means of controlling the temperature in different areas is by use of thermostatically activated zone valves to regulate the temperature of individual circuits.

Three-port thermostatic valves may be either mixing or diverting. The mixing valve has two inlets and one outlet. The diverting valve has one inlet and two outlets. Selection will depend on the design criteria, as shown in the illustrations.

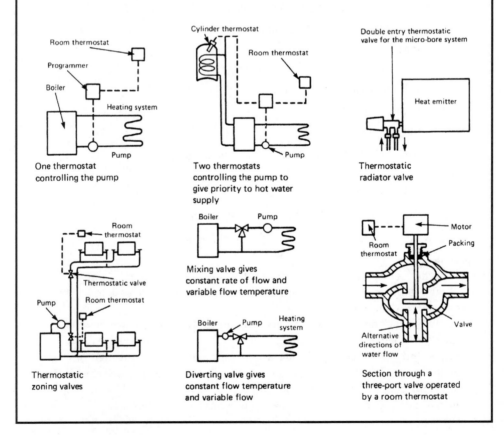

The diverter valve may be used to close the heating circuit to direct hot water from the boiler to the hot water cylinder. The reverse is also possible, depending on whether hot water or heating is considered a priority. With either, when the thermostat on the priority circuit is satisfied it effects a change in the motorised diverter valve to direct hot water to the other circuit.

A rod-type thermostat may be fitted into the hot water storage cylinder, or a surface contact thermostat applied below the insulation. At the pre-set temperature (about 60°C) a brass and invar steel strip expands to break contact with the electricity supply. A room thermostat also operates on the principle of differential expansion of brass and invar steel. Thermostatic radiator valves have a sensitive element which expands in response to a rise in air temperature to close the valve at a preset temperature, normally in range settings 5–27°C. Sensors are either a thermostatic coil or a wax or liquid charged compartment which is insulated from the valve body.

A clock controller sets the time at which the heating and hot water supply will operate. Programmers are generally more sophisticated, possibly incorporating 7 or 28-day settings, bypass facilities and numerous on/off functions throughout the days.

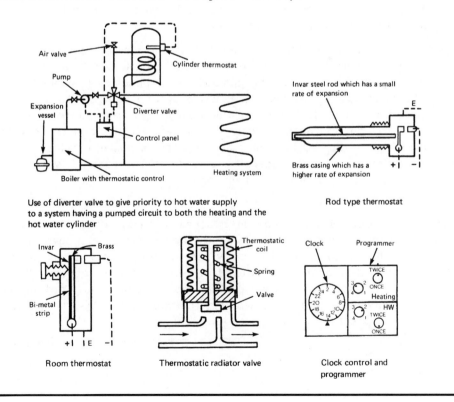

Use of diverter valve to give priority to hot water supply to a system having a pumped circuit to both the heating and the hot water cylinder

Rod type thermostat

Room thermostat

Thermostatic radiator valve

Clock control and programmer

Ref. Building Regulations, Approved Document L1: Conservation of fuel and power in dwellings –

From 2002 it has been mandatory in the UK to provide a higher standard of controls for hot water and heating installations. This is to limit consumption of finite fuel resources and to reduce the emission of atmospheric pollutants. All new installations and existing systems undergoing replacement components are affected.

Requirements for 'wet' systems –

- Only boilers of a minimum efficiency can be installed. See SEDBUK values on page 59.

- Hot water storage cylinders must be to a minimum acceptable standard, i.e. BSs 1566 and 3198: Copper indirect cylinders and hot water storage combination units for domestic purposes, respectively for vented systems. BS 7206: Specification for unvented hot water storage units and packages, for sealed systems. Vessels for unvented systems may also be approved by the BBA, the WRc or other accredited European standards authority. See pages 483 and 484.

- New systems to be fully pumped. If it is impractical to convert an existing gravity (convection) hot water circulation system, the heating system must still be pumped, i.e. it becomes a semi-gravity system, see pages 90 and 94. Existing system controls to be upgraded to include a cylinder thermostat and zone (motorised) valve to control the hot water circuit temperature and to provide a boiler interlock. Other controls are a programmer or clock controller, a room thermostat and thermostatic radiator valves (TRVs to BS EN 215) in the bedrooms.

Note: The boiler is said to be 'interlocked' when switched on or off by the room or cylinder thermostat (or boiler energy management system). The wiring circuit to and within the boiler and to the pump must ensure that both are switched off when there is no demand from the hot water or heating system, i.e. the boiler must not fire unnecessarily even though its working thermostat detects the water content temperature to be below its setting.

continued . . . .

Requirements for 'wet' systems (continued) –

- Independent/separate time controls for hot water and space heating. The exceptions are:

  (1) combination boilers which produce instantaneous hot water, and

  (2) solid fuel systems.

- Boiler interlock to be included to prevent the boiler firing when no demand for hot water or heating exists.

- Automatic by-pass valve to be fitted where the boiler manufacturer specifies a by-pass circuit.

  Note: A circuit by-pass and automatic control valve is specified by some boiler manufacturers to ensure a minimum flow rate whilst the boiler is firing. This is particularly useful where TRVs are used, as when these begin to close, a by-pass valve opens to maintain a steady flow of water through the boiler. An uncontrolled open by-pass or manually set by-pass valve is not acceptable as this would allow the boiler to operate at a higher temperature, with less efficient use of fuel.

- Independent temperature control in living and sleeping areas (TRVs could be used for bedroom radiators).

- Installations to be inspected and commissioned to ensure efficient use by the local authority Building Control Department or self-certified by a 'competent person', i.e. CORGI, OFTEC or HETAS approved (see page 60).

- System owners/users to be provided with equipment operating guides and maintenance instructions. This 'log-book' must be completed by a 'competent person'.

- Dwellings with over 150 m² living space/floor area to have the heating circuits divided into at least two zones. Each to have independent time and temperature control and to be included in the boiler interlock arrangement. A separate control system is also required for the hot water.

continued . . . .

Requirements for 'dry' systems –

- Warm air or dry systems (see page 98) should also benefit fully from central heating controls. Although gas-fired air heaters are not covered by SEDBUK requirements, these units should satisfy the following standards:
  BS EN 778: Domestic gas-fired forced convection air heaters for space heating not exceeding a net heat input of 70 kW, without a fan to assist transportation of combustion air and/or combustion products, or
  BS EN 1319: Domestic gas-fired forced convection air heaters for space heating, with fan-assisted burners not exceeding a net heat input of 70 kW.

- Replacement warm air heat exchanger units can only be fitted by a 'competent person'. All newly installed ducting should be fully insulated.

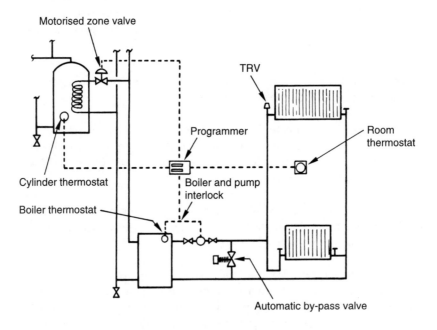

Typical semi-gravity system of hot water and heating controls

Schematic of control systems –

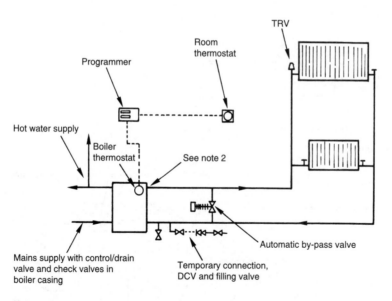

Cylinder thermostat

TRV

Room thermostat

Programmer

Boiler and pump interlock

Expansion valve and tundish

Boiler thermostat

Expansion vessel

Fill valve, double check valve and temporary connection

Automatic by-pass valve

3 port motorised zone valve

Typical fully pumped system of hot water and heating

TRV

Room thermostat

Programmer

Hot water supply

Boiler thermostat

See note 2

Automatic by-pass valve

Mains supply with control/drain valve and check valves in boiler casing

Temporary connection, DCV and filling valve

Notes:
1. Hot water draw off taps supplied direct from mains, through instantaneous water heater
2. Heating water is sealed. Additional components include heating pump and expansion vessel in boiler casing, with expansion valve and tundish (see upper diagram)

Typical combination boiler (see also page 43)

95

Optimum Start Controls – these have a control centre which computes the building internal temperature and the external air temperature. This is used to programme the most fuel efficient time for the boiler and associated plant to commence each morning and bring the building up to temperature ready for occupation. The system may also have the additional function of optimising the system shutdown time.

Compensated Circuit – this system also has a control centre to compute data. Information is processed from an external thermostat/ sensor and a heating pipework immersion sensor. The principle is that the boiler water delivery temperature is varied relative to outside air temperature. The warmer the external air, the cooler the system water and vice versa.

The capital cost of equipment for these systems can only be justified by substantial fuel savings. For large commercial and industrial buildings of variable occupancy the expenditure is worthwhile, particularly in the intermediate seasons of autumn and spring, when temperatures can vary considerably from day to day.

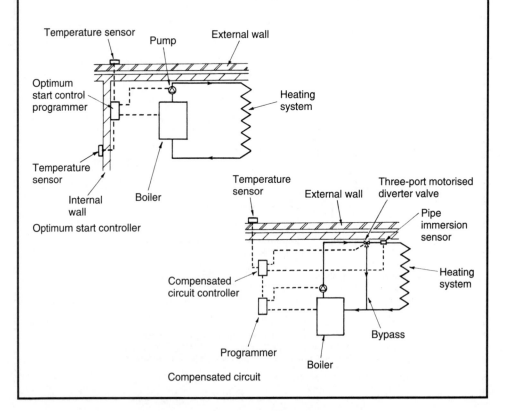

Optimum start controller

Compensated circuit

Energy management systems can vary considerably in complexity and degree of sophistication. The simplest timing mechanism to switch systems on and off at predetermined intervals on a routine basis could be considered as an energy management system. This progresses to include additional features such as programmers, thermostatic controls, motorised valves, zoning, optimum start controllers and compensated circuits. The most complex of energy management systems have a computerised central controller linked to numerous sensors and information sources. These could include the basic internal and external range shown schematically below, along with further processed data to include: the time, the day of the week, time of year, percentage occupancy of a building, meteorological data, system state feedback factors for plant efficiency at any one time and energy gain data from the sun, lighting, machinery and people.

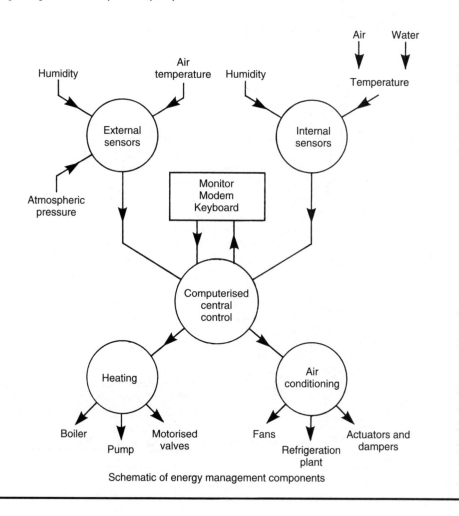

Schematic of energy management components

If there is sufficient space within floors and ceilings to accommodate ducting, warm air can be used as an alternative to hot water in pipes. There are no obtrusive emitters such as radiators. Air diffusers or grilles with adjustable louvres finish flush with the ceiling or floor. The heat source may be from a gas, oil or solid fuel boiler with a pumped supply of hot water to a heat exchanger within the air distribution unit. The same boiler can also be used for the domestic hot water supply. Alternatively, the unit may burn fuel directly, with air delivered around the burner casing. Control is simple, using a room thermostat to regulate heat exchanger and fan. The risk of water leakage or freezing is minimal, but air ducts should be well insulated to reduce heat losses. Positioning grilles in doors is an inexpensive means for returning air to the heater, but a return duct is preferred. Fresh air can be supplied to rooms through openable windows or trickle ventilators in the window frames. If rooms are completely sealed, fresh air should be drawn into the heating unit. The minimum ratio of fresh to recirculated air is 1:3.

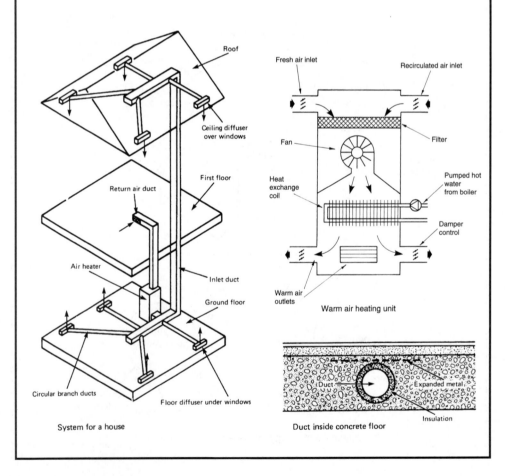

Roof

Ceiling diffuser over windows

First floor

Return air duct

Air heater

Inlet duct

Ground floor

Circular branch ducts

Floor diffuser under windows

System for a house

Fresh air inlet

Recirculated air inlet

Fan

Filter

Heat exchange coil

Pumped hot water from boiler

Damper control

Warm air outlets

Warm air heating unit

Duct

Expanded metal

Insulation

Duct inside concrete floor

The thermal transmittance rate from the inside to the outside of a building, through the intermediate elements of construction, is known as the 'U' value. It is defined as the energy in watts per square metre of construction for each degree Kelvin temperature difference between inside and outside of the building, i.e. W/m^2 K. The maximum acceptable 'U' values vary with building type and can be found listed in Approved Documents L1 and L2 to the Building Regulations.

Typical maximum 'U' values:

External walls . . . . 0·35

Pitched roof . . . . 0·16

Pitched roof containing a room . . . . 0·20

Flat roof . . . . 0·25

External floor . . . . 0·25

Windows, doors and rooflights . . . . 2·00 (ave.) Wood/uPVC

Windows, doors and rooflights . . . . 2·20 (ave.) Metal

Note: Windows, doors and rooflights, maximum 25% of floor area.

Non-domestic buildings also have a maximum 'U' value of 0·7 for vehicle access doors, along with the following requirements for windows and personnel doors:

Residential buildings – maximum 30% of exposed wall area.

Industrial and storage buildings – max. 15% of exposed wall area.

Places of assembly, offices and shops – maximum 40% of exposed wall area.

Rooflights – maximum 20% of rooflight to roof area.

E.g. A room in a dwelling house constructed to have maximum 'U' values has an external wall area of 30 m^2 to include 3 m^2 of double glazed window. Given internal and external design temperatures of 22°C and –2°C respectively, the heat loss through this wall will be:

Area × 'U' × temperature difference

Wall: 27 × 0·35 × 24 = 226·80

Window: 3 × 2·00 × 24 = 144·00
                            ———————
                            370·80 watts

A heat emitter should be capable of providing sufficient warmth to maintain a room at a comfortable temperature. It would be uneconomical to specify radiators for the rare occasions when external temperatures are extremely low, therefore an acceptable design external temperature for most of the UK is –1°C. Regional variations will occur, with a figure as low as –4°C in the north. The following internal design temperatures and air infiltration rates are generally acceptable:

Room	Temperature 0°C	Air changes per hour
Living	21	1·5
Dining	21	1·5
Bed/sitting	21	1·5
Bedroom	18	1·0
Hall/landing	18	1·5
Bathroom	22	2·0
Toilet	18	2·0
Kitchen	18	2·0

The study in the part plan shown below can be used to illustrate the procedure for determining heat losses from a room.

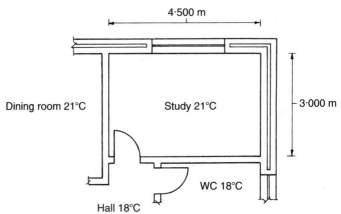

External design temperature –1°C
Room height = 2·3 m
Door area = 2 m^2
Window area = 1·5 m^2
Ventilation rate = 1·5 a/c per hour
Bedrooms above at 18°C

To determine the total heat loss or heating requirement for a room, it is necessary to obtain the thermal insulation properties of construction. For the room shown on the previous page, the 'U' values can be taken as:

External wall . . . . 0·35 W/m² K

Window . . . . 2·00

Internal wall. . . . . 2·00

Door. . . . . 4·00

Floor. . . . . 0·25

Ceiling. . . . . 2·50

Heat is also lost by air infiltration or ventilation. This can be calculated and added to the heat loss through the structure, to obtain an estimate of the total heating requirement.

Heat loss by ventilation may be calculated using the following formula:

$$\text{Watts} = \frac{\text{Room volume} \times \text{A/c per hour} \times \text{Temp. diff. (int. – ext.)}}{3}$$

Note: The lower denomination 3, is derived from density of air (1·2 kg/m³) × s.h.c. of air (1000 J/kg K) divided by 3600 seconds.

For the study shown on the previous page:

(4·5 × 3 × 2·3) × 1·5 × (21 – –1) divided by 3 = 341·55 watts

Heat loss through the structure is obtained by summating the elemental losses:

Element	Area (m²)		'U' value		Temp. diff. (int. – ext.)		Watts
External wall	15·75	×	0·35	×	22	=	121·28
Window	1·5		2·00		22		66
Internal wall	4·9		2·00		3		29·4
Door	2		4·00		3		24
Floor	13·5		0·25		22		74·25
Ceiling	13·5		2·50		3		101·25
							416·18

Total heat loss from the study = 341·55 + 416·18 = 757·73, i.e. 758 watts

Radiators are specified by length and height, number of sections, output in watts and number of panels. Sections refer to the number of columns or verticals in cast iron radiators and the number of corrugations in steel panel radiators. Panels can be single, double or triple. Design of radiators and corresponding output will vary between manufacturers. Their catalogues should be consulted to determine exact requirements. The following extract shows that a suitable single panel radiator for the previous example of 758 watts, could be:

> 400 mm high × 800 mm long × 20 sections (769 watts), or
> 500 mm high × 640 mm long × 16 sections (749 watts).

Selection will depend on space available. Over-rating is usual to allow for decrease in efficiency with age and effects of painting.

Height (mm)	Length (mm)	Sections	Watts (single)	Watts (double)
400	480	12	472	903
	640	16	621	1194
	800	20	769	1484
	960	24	915	1773
	1120	28	1059	2060
	1280	32	1203	2346
	1440	36	1346	2632
	1600	40	1488	2916
	1760	44	1630	3200
	1920	48	1770	3483
	2080	52	1911	3765
	2240	56	2051	4047
500	480	12	569	1063
	640	16	749	1406
	800	20	926	1748
	960	24	1102	2088
	1120	28	1276	2426
	1280	32	1449	2763
	1440	36	1621	3099
	1600	40	1792	3434
	1760	44	1963	3768
	1920	48	2133	4101
	2080	52	2302	4433
	2240	56	2470	4765

Note: Radiators are also manufactured in 300, 600, 750 and 900 mm standard heights.

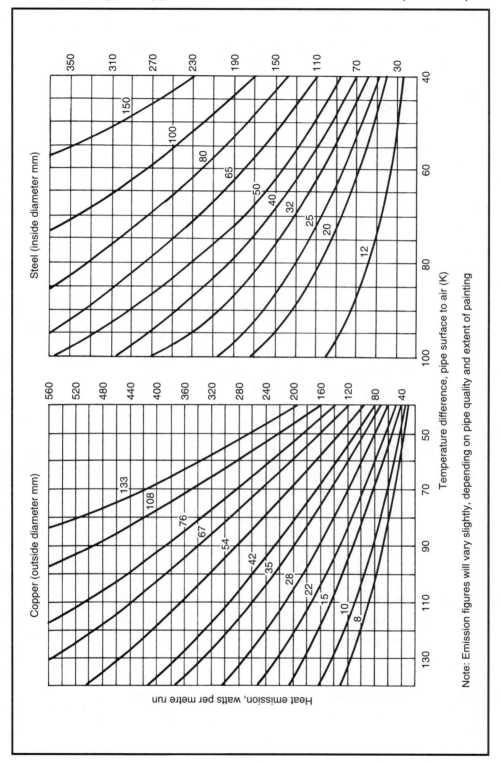

Steel (inside diameter mm)

Copper (outside diameter mm)

Heat emission, watts per metre run

Temperature difference, pipe surface to air (K)

Note: Emission figures will vary slightly, depending on pipe quality and extent of painting

To determine the overall boiler rating, the requirement for hot water (see Chapter 2) is added to that necessary for heating. Heating requirements are established by summating the radiator specifications for each of the rooms. To this figure can be added a nominal percentage for pipework heat losses, the amount depending on the extent of insulation.

E.g. if the total radiator output in a house is 18 kW and an additional 5% is added for pipework losses, the total heating requirement is:

$$18 + (18 \times 5/100) = 18 \cdot 9 \text{ kW.}$$

Given the manufacturer's data of 80% boiler efficiency, the boiler gross heat input will be:

$$18 \cdot 9 \times 100/80 = 23 \cdot 63 \text{ kW.}$$

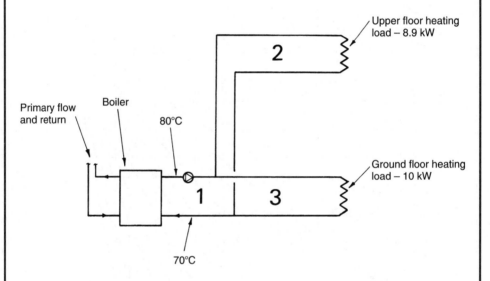

Pipes 1 – Heating flow and return at boiler

Pipes 2 – to upper floor

Pipes 3 – to ground floor

Schematic illustration, assuming a heating load of 8·9 kW on the upper floor and 10 kW on the ground floor, i.e. 18·9 kW total.

The size of pipework can be calculated for each sub-circuit and for the branches to each emitter. Unless emitters are very large, 15 mm o.d. copper tube or the equivalent is standard for connections to radiators in small bore installations. To illustrate the procedure, the drawing on the previous page allows for calculation of heating flow and return pipes at the boiler, and the supply pipes to each area of a house.

Pipes 1 supply the total heating requirement, 18·9 kW.
Pipes 2 supply the upper floor heating requirement, 8·9 kW.
Pipes 3 supply the lower floor heating requirement, 10 kW.

For each pair of pipes (flow and return) the mass flow rate is calculated from:

$$kg/s = \frac{kW}{S.h.c. \times temp. \, diff. \, (flow - return)}$$

Specific heat capacity (s.h.c.) can be taken as 4·2 kJ/kg K. The temperature differential between pumped heating flow and return will be about 10 K, i.e. 80°C – 70°C.

Therefore, the mass flow rate for:

$$Pipes \, 1 = \frac{18·9}{4·2 \times 10} = 0·45 \, kg/s$$

$$Pipes \, 2 = \frac{8·9}{4·2 \times 10} = 0·21 \, kg/s$$

$$Pipes \, 3 = \frac{10·0}{4·2 \times 10} = 0·24 \, kg/s$$

Selecting a pumped water velocity of 0·8 m/s (see page 55) and copper tube, the design chart on page 108 indicates:

Pipes 1 = 35 mm o.d.
Pipes 2 = 22 mm o.d.
Pipes 3 = 22 mm o.d.

The specification for a pump is very much dependent on the total length of pipework, summated for each section within a system. In existing buildings this can be established by taking site measurements. For new buildings at design stage, estimates can be taken from the architects' working drawings. Actual pipe lengths plus an allowance for resistance due to bends, tees and other fittings (see page 25), provides an effective length of pipework for calculation purposes.

Using the previous example, given that pipes 1, 2 and 3 are 6 m, 10 m and 12 m effective lengths respectively, the design chart shown on the following page can be used to determine resistance to water flow in each of the three sections shown:

Pressure drop in pipes 1 = 200 N/m^2 per metre (or pascals per metre).

Pressure drop in pipes 2 and 3 = 360 N/m^2 per metre (Pa per m).

Therefore: Pipes 1 @ 6 m × 200 Pa = 1200
               Pipes 2 @ 10 m × 360 Pa = 3600
               Pipes 3 @ 12 m × 360 Pa = 4320
                                    9120 Pa or 9·12 kPa

From this calculation, the pump specification is 0·45 kg/s at 9·12 kPa.

However, a higher figure for pump pressure will be necessary as the resistances in branch pipes to individual emitters will also need to be included. Pump selection is from manufacturer's pump performance charts similar to that shown on page 57.

Note: The smaller the pipe diameter, the greater the pressure drop or resistance to flow.

A simple and reasonably accurate estimate for determining boiler size.

Procedure –

- Establish dwelling dimensions and factor for location –

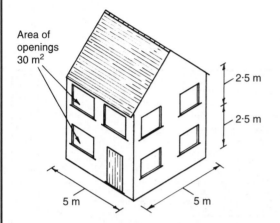

Area of openings 30 m²

2·5 m

2·5 m

5 m   5 m

UK location	Factor
North & Midlands	29
Scotland	28·5
South east	27
Wales	27
Northern Ireland	26·5
South west	25

Detached house, location south east

- Approximate heat losses:

  Openings area (30 m²) × Openings 'U' value (2·00 ave.)[*] = 60 (A).

  Gross wall area (100 m²) – Openings area (30 m² × Wall 'U' value (0·35)[*] – 24·5 (B).

  Roof length (5 m) × Roof width (5 m) × Roof 'U' value (0·16)[*] = 4 (C).

  Floor length (5 m) × Floor width (5 m) × Standard correction factor (0·7) = 17·5 (D).

  (For ceiling and floors in a mid-position flat, use zero where not exposed.)

- Summate fabric losses: A + B + C + D = 106.

- Multiply by location factor: 106 × 27 = 2862 watts.

- Calculate ventilation losses:

  Floor area (25 m²) × Room height (2.5 m) × No. of floors (2) = Volume (125 m³ × Standard ventilation correction factor (0·25) × Location factor (27) = 843·75 watts.

- Boiler input (net) rating = 2862 + 843·75 + 2000 (watts for hot water) + calcs. for any extension to building = 5706 watts or 5·71 kW.

[*]See page 99 for 'U' values.

Water Flow Resistance Through Copper Tube

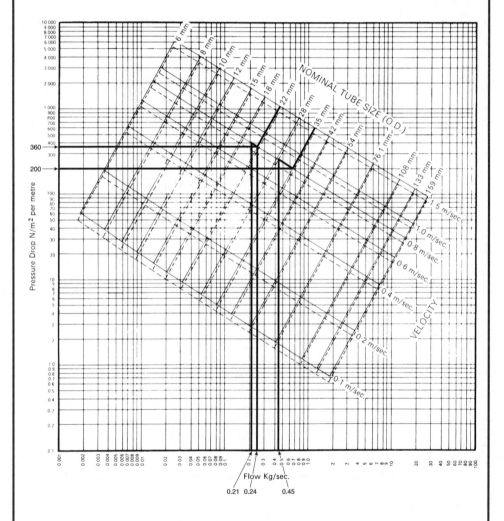

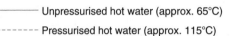 Unpressurised hot water (approx. 65°C)

------ Pressurised hot water (approx. 115°C)

Reproduced with the kind permission of the Copper Development Association.

4 FUEL CHARACTERISTICS AND STORAGE

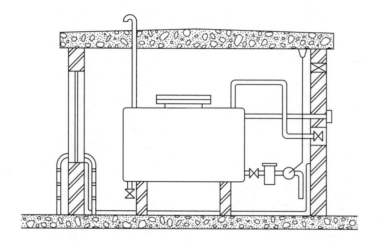

FUELS – FACTORS AFFECTING CHOICE

SOLID FUEL – PROPERTIES AND STORAGE

DOMESTIC SOLID FUEL BOILERS

SOLID FUEL – FLUES

OIL – PROPERTIES

OIL – STORAGE

OIL-FIRED BURNERS

OIL – FLUES

NATURAL GAS AND SUPPLY PROPERTIES

LPG – PROPERTIES AND STORAGE

ELECTRIC BOILER

ELECTRICITY – ELECTRODE BOILER

One of the most important considerations for providing an effective means of heating water is selection of an appropriate fuel. Choice and selection is a relatively new concept, as until the 1960s mains gas was rarely available outside of large towns and cities. Also, the cost of fuel oil was prohibitive for most people. The majority of domestic premises were heated by solid fuel for open fires with a back boiler for hot water. Solid fuel boilers for hot water and central heating were available, but the associated technology of pumps and thermostatic controls were rudimentary by today's standards. Systems of the time required considerable attention, not least frequent replenishment of fuel and disposal of ash. The post-1960s era led to much higher expectations in domestic comfort and convenience standards. This coincided with considerable developments in fuel burning appliances to complement the availability of new gas and oil resources from off-shore sources.

Practical factors and amenity issues may still limit or simplify choice, e.g. in some areas mains gas is not available and some buildings may have very limited space for fuel storage, or none at all. Personal preference as a result of previous experience, sales presentations or promotions may also have an important influence.

Amenity factors:

Facility to control the fuel, i.e. response to thermostatic and programmed automation.
Space for fuel storage.
Space for a boiler or special facilities to accommodate it.
Accessibility for fuel delivery.
Planning issues: chimneys and flue arrangements.
Location – conformity with Clean Air Act and exhaust emissions.
Maintenance requirements and after-care programme.
Availability.

Economic factors:

Capital cost of installation.
Cost of fuel storage facility.
Cost of special equipment.
Cost of equipment accommodation/plant room.
Cost of constructing a service area/access road.
Fuel costs – current and projected.
Flexibility of boiler, i.e. facility to change to another fuel.

Appropriate as logs of wood or as a coal product for open fires, stoves and boilers. A considerable amount of space is required for storage and manual handling is very much a feature. Arrangements must be made for fuel deliveries and disposal of ashes. Although the combustion efficiency is generally lower than oil or gas, some degree of automation is possible with the more efficient slow burning anthracites. Domestic boilers have several days' burning potential by gravity fed integral hopper. Instantaneous control is not possible and skilful operation is required to maintain boilers at low output.

Chimney construction and flue requirements must comply with Approved Document J to the Building Regulations. These are generally much larger and more visual than that required for use with other fuels. The sulphur content from burnt coal products is corrosive to many materials, therefore flue construction must not contain stainless steel linings or other materials which could be affected. The sulphur also contributes to atmospheric pollution.

Properties:

Fuel type	Calorific value MJ/kg	Sulphur content %	Bulk density* kg/m³
Anthracite[†]	33	1.0	750–800
Coking coal	30	1.0	
Dry steam coal[†]	30	1.1	
Strong caking coal	29	1.9	600–800
Medium caking coal	27	1.9	
Weak caking coal	26	1.9	
Non-caking coal	24	1.8	
Manufactured coke[†]	28	N/A	400–500
Wood	19	N/A	300–800

Notes:

* Variation depending on granular size. Unit size and species for wood.

[†] Smokeless fuels.

When solid fuel is to be used it is essential to consider
accommodation for fuel storage and facilities available. For domestic
and small buildings where requirements are minimal, a brick or
concrete bunker of nominal size is adequate. Industrial and
commercial premises will require a fuel bunker or hopper above the
boiler to reduce manual handling. Motorised feed mechanisms can be
used to regulate fuel delivery to the boilers and vacuum pumps can
effect extraction of ashes.

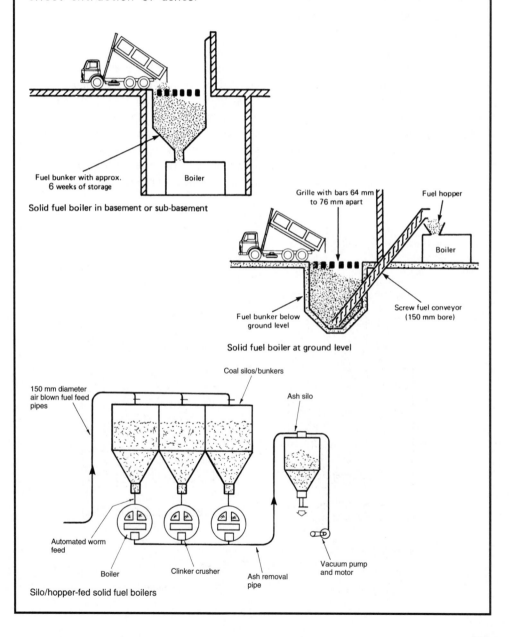

Fuel bunker with approx.
6 weeks of storage

Boiler

Solid fuel boiler in basement or sub-basement

Grille with bars 64 mm
to 76 mm apart

Fuel hopper

Boiler

Fuel bunker below
ground level

Screw fuel conveyor
(150 mm bore)

Solid fuel boiler at ground level

Coal silos/bunkers

150 mm diameter
air blown fuel feed
pipes

Ash silo

Automated worm
feed

Boiler

Clinker crusher

Ash removal
pipe

Vacuum pump
and motor

Silo/hopper-fed solid fuel boilers

Domestic Solid Fuel Boilers

Back boilers situated behind a fireplace are limited to providing room heat from the fire, hot water by gravity circulation to a storage cylinder and perhaps a couple of radiators or a towel rail off the primary flow and return. They were standard installations in many 1930s houses, but are now virtually obsolete. The combined room heater and boiler shown below is an improvement, having an enclosed fire and a convected outlet to heat the room in which it is installed. The water jacket is of sufficient capacity to provide hot water for storage and for several radiators. These appliances will require re-stoking every few hours.

Independent boilers are free standing, automatically fed by hopper and require only a flue. A chimney structure is not necessary, provided the flue satisfies Approved Document J to the Building Regulations. The integral fuel store contains small granules or 'peas' of anthracite and will require minimal attention with a burning capacity of several days. Automatic control is by thermostat in the water way to regulate a fan assisted air supply for complete combustion. These boilers are designed with sufficient capacity to provide hot water and central heating for most domestic situations.

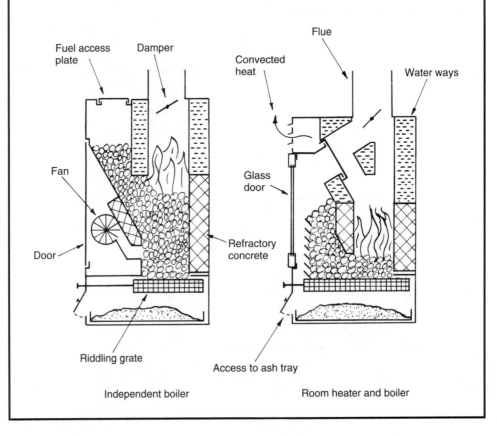

Independent boiler

Room heater and boiler

Flue pipes may be used to connect a solid fuel burning appliance to a chimney. They must not pass through a roof space, partition, internal wall or floor. Acceptable connecting flue pipe materials are:

- Cast iron to BS 41: Specification for cast iron spigot and socket flue or smoke pipes and fittings.
- Mild steel with a flue wall thickness of at least 3 mm, complying with BS 1449-1: Steel plate, sheet and strip.
- Stainless steel with a flue wall thickness of at least 1 mm, complying with BS EN 10 0088-1: Stainless steels, grades 1·4401, 1·4404, 1·4432 or 1·4436.
- Vitreous enamelled steel pipe complying with BS 6999: Specification for vitreous-enamelled low-carbon-steel flue pipes for solid-fuel-burning appliances with a maximum rated output of 45 kW.

All spigot and socket jointed pipes to be fitted socket uppermost and sealed with a non-combustible rope and fire cement or proprietory equivalent.

Any combustible material used in construction must be at least 200 mm from the inside surface of the flue. Where any metal fixings are in contact with combustible materials they must be at least 50 mm from the inside surface of a flue.

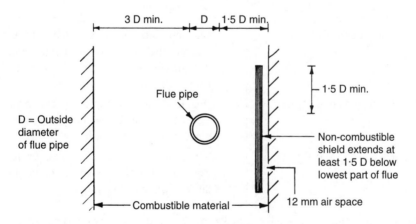

3 D min.   D   1·5 D min.

Flue pipe

D = Outside diameter of flue pipe

1·5 D min.

Non-combustible shield extends at least 1·5 D below lowest part of flue

Combustible material

12 mm air space

Other requirements for closeness of combustible material

Provisions for Solid Fuel Appliance Flues

Flue outlets must be above the roof line to effect clear, unhindered dispersal of combustion products without creating a fire hazard. See page 662 – 'Open Fire Places and Flues' in the *Building Construction Handbook*.

Flue length and height must be sufficient to encourage adequate draught and efflux (discharge) velocity at the terminal, with regard to limiting the possibility of condensation occurring in the flue. Flue gases cool relative to the flue pipe and surrounding structure temperature, until dew point of water occurs at about 60°C. Flue linings must therefore be impervious and resistant to corrosion. If condensation is a problem, a small diameter vertical drain can be located at the base of the flue.

Flue direction should be straight and vertical wherever possible. Horizontal runs are to be avoided. If the appliance has a back outlet connection an exception is made, but the horizontal flue length must not exceed 150 mm before connecting to a chimney or vertical flue. Bends should not exceed 45°C to the vertical to maintain a natural draught and to ease cleaning.

Flue size is never less than that provided on the appliance outlet.

Boiler, cooker or stove	Min. flue size
< 20 kW rated output	125 mm dia. or square/rectangular equivalent area, with a minimum dimension of 100 mm in straight flues and 125 mm in bends
20–30 kW rated output	150 mm dia. or square/rectangular equivalent area, with a minimum dimension of 125 mm
30–50 kW rated output	175 mm dia. or square/rectangular equivalent area, with a minimum dimension of 150 mm

Flue size in chimneys varies between 125 and 200 mm diameter (or square/rectangular equivalent) depending on application and appliance rating.

Appliances require air (oxygen) for efficient combustion of fuel. This requires purpose made ventilation openings in the structure, size depending on the appliance type and rating.

Appliance type	Permanently open ventilation
Boiler, cooker or stove with a flue draught stabiliser	300 mm^2/kW for the first 5 kW of rated output, 850 mm^2/kW thereafter
As above, without a flue draught stabiliser	550 mm^2/kW of rated output above 5 kW

E.g. A 20 kW boiler attached to a draught stabilised flue.

$$(300 \times 5) + (850 \times 15) = 14250 \text{ mm}^2$$

Taking the square root of 14250, indicates an open draught of at least 120 × 120 mm.

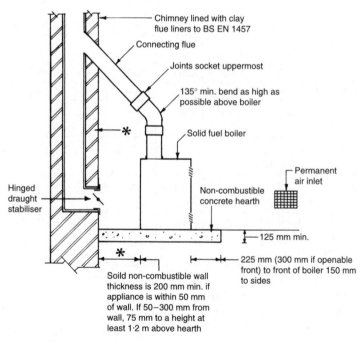

Chimney lined with clay flue liners to BS EN 1457

Connecting flue

Joints socket uppermost

135° min. bend as high as possible above boiler

Solid fuel boiler

Permanent air inlet

Non-combustible concrete hearth

Hinged draught stabiliser

125 mm min.

225 mm (300 mm if openable front) to front of boiler 150 mm to sides

Soild non-combustible wall thickness is 200 mm min. if appliance is within 50 mm of wall. If 50–300 mm from wall, 75 mm to a height at least 1·2 m above hearth

Solid fuel boiler and flue connections

Refs. Building Regulations, Approved Document J: Combustion appliances and fuel storage systems. Section 2.
BS 5854: Code of practice for flues and flue structures in buildings.

Fuel for boilers is manufactured by processing crude oil. The crude is distilled and condensed to produce a variety of commercial brands including gasolenes, kerosenes and gas oils. Distillates are blended to create several grades suitable as boiler fuels.

Kerosene (known commercially as Class C2) is an unblended relatively expensive light distillate suitable for domestic vaporising or atomising oil-fired boilers. Gas oil (Class D) is a heavier and less expensive distillate suitable for larger atomising burners in domestic and industrial applications. Fuel oils (Classes E, F, G and H) are a blend of residual oils with distillates that are considerably cheaper than the other classes. They are also heavier and generally require storage and handling plant with heating facilities. They require pre-heating before pumping and atomising for burning. These oils are limited to large-scale plant that has high level chimneys to discharge the pollutants and dirty flue gases characteristic of their high sulphur content.

Characteristics:

	Kerosene	Gas oil	Residue-containing burner fuels			
Class	C2	D	E	F	G	H
Density	790	840	930	950	970	990 kg/m^3
Flash point	38	56	66	66	66	66°C
Calorific value	46·4	45·5	43·4	42·9	42·5	42·2 MJ/kg
Sulphur content	0·2	0·2	3·5	3·5	3·5	3·5%
Kinematic viscosity	2·0	5·5	8·2	20	40	56*
Minimum storage temp.	N/A	N/A	10	25	40	45°C

Note: * Class C2 and D at 40°C.
         Classes E, F, G and H at 100°C.

Ref: BS 2869: Specification for fuel oils

An oil storage tank may be located externally. Unless the tank is underground or provided with a fire resistant barrier, it must be sited at least 1.8 m from the building. A plant room may be used if constructed in accordance with the local fire regulations. It must be built of fire resistant materials, with the base and walls to flood level rendered with cement mortar to contain all the oil in the tank plus 10% in the event of a leakage. Where the oil storage room is within a building, it should be totally enclosed with walls and floors of at least 4 hours' fire resistance.

As a guide to tank capacity, it should equate to the normal delivery plus 2 weeks' supply at maximum consumption or 3 weeks' average supply – take the greater. Supply pipelines can be as little as 8 or 10 mm o.d. annealed copper in coils to eliminate jointing. They can also be of steel for larger installations. Industrial supplies have the pipes insulated and trace wired to keep the oil warm. The tank should be elevated to provide at least 0·5 m head to effect the level controller or metering valve. If this is impractical, the supply can be pumped. The maximum head is 4 m.

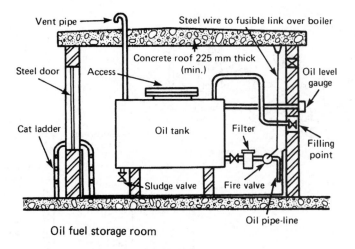

Oil fuel storage room

Refs: Environment Agency publication PPG 2: Above ground oil storage tanks.
BS 5410-1: Code of practice for oil firing – Installations up to 44 kW output capacity for space heating and hot water supply purposes.

Installation of Oil Tank and Oil Supply

An oil storage tank is usually rectangular with a raised top designed to shed water. Tanks for domestic application have a standard capacity of 2275 litres (2·275 m³) for economic deliveries of 2 m³. A vertical sight glass attached to the side provides for easy visual indication of the level. Tanks are made from ungalvanised welded carbon steel or sectional pressed ungalvanised carbon steel with internal strutting to prevent deformity when full. They are also produced in plastic. Brick piers or a structural steel framework is used to raise the tank above the ground. This is necessary to avoid corrosion from ground contact and to create sufficient head or pressure (0·5 m min.) from the outlet to the burner equipment. Location must be within 30 m of the oil tanker vehicle access point, otherwise an extended fill line must be provided.

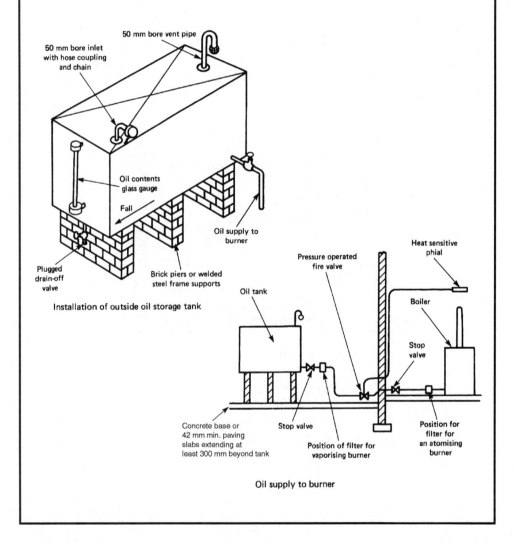

50 mm bore vent pipe

50 mm bore inlet with hose coupling and chain

Oil contents glass gauge

Fall

Plugged drain-off valve

Brick piers or welded steel frame supports

Installation of outside oil storage tank

Oil supply to burner

Pressure operated fire valve

Heat sensitive phial

Oil tank

Boiler

Stop valve

Stop valve

Concrete base or 42 mm min. paving slabs extending at least 300 mm beyond tank

Position of filter for vaporising burner

Position for filter for an atomising burner

Oil supply to burner

There are two types of oil burner: 1. vaporising; 2. atomising.

1. The natural draught vaporising burner consists of a cylindrical pot which is fed with oil at its base from a constant oil level controller. When the burner is lit, a thin film of oil burns in the bottom. Heat is generated and the oil is vaporised. When the vapour comes into contact with air entering the lowest holes, it mixes with the air and ignites. At full firing rate more air and oil mix until a flame burns out of the top of the burner.

2. The pressure jet atomising burner has an atomising nozzle. This produces a fine spray of oil which is mixed with air forced into the burner by a fan. Ignition electrodes produce a spark to fire this air/oil mixture.

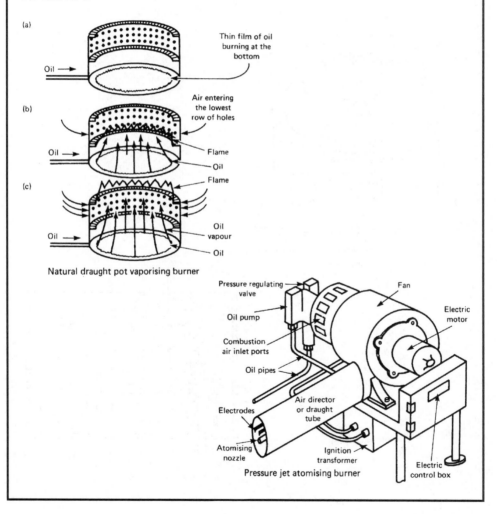

Natural draught pot vaporising burner

Pressure jet atomising burner

Wall-flame Oil Burner/Oil-level Controller

The wall-flame burner consists of a steel base plate securing a centrally placed electric motor. The armature of this motor is wound on a hollow metal shroud which dips into an oil well. A constant oil-level controller feeds the well, just covering the edge of the shroud. The shroud is circular with its internal diameter increasing towards the top, from which two holes connect with a pair of oil pipes. When the motor is engaged, oil is drawn up to the pipes and thrown onto the flame ring. Simultaneously, air is forced onto the rings by the fan. This air/oil mixture is ignited by the electrodes.

The constant oil-level controller is used to feed vaporising burners. If the inlet valve fails to close, oil flows into the trip chamber. The trip float rises and operates the trip mechanism, thus closing the valve.

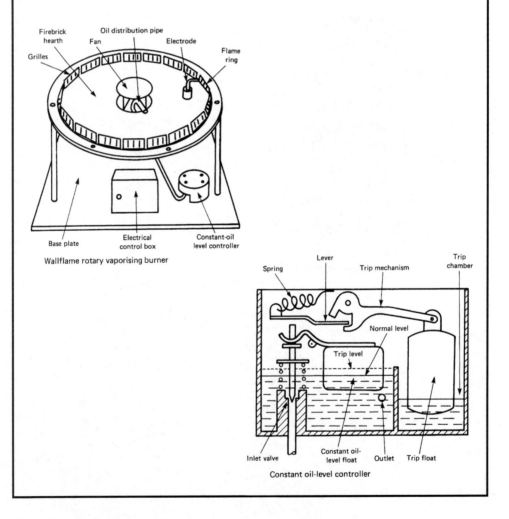

Wallflame rotary vaporising burner

Constant oil-level controller

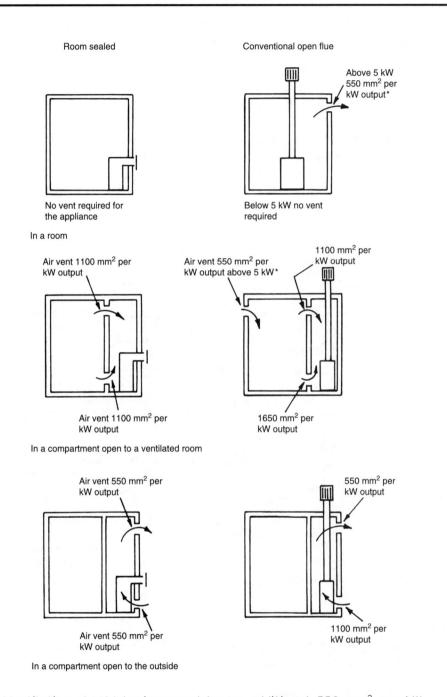

Room sealed

No vent required for
the appliance

In a room

Conventional open flue

Above 5 kW
550 mm^2 per
kW output*

Below 5 kW no vent
required

Air vent 1100 mm^2 per
kW output

Air vent 550 mm^2 per
kW output above 5 kW*

1100 mm^2 per
kW output

Air vent 1100 mm^2 per
kW output

1650 mm^2 per
kW output

In a compartment open to a ventilated room

Air vent 550 mm^2 per
kW output

550 mm^2 per
kW output

Air vent 550 mm^2 per
kW output

1100 mm^2 per
kW output

In a compartment open to the outside

* Ventilation should be increased by an additional 550 mm^2 per kW
output where the appliance has a draught break, i.e. a draught
stabiliser or draught diverter.

Outlets from flues serving oil-fired appliances, rated up to 45 kW output, must be carefully located to ensure:

- natural draught for fuel combustion
- efficient and safe dispersal of combusted fuel products
- adequate air intake if combined with a balanced flue.

In conjunction with the air inlet provisions shown on the previous page, the following guidance should ensure efficient combustion and burnt fuel gas dispersal.

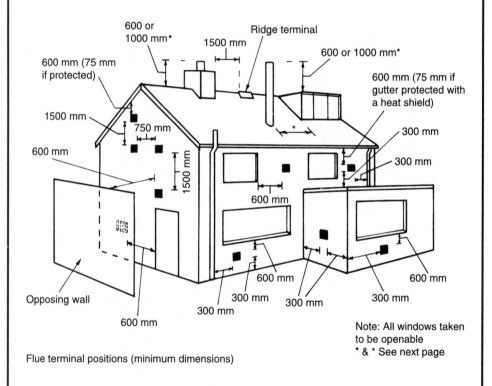

Flue terminal positions (minimum dimensions)

Note: All windows taken to be openable
• & * See next page

Ref. Building Regulations, Approved Document J: Combustion appliances and fuel storage systems. Section 4.

The following guidance provides minimum acceptable dimensions with regard to appliance efficiency, personnel and fire safety. The listing should be read with the illustration on the previous page. Local conditions such as wind patterns may also influence location of terminals. Flue terminal guards may be used as a protective barrier where direct contact could occur.

Location of terminal	Pressure jet atomising burner	Vaporising burner
Directly under an openable window or a ventilator	600	
Horizontally to an openable window or a ventilator	600	
Under eaves, guttering or drainage pipework	600	
As above, with a 750 mm wide heat shield	75	
Horizontally from vertical drain or discharge pipes	300	
Horizontally from internal or external corners	300	
Horizontally from a boundary	300	
Above ground or balcony	300	
From an opposing wall or other surface	600	
Opposite another terminal	1200	
Vertically from a terminal on the same wall	1500	
Horizontally from a terminal on the same wall	750	
From a ridge terminal to a vertical structure	1500	
Above the intersection with a roof	600	1000 •
Horizontally to a vertical structure	750	2300 ˙
Above a vertical structure < 750 mm (pressure jet burner) or < 2300 mm (vaporising burner) horizontally from a terminal	600	1000 ˙

Note: The vaporising burner column is annotated with an arrow reading "Not to be used in these situations".

Notes:
Dimensions in mm.
No terminal to be within 300 mm of combustible material.
Where a vaporising burner is used, the terminal should be at least 2300 mm horizontally from a roof.
See previous page for • and ˙.

Natural Gas – Properties

UK gas supplies originate from decaying organic matter found at depths up to 3 km below the North Sea. Extract is by drilling rigs and pipelines to the shore. On shore it is pressurised to about 5 kPa throughout a national pipe network.

Properties of natural gas:

Methane	89·5%
Ethane	4·5%
Propane	1·0%
Pentane	0·5%
Butane	0·5%
Nitrogen	3·5%
Carbon dioxide	0·5%

The composition shown will vary slightly according to source location. All the gases above are combustible except for nitrogen. Natural gas is not toxic, but incomplete combustion will produce carbon monoxide, hence the importance of correct burner and flue installations. A distinctive odour is added to the gas, as in its natural state it has no detectable smell. Natural gas is lighter than air with a specific gravity of about 0.6, relative to 1.0 for air.

Characteristics:

Calorific value	36–40 MJ/m^3
Specific gravity	0·5–0·7
Wobbe No.	approx. 50%
Sulphur	approx. 20 mg/m^3

Note: The Wobbe No. is sometimes used to represent the thermal input of an appliance for a given pressure and burner orifice. It is calculated from:

$$\frac{\text{Calorific value}}{\sqrt{\text{Specific gravity}}} \quad e.g. \quad \frac{40}{\sqrt{0\cdot6}} = 51\%$$

Natural gas has many advantages over other fuels, including: clean and efficient burning, no storage, less maintenance, relatively economic and a minimum of ancillaries.

LPGs are a by-product of the oil refining process. They are also found naturally in the north sea and other oil fields. These gases are liquefied in containers to about 1/200 of their volume as a gas by application of moderate pressure for convenience in transportation and storage. They are marketed as two grades, propane and butane, under various brand names. Both grades are heavier than air, therefore periphery walls around storage containers are unacceptable. If there were a leakage, the vapour would be trapped at low level and be unable to disperse. Calorific values differ considerably from natural gas, therefore appliances are not interchangeable. Siting of storage vessels should be away from buildings, boundaries and fixed sources of ignition as a precaution in event of fire.

Storage tank capacity (m³)	Min. distance from building or boundary (m)
<0·45	–
0·45–2·25	3·0
2·25–9·00	7·5
>9·00	15·0

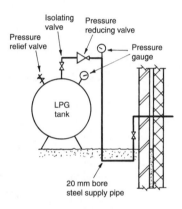

Containerised LPG

Characteristics:

Propane:

Calorific value	96 MJ/m³ (dry) 50 MJ/kg
Specific gravity	1·4–1·55
Sulphur content	0·02%
Air for combustion	24 m³ per m³ of gas

Butane:

Calorific value	122 MJ/m³ (dry) 50 MJ/kg
Specific gravity	1·9–2·1
Sulphur content	0·02%
Air for combustion	30 m³ per m³ of gas

Refs. Building Regulations, Approved Document J, Section 5: Provisions for liquid fuel storage and supply.
BS 5588-0: Fire precautions in the design, construction and use of buildings. Guide to fire safety codes of practice for particular premises/applications.

LPG may be stored below or above ground in tanks and above ground in cylinders. Tanks are provided in a standard volume of 2 or 4 m³ (2000 or 4000 litres capacity), sited no more than 25 m from a road or driveway for hose connection to the replenishment tanker. Cylinder location is less critical, these are in a set of 4 (47 kg each) for use two at a time, with a simple change over facility as required. Tanks and cylinders must not obstruct exit routes. Where a tank is located in the ground, it is fitted with sacrificial anodes to prevent decay by electrolytic activity.

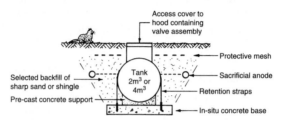

Below ground installation (section)

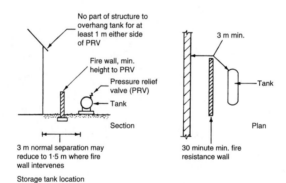

Storage tank location

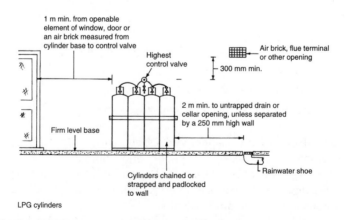

LPG cylinders

Electrically powered boilers have the advantage of no maintenance, no flue, over 99% efficiency* and no direct discharge of noxious gases.

* Energy loss is at the power station where conversion of fuel energy into electricity can be as little as 50% efficient.

Primary thermal store (> 15 litres capacity) – these use off-peak electricity, normally through a 3 kW immersion heater as an economic means for creating a store of hot water. They have the option of supplementary power at standard tariff through higher rated immersion heaters to satisfy greater demand.

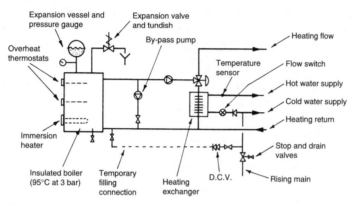

Primary store boiler

Instantaneous (< 15 litres capacity) – these low water content, high powered (6–12 kW) units provide direct heat energy at standard tariff in response to programmed demand. They are very compact, generally about 100 mm square x 1 m in height. Integral controls include a thermal safety cut-out and 'soft' switching to regulate power supply as the unit is engaged.

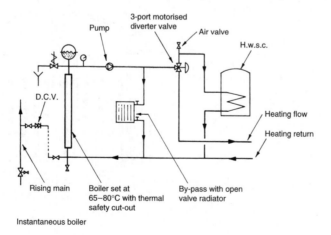

Instantaneous boiler

Electricity – Electrode boiler

Electricity can be used directly in convectors, fan heaters, element fires, etc., or indirectly as shown below as hot water thermal storage heating. It is an alternative use of off-peak electricity to storage in concrete floors or thermal block space heaters and has the advantage of more effective thermostatic control.

Electricity is converted to heat energy in water by an electrode boiler and stored in a pressurised insulated cylinder at about 180°C. The water is circulated by a pump programmed for daytime use to heat emitters in the building. Careful design of the storage vessel is essential to maintain sufficient thermal capacity for the heating requirements. An assessment of demand will need to be presented to the supply authority and a reduced rate of electricity tariff may be negotiated, possibly between 1900 and 0700 hours.

Calorific value of electricity . . . . 3·6 MJ/kWh

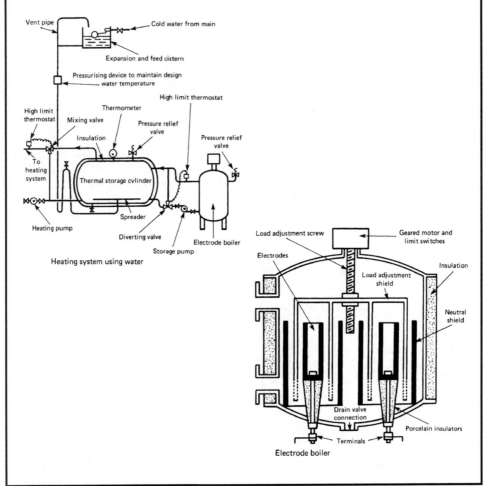

Heating system using water

Electrode boiler

5 VENTILATION SYSTEMS

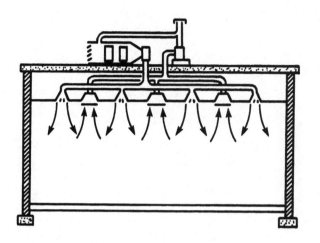

VENTILATION REQUIREMENTS

GUIDE TO VENTILATION RATES

DOMESTIC ACCOMMODATION

NON-DOMESTIC BUILDINGS

NATURAL VENTILATION

PASSIVE STACK VENTILATION

MECHANICAL VENTILATION

TYPES OF FAN

FAN LAWS

SOUND ATTENUATION IN DUCTWORK

AIR FILTERS

LOW VELOCITY AIR FLOW IN DUCTS

AIR DIFFUSION

VENTILATION DESIGN

DUCT SIZING

RESISTANCES TO AIR FLOW

Ventilation – a means of changing the air in an enclosed space to:

- Provide fresh air for respiration – approx. 0·1 to 0·2 l/s per person.
- Preserve the correct level of oxygen in the air – approx. 21%.
- Control carbon dioxide content to no more than 0·1%. Concentrations above 2% are unacceptable as carbon dioxide is poisonous to humans and can be fatal.
- Control moisture – relative humidity of 30% to 70% is acceptable.
- Remove excess heat from machinery, people, lighting, etc.
- Dispose of odours, smoke, dust and other atmospheric contaminants.
- Relieve stagnation and provide a sense of freshness – air movement of 0·15 to 0·5 m/s is adequate.

Measures for control:

Health and Safety at Work, etc. Act.
The Factories Act.
Offices, Shops and Railway Premises Act.
Building Regulations, Approved Document F – Means of Ventilation.
BS 5720 Code in Practice for Mechanical Ventilation and Air Conditioning in Buildings.

The statutes provide the Health and Safety Executive with authority to ensure buildings have suitably controlled internal environments. The Building Regulations and the British Standard provide measures for application.

Requirements for an acceptable amount of fresh air supply in buildings will vary depending on the nature of occupation and activity. As a guide. between 8 and 32 l/s per person of outdoor air supply can be applied between the extremes of non-smoking to very heavy smoking environments. Converting this to m^3/h (divide by 1000, multiply by 3600), equates to 29 to 115 m^3/h per person.

Air changes per hour or ventilation rate is the preferred criteria for system design. This is calculated by dividing the quantity of air by the room volume and multiplying by the occupancy.

E.g. 50 m^3/h, 100 m^3 office for five persons: 50/100 × 5 = 2·5 a/c per h.

Guide to Ventilation Rates

Room/building/accommodation	Air changes per hour
Assembly/entrance halls	3-6
Bathrooms (public)	6*
Boiler plant rooms	10-30[†]
Canteens	8-12
Cinema/theatre	6-10
Classrooms	3-4
Dance halls	10-12
Dining hall/restaurants	10-15
Domestic habitable rooms	approx. 1*
Factories/garages/industrial units	6-10
Factories – fabric processing	10-20
Factories (open plan/spacious)	1-4
Factories with unhealthy fumes	20-30
Foundries	10-15
Hospital wards	6-10
Hospital operating theatres	10-20
Kitchens (commercial)	20-60*
Laboratories	6-12
Laundries	10-15
Lavatories (public)	6-12*
Libraries	2-4
Lobbies/corridors	3-4
Offices	2-6
Smoking rooms	10-15
Warehousing	1-2

Notes:

* For domestic applications see pages 135 and 136.

[†] 18 air changes per hour is generally acceptable, plus an allowance of 0·5 l/s (1·8 m^3/h) per kW boiler rating for combustion air. Double the combustion allowance for gas boilers with a diverter flue.

See also: BS 5720: Code of practice for mechanical ventilation and air conditioning of buildings.

Approved Document F (Ventilation) provides the minimum requirements for comfortable background ventilation and for preventing the occurrence of condensation. It is effected without significantly reducing the high standards of thermal insulation necessary in modern buildings.

Definitions:

* Habitable room – any room used for dwelling purposes, not including a kitchen.
* Bathroom – any room with a bath and/or shower.
* Sanitary accommodation – any room with a WC.
* Ventilation opening – a means of ventilation, permanent or variable (open or closed) providing access to external air, e.g. door, window, louvre, air brick or PSV.
* PSV – passive stack ventilation is a system of vertical ducting from room ceilings to roof outlets providing ventilation by stack effect and wind passing over the roof.
* Rapid ventilation – openable window or mechanical fan system.
* Background ventilation – permanent vents, usually trickle ventilators set in a window frame (see below). An air brick with a sliding 'hit and miss' ventilator could also be used.

Note: With rapid and background ventilation, some part of the ventilation opening should be at least 1·75 m above the floor.

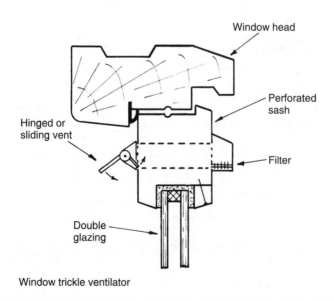

Window trickle ventilator

Habitable rooms – rapid ventilation by openable window at least equivalent to 1/20 floor area and background ventilation of 8000 mm² minimum.

Kitchen – rapid ventilation by a window opening (no minimum size) and mechanical extract ventilation of at least 60 l/s (30 l/s if incorporated in cooker hood) or PSV, plus background ventilation of 4000 mm² minimum.

Bathroom – rapid ventilation by a window opening (no minimum size) and mechanical extract ventilation of at least 15 l/s or PSV, plus background ventilation of 4000 mm² minimum.

Sanitary accommodation – rapid ventilation by openable window at least equivalent to 1/20 floor area or mechanical extract ventilation of at least 6 l/s. Background ventilation of 4000 mm² minimum.

Utility room – rapid ventilation by window opening (no minimum size) and mechanical extract ventilation of at least 30 l/s or PSV, plus background ventilation of 4000 mm² minimum.

Note: PSV may also be used in sanitary accommodation occupying an internal room, i.e. no external walls. Wherever a fan is fitted to an internal room, it must have a 15-minute overrun and an air inlet of, or equvalent to, a 10 mm gap under the door.

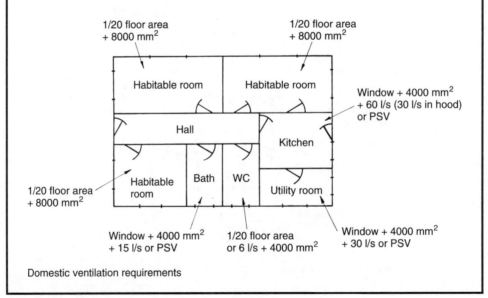

Domestic ventilation requirements

Approved Document F to the Building Regulations includes provisions for buildings other than domestic dwellings.

Definitions:
- Occupiable room - office, workroom, classroom, hotel bedroom, etc. Does not include bathrooms, sanitary accommodation, utility rooms, storage rooms, plant rooms and corridors.
- Kitchen - in this instance a facility with domestic-type appliances, not a kitchen designed for commercial purposes. See page 134 for guidance on commercial kitchens.

Occupiable room - rapid ventilation by openable window at least equivalent to 1/20 floor area. Background ventilation of 4000 mm^2 for floor areas up to 10 m^2 with an additional 400 mm^2 thereafter for every 1 m^2 of floor. Where a mechanical supply of fresh air is used, a minimum of 8 l/s is acceptable for non-smoking situations. For rooms with provision for smoking, see guidance on page 133.

Kitchen - rapid ventilation by a window opening (no minimum size) and mechanical extract ventilation of at least 30 l/s adjacent to a hob, or 60 l/s if provided elsewhere. Background ventilation of 4000 mm^2 minimum.

Bathroom and shower rooms - Rapid ventilation by window opening (no minimum size) and mechanical extract ventilation of at least 15 l/s per bath or shower.

Sanitary accommodation and general washing facilities - rapid ventilation by openable window at least equivalent to 1/20 floor area, or mechanical ventilation at 6 l/s per WC or three air changes per hour. Background ventilation of at least 4000 mm^2 per WC.

Notes: Kitchens and bathrooms with domestic-type facilities may use PSV as an alternative to mechanical extract. Requirements for internal kitchens, bathrooms and sanitary accommodation without windows can be satisfied with mechanical fan extract ventilation. Control is by light switch or sensor to include a 15-minute overrun. A room air inlet of, or equivalent to, a 10 mm gap under the door is also necessary.

Natural ventilation is an economic means of providing air changes in a building. It uses components integral with construction such as air bricks and louvres, or openable windows. The sources for natural ventilation are wind effect/pressure and stack effect/pressure.

Stack effect is an application of convected air currents. Cool air is encouraged to enter a building at low level. Here it is warmed by the occupancy, lighting, machinery and/or purposely located heat emitters. A column of warm air rises within the building to discharge through vents at high level, as shown on the following page. This can be very effective in tall office-type buildings and shopping malls, but has limited effect during the summer months due to warm external temperatures. A temperature differential of at least 10 K is needed to effect movement of air, therefore a supplementary system of mechanical air movement should be considered for use during the warmer seasons.

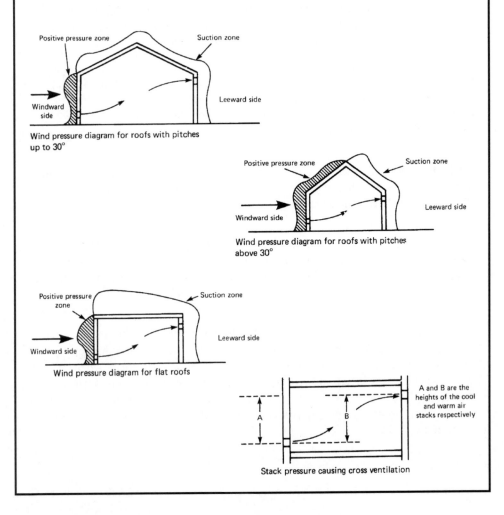

Wind pressure diagram for roofs with pitches up to 30°

Wind pressure diagram for roofs with pitches above 30°

Wind pressure diagram for flat roofs

A and B are the heights of the cool and warm air stacks respectively

Stack pressure causing cross ventilation

The rates of air change are determined by the building purpose and occupancy, and local interpretation of public health legislation. Public buildings usually require a ventilation rate of 30 m^3 per person per hour.

Wind passing the walls of a building creates a slight vacuum. With provision of controlled openings this can be used to draw air from a room to effect air changes. In tall buildings, during the winter months, the cool more dense outside air will tend to displace the warmer lighter inside air through windows or louvres on the upper floors. This is known as stack effect. It must be regulated otherwise it can produce draughts at low levels and excessive warmth on the upper floors.

Ventilation and heating for an assembly hall or similar building may be achieved by admitting cool external air through low level convectors. The warmed air rises to high level extract ducts. The cool air intake is regulated through dampers integral with the convectors.

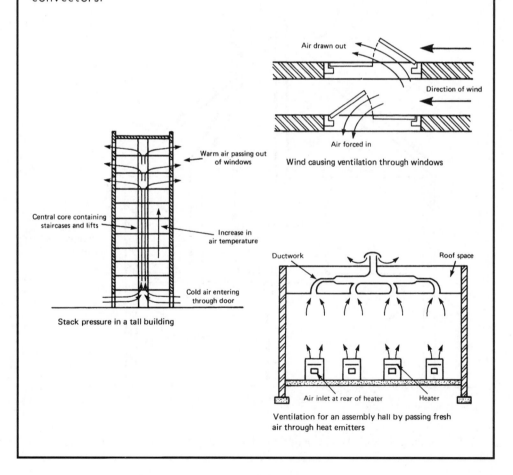

Air drawn out

Direction of wind

Air forced in

Wind causing ventilation through windows

Warm air passing out of windows

Central core containing staircases and lifts

Increase in air temperature

Cold air entering through door

Stack pressure in a tall building

Ductwork

Roof space

Air inlet at rear of heater

Heater

Ventilation for an assembly hall by passing fresh air through heat emitters

PSV consists of vertical or near vertical ducts of 100 to 150 mm diameter, extending from grilles set at ceiling level to terminals above the ridge of a roof. Systems can be applied to kitchens, bathrooms, utility rooms and sometimes sanitary accommodation, in buildings up to four storeys requiring up to three stacks/ducts. More complex situations are better ventilated by a Mechanical Assisted Ventilation System (MAVS), see next page.

PSV is energy efficient and environmentally friendly with no running costs. It works by combining stack effect with air movement and wind passing over the roof. It is self-regulating, responding to a temperature differential when internal and external temperatures vary.

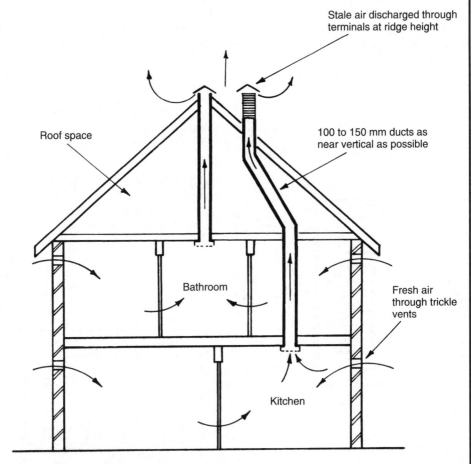

Stale air discharged through terminals at ridge height

Roof space

100 to 150 mm ducts as near vertical as possible

Bathroom

Fresh air through trickle vents

Kitchen

PSV to a dwelling house

Ref.: Building Regulations, Approved Document F1.

MAVS may be applied to dwellings and commercial premises where PSV is considered inadequate or impractical. This may be because the number of individual ducts would be excessive, i.e. too space consuming and obtrusive with several roof terminals. A low powered (40 W) silent running fan is normally located within the roof structure. It runs continuously and may be boosted by manual control when the level of cooking or bathing activity increases. Humidity sensors can also be used to automatically increase air flow.

MAVS are acceptable to Approved Document F1 of the Building Regulations as an alternative to the use of mechanical fans in each room. However, both PSV and MAVS are subject to the spread of fire regulations (Approved Document B). Ducting passing through a fire resistant wall, floor or ceiling must be fire protected with fire resistant materials and be fitted with a fusible link automatic damper.

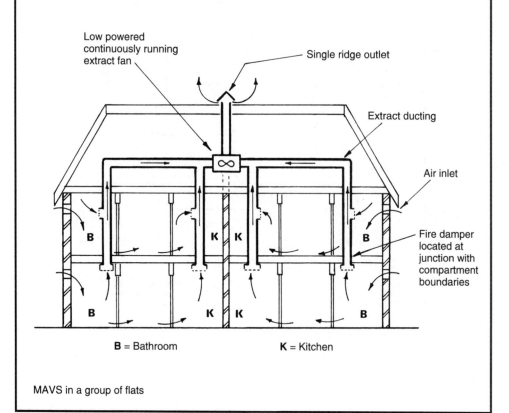

Low powered continuously running extract fan

Single ridge outlet

Extract ducting

Air inlet

Fire damper located at junction with compartment boundaries

B = Bathroom   **K** = Kitchen

MAVS in a group of flats

Mechanical Ventilation with Heat Recovery (MVHR)

MVHR is a development of MAVS to include energy recovery from the warmth in fan extracted moist air from bathrooms and kitchens. The heat recovery unit contains an extract fan for the stale air, a fresh air supply fan and a heat exchanger. This provides a balanced continuous ventilation system, obviating the need for ventilation openings such as trickle ventilators. Apart from natural leakage through the building and air movement from people opening and closing external doors, the building is sealed to maximise energy efficiency. Up to 70% of the heat energy in stale air can be recovered, but this system is not an alternative to central heating. A space heating system is required and MVHR can be expected to contribute significantly to its economic use. MVHR complies with the 'alternative approaches' to ventilation of dwellings, as defined in Approved Document F1 to the Building Regulations.

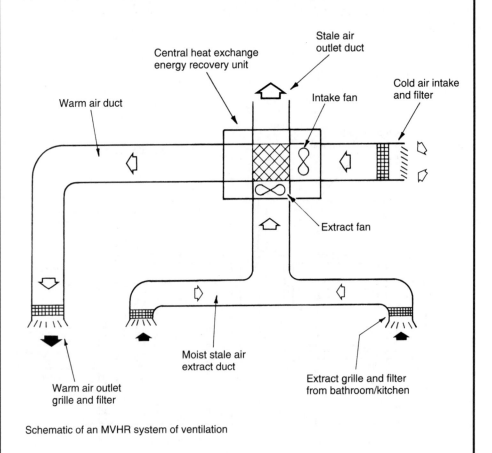

Schematic of an MVHR system of ventilation

Mechanical ventilation systems are frequently applied to commercial buildings, workshops, factories, etc., where the air change requirements are defined for health and welfare provision. There are three categories of system:

1. Natural inlet and mechanical extract
2. Mechanical inlet and natural extract
3. Mechanical inlet and mechanical extract

The capital cost of installing mechanical systems is greater than natural systems of air movement, but whether using one or more fans, system design provides for more reliable air change and air movement. Some noise will be apparent from the fan and air turbulence in ducting. This can be reduced by fitting sound attenuators and splitters as shown on page 147. Page 152 provides guidance on acceptable noise levels.

Internal sanitary accommodation must be provided with a shunt duct to prevent smoke or smells passing between rooms. In public buildings, duplicated fans with automatic changeover are also required in event of failure of the duty fan. Basement car parks require about 20 air changes per hour and should also be provided with duplicate fans with a fan failure automatic changeover facility.

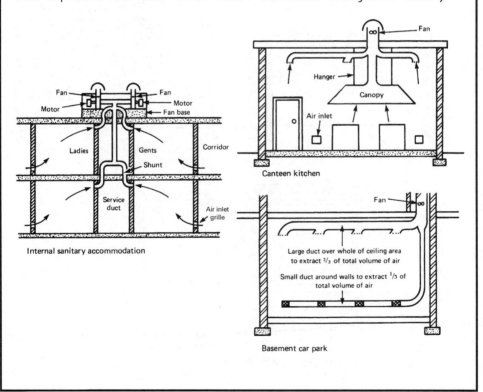

Internal sanitary accommodation

Canteen kitchen

Basement car park

Fan assisted ventilation systems supplying external air to habitable rooms must have a facility to pre-heat the air. They must also have control over the amount of air extracted, otherwise there will be excessive heat loss. A mechanical inlet and mechanical extract system can be used to regulate and balance supply and emission of air by designing the duct size and fan rating specifically for the situation.

Air may be extracted through specially made light fittings. These permit the heat enhanced air to be recirculated back to the heating unit. This not only provides a simple form of energy recovery, but also improves the light output by about 10%. With any form of recirculated air ventilation system, the ratio of fresh to recirculated air should be at least 1:3. i.e. min. 25% fresh, max. 75% recirculated.

In large buildings where smoking is not permitted, such as a theatre, a downward air distribution system may be used. This provides a uniform supply of warm filtered air.

Ductwork in all systems should be insulated to prevent heat losses from processed air and to prevent surface condensation.

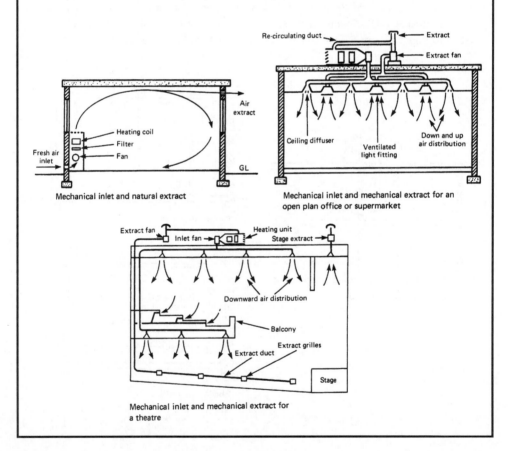

Mechanical inlet and natural extract

Mechanical inlet and mechanical extract for an open plan office or supermarket

Mechanical inlet and mechanical extract for a theatre

Propeller fan – does not create much air pressure and has limited effect in ductwork. Ideal for use at air openings in windows and walls.

Axial flow fan – can develop high pressure and is used for moving air through long sections of ductwork. The fan is integral with the run of ducting and does not require a base.

Bifurcated axial flow fan – used for moving hot gases, e.g. flue gases, and greasy air from commercial cooker hoods.

Cross-flow or tangential fan – used in fan convector units.

Centrifugal fan – can produce high pressure and has the capacity for large volumes of air. Most suited to larger installations such as air conditioning systems. It may have one or two inlets. Various forms of impeller can be selected depending on the air condition. Variable impellers and pulley ratios from the detached drive motor make this the most versatile of fans.

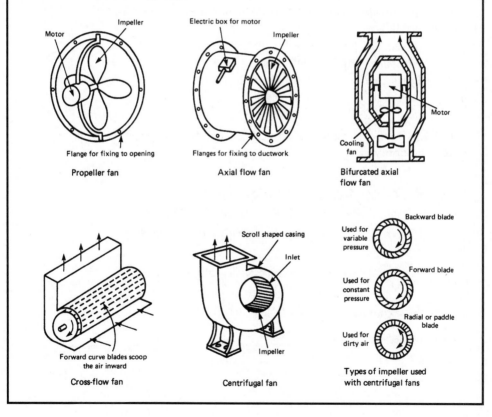

Fan Laws

Fan performance depends very much on characteristics such as type and configuration of components. Given a standard set of criteria against which a fan's performance is measured, i.e. 20°C dry bulb temperature, 101.325 kPa (1013 mb) atmospheric pressure, 50% relative humidity and 1.2 kg/m^3 air density, any variation in performance can be predicted according to the following fan laws:

- Discharge (volumetric air flow) varies directly with the fan speed.

$$Q2 = Q1(N2/N1)$$

- Fan pressure is proportional to the fan speed squared.

$$P2 = P1(N2/N1)^2$$

- Fan power is proportional to the fan speed cubed.

$$W2 = W1(N2/N1)^3$$

where: Q = air volume in m^3/s
       N = fan speed in rpm
       P = pressure in pascals (Pa)
       W = power in watts or kilowatts.

E.g. a mechanical ventilation system has the following fan characteristics:

Discharge (Q1) = 6 m^3/s
Pressure (P1)   = 400 Pa
Power (W1)   = 3 kW
Speed (N1)   = 1500 rpm

If the fan speed is reduced to 1000 rpm, the revised performance data will apply:

Discharge (Q2) = 6(1000/1500) = 4 m^3/s
Pressure (P2)   = 400(1000/1500)^2 = 178 Pa
Power (W2)   = 3000(1000/1500)^3 = 890 W

$$\text{Fan efficiency} = \frac{\text{Total fan pressure} \times \text{Air volume}}{\text{Power}} \times \frac{100}{1}$$

So, for this example: $\dfrac{178 \times 4}{890} \times \dfrac{100}{1} = 80\%$

Fans and air turbulence can be a significant noise source in air distribution systems. System accessories and fittings such as ductwork material, grilles/diffusers, mixing boxes, tee junctions and bends can compound the effect of dynamic air. Ducts of large surface area may need to be stiffened to prevent reverberation.

Fans may be mounted on a concrete base, with either cork, rubber or fibre pad inserts. Strong springs are an alternative. Duct connections to a fan should have a flexible adaptor of reinforced PVC.

Sound attenuation in ducting can be achieved by continuously lining the duct with a fire resistant, sound absorbing material. Where this is impractical, strategically located attenuators/silencers composed of perforated metal inserts or a honeycomb of sound absorbent material can be very effective. These have a dual function as system sound absorbers and as absorbers of airborne sound transmission from adjacent rooms sharing the ventilation system.

To prevent air impacting at bends, a streamlining effect can be achieved by fixing vanes or splitters to give the air direction.

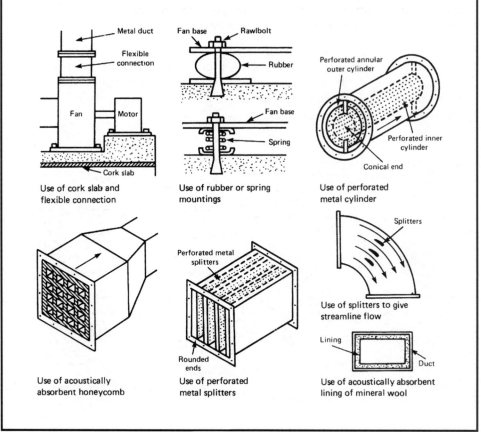

Use of cork slab and flexible connection

Use of rubber or spring mountings

Use of perforated metal cylinder

Use of acoustically absorbent honeycomb

Use of perforated metal splitters

Use of splitters to give streamline flow

Use of acoustically absorbent lining of mineral wool

Cell or panel – flat or in a vee formation to increase the surface contact area. Available in dry or wet (viscous) composition in disposable format for simple fitting within the ductwork. A rigid outer frame is necessary to prevent flanking leakage of dirty air. The dry type can be manufactured from dense glass paper in deep pleats arranged parallel to the air flow. Dry filters can be vacuum cleaned to extend their life, but in time will be replaced. The viscous filter is coated with an odourless, non-toxic, non-flammable oil. These can be cleaned in hot soapy water and recoated with oil.

Bag – a form of filtration material providing a large air contact area. When the fan is inactive the bag will hang limply unless wire reinforced. It will resume a horizontal profile during normal system operation. Fabric bags can be washed periodically and replaced.

Roller – operated manually or by pressure sensitive switch. As the filter becomes less efficient, resistance to air flow increases. The pressure effects a detector which engages a motor to bring down clean fabric from the top spool. Several perforated rollers can be used to vee format and increase the fabric contact area.

Activated carbon – disposable filter composed of carbon particles arranged to provide a large surface area. Used specifically above commercial cooker hoods as this material has very effective adsorption of hot greasy fumes.

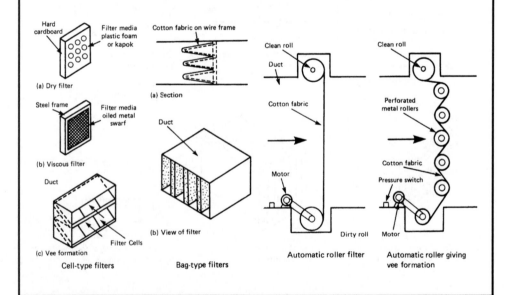

Viscous – these have a high dust retention capacity and are often specified for application to industrial situations. An improvement on the panel type has close spaced corrugated metal plates continuously sprayed with oil. A rotating variation has filter plates hung from chains. The lower plates in the cycle pass through a bath of oil which removes attached particles and resurfaces the plates with clean oil.

Electrostatic unit – this has an ionising area which gives suspended dust particles a positive electrostatic charge. These are conveyed in the air stream through metal plates which are alternately charged positive and earthed negative. Positively charged particles are repelled by the positive plates and attracted to the negative plates. The negative plates can also be coated with a thin layer of oil or gel for greater retention of dust. The unit can have supplementary, preliminary and final filters as shown below, giving an overall efficiency of about 99%.

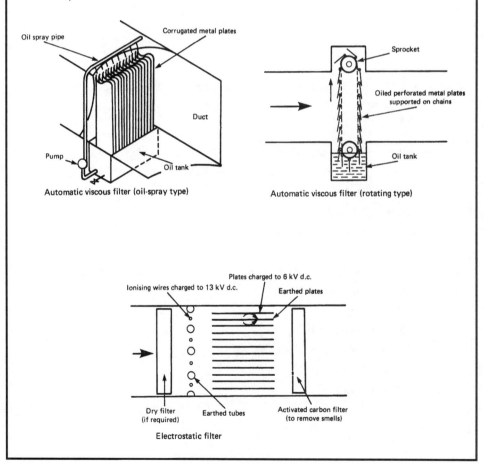

Oil spray pipe

Corrugated metal plates

Duct

Pump

Oil tank

Automatic viscous filter (oil-spray type)

Sprocket

Oiled perforated metal plates supported on chains

Oil tank

Automatic viscous filter (rotating type)

Plates charged to 6 kV d.c.

Ionising wires charged to 13 kV d.c.

Earthed plates

Dry filter (if required)

Earthed tubes

Activated carbon filter (to remove smells)

Electrostatic filter

Simple ducted air systems, typical of those serving internal WCs and bathrooms, operate at relatively low air velocity with little frictional resistance or pressure drop. In these situations the relationship between air flow and duct diameter can be expressed as:

$$Q = 6{\cdot}3 \times 10^{-7} \times \sqrt{d^5 \times h \div L}$$

where: Q = air flow rate in m^3/sec.
d = duct diameter in mm.
h = pressure drop in mm water gauge.
L = length of duct in metres.

To determine duct diameter from design input data, the formula is represented:

$$d = 305 \times \sqrt[5]{Q^2 \times L \div h}$$

E.g.   A 10 m long ventilation duct is required to provide air at $0{\cdot}10\,m^3$ at a pressure drop of $0{\cdot}15$ mm wg.

$0{\cdot}15$ mm = $1{\cdot}5$ pascals (Pa) (over 10 m of ducting)
    = $0{\cdot}015$ mm per m, or $0{\cdot}15$ Pa per m.
$d = 305 \times \sqrt[5]{(0{\cdot}10)^2 \times 10 \div 0{\cdot}15}$
$d = 305 \times \sqrt[5]{0{\cdot}6667}$
$d = 305 \times 0{\cdot}922 = 281$ mm diameter.

To check that the calculated diameter of 281 mm correlates with the given flow rate (Q) of $0{\cdot}10\ m^3$/sec:

$Q = 6{\cdot}3 \times 10^{-7} \times \sqrt{d^5 \times h \div L}$
$Q = 6{\cdot}3 \times 10^{-7} \times \sqrt{(281)^5 \times 0{\cdot}15 \div 10}$
$Q = 6{\cdot}3 \times 10^{-7} \times 162110$
$Q = 0{\cdot}102\ m^3$/sec

Diffusers – these vary considerably in design from standard manufactured slatted grilles to purpose-made hi-tech profiled shapes and forms compatible with modern interiors. The principal objective of air distribution and throw must not be lost in these designs.

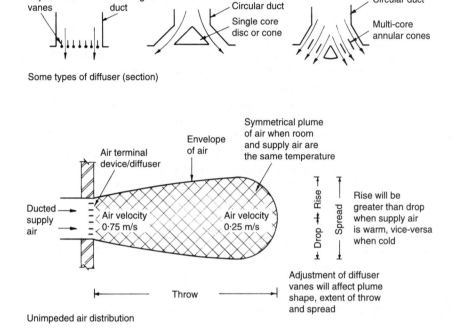

Some types of diffuser (section)

Unimpeded air distribution

Coanda effect – diffuser location must be selected to avoid unwanted draughts, air delivery impacting on beams, columns and other air deliveries. Where structural elements are adjacent, such as a wall and ceiling, the air delivery may become entrained and drawn to the adjacent surface. This can be advantageous as the plume of air throw, although distorted, may extend to run down the far wall as well.

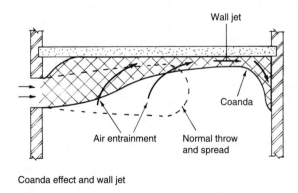

Coanda effect and wall jet

Ventilation Design – Air Velocity

Air velocity within a room or workplace should be between 0·15 and 0·50 m/s, depending on the amount of activity. Sedentary tasks such as desk work will fall into the range of 0·15 to 0·30 m/s, whilst more active assembly work, shopwork and manufacturing, between 0·30 and 0·50 m/s. These figures are designed to provide a feeling of freshness, to relieve stagnation without noise distraction from air movement equipment.

Conveyance of air and discharge through ducting and outlet diffusers will produce some noise. This should not be distracting and must be maintained at an unobtrusive level. As the extent of occupancy activity and/or machinery and equipment noise increases, so may the ducted air velocity, as background noise will render sound from air movement unnoticeable. For design purposes, the greater the ducted air velocity, the smaller the duct size and the less space consuming the ducting. However, some regard must be made for acceptable ducted air noise levels and the following table provides some guidance:

Situation	Ducted air velocity (m/s)
Very quiet, e.g. sound studio, library, study, operating theatres	1·5–2·5
Fairly quiet, e.g. private office, habitable room, hospital ward	2·5–4·0
Less quiet, e.g. shops, restaurant, classroom, general office	4·0–5·5
Non-critical, e.g. gyms, warehouse, factory, department store	5·5–7·5

Estimation of duct size and fan rating can be achieved by simple calculations and application to design charts. The example below is a graphical representation of the quantity of air (m³/s), friction or pressure reduction (N/m² per m) or (Pa per m) and air velocity (m/s) in circular ductwork. Conversion to equivalent size square or rectangular ductwork is shown on pages 157 and 158.

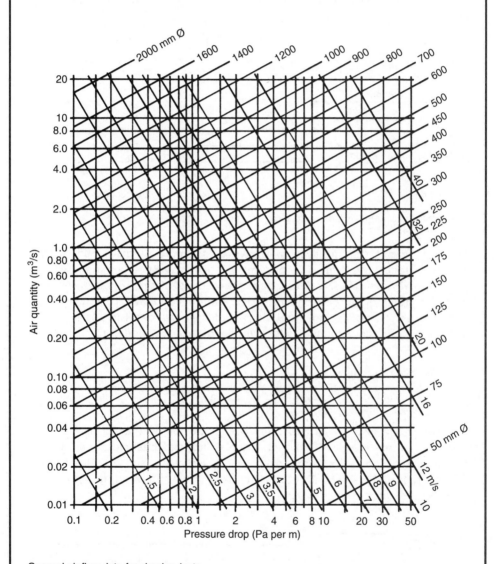

General air flow data for circular ducts

For mechanical supply and extract systems, the air volume flow rate or quantity of air can be calculated from the following formula:

$$Q(m^3/s) = \frac{Room\ volume \times Air\ changes\ per\ hour}{Time\ in\ seconds}$$

Air changes per hour can be obtained from appropriate legislative standards for the situation or the guidance given on pages 133 and 134.

E.g.

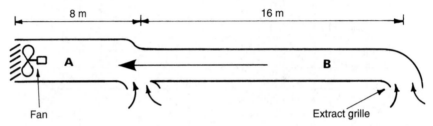

Room volume = 1800 m³, requiring six air changes per hour

The ducted extract air system shown is a simple straight run, with duct A effectively 8 m long and duct B effectively 16 m long. Where additional bends, tees, offsets and other resistances to air flow occur, a nominal percentage increase should be added to the actual duct length. Some design manuals include 'k' factors for these deviations and an example is shown on pages 159 and 160.

For the example given:

$$Q = \frac{1800 \times 6}{3600} = 3\ m^3/s$$

Disposition of extract grilles and room function will determine the quantity of air removed through each grille and associated duct. In this example the grilles are taken to be equally disposed, therefore each extracts 1·5 m³/s. Duct A therefore must have capacity for 3 m³/s and duct B, 1·5 m³/s.

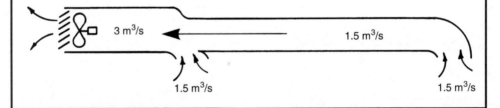

There are several methods which may be used to establish ventilation duct sizes, each having its own priority. The following shows three of the more popular, as applied to the design chart on page 153.

- Equal velocity – applied mainly to simple systems where the same air velocity is used throughout. For example, selected velocity is 7 m/s (see page 152), therefore the design chart indicates:

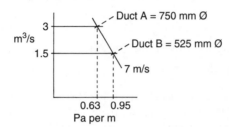

- Velocity reduction – air velocity is selected for the main section of ductwork and reduced for each branch. For example, selected air velocities for ducts A and B are 8 m/s and 5 m/s respectively:

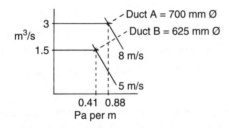

- Equal friction/constant pressure drop – air velocity is selected for the main section of ductwork. From this, the friction is determined and the same figure applied to all other sections. For example, selected air velocity through duct A is 7 m/s:

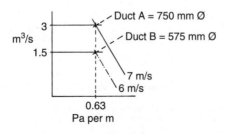

Using the example on page 154 with the equal velocity method of duct sizing shown on page 155, the fan will be required to extract $3\,m^3$ of air per second at a pressure of:

Duct (A) = 8 m × 0·63 Pa per m = 5·04 Pa
Duct (B) = 16 m × 0·95 Pa per m = 15·20 Pa

$$\overline{}$$

20·24 Pa (i.e. 20·25)

System pressure loss is calculated from: $k = P/Q^2$

where: k = pressure loss coefficient
P = pressure loss (Pa)
Q = air volume flow rate (m^3/s)

Therefore: $k = 20·25/3^2 = 2·25$

Using this coefficient, the system characteristic curve may be drawn between the operating air volume flow rate of $3\,m^3/s$ down to a nominal low operating figure of, say, $0.5\,m^3/s$. By substituting figures in this range in the above transposed formula, $P = k \times Q^2$ we have:

$P = 2·25 \times (0·5)^2 = 0·56\ Pa$ [$0·5\ m^3/s$ @ $0·56$ Pa]
$P = 2·25 \times (1.0)^2 = 2·25\ Pa$ [$1·0\ m^3/s$ @ $2·25$ Pa]
$P = 2·25 \times (1.5)^2 = 5·06\ Pa$ [$1·5\ m^3/s$ @ $5·06$ Pa]
$P = 2·25 \times (2.0)^2 = 9·00\ Pa$ [$2·0\ m^3/s$ @ $9·00$ Pa]
$P = 2·25 \times (2.5)^2 = 14·06\ Pa$ [$2·5\ m^3/s$ @ $14·06$ Pa]
$P = 2·25 \times (3.0)^2 = 20·25\ Pa$ [$3·0\ m^3/s$ @ $20·25$ Pa]

Plotting these figures graphically against fan manufacturers data will provide an indication of the most suitable fan for the situation:

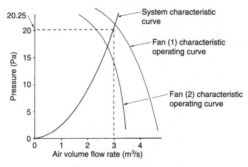

Select fan (1), as with variable settings this would adequately cover the system design characteristics.

Some ventilation design manuals limit data presentation to circular profile ductwork only. It is often more convenient for manufacturers and installers if square or rectangular ductwork can be used. This is particularly apparent where a high aspect ratio profile will allow ducting to be accommodated in depth restricted spaces such as suspended ceilings and raised floors.

Aspect ratio:

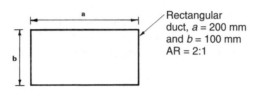

Rectangular duct, $a = 200$ mm and $b = 100$ mm AR = 2:1

The numerical relationship between dimension a to b. Square = 1:1.

Conversion of circular ductwork to square or rectangular (or vice versa) using the equal velocity of flow formula:

$$d = \frac{2ab}{a + b}$$

where: d = duct diameter
    a = longest dimension of rectangular duct
    b = shortest dimension of rectangular duct.

E.g. a 400 mm diameter duct to be converted to a rectangular profile of aspect ratio 3:1.

$$a = 3b$$

Substituting in the above formula:

$$400 = \frac{2 \times 3b \times b}{3b + b} = \frac{6b^2}{4b} = \frac{6b}{4}$$

Therefore:

$$b = \frac{4 \times 400}{6} = 267 \text{ mm}$$

$$a = 3b = 800 \text{ mm}$$

For equal volume of flow and pressure drop there are two possible formulae:

1.
$$d = 1.265 \times \left[\frac{(a \times b)^3}{a + b}\right]^{0.2}$$

2.
$$d = \left[\frac{32(a \times b)^3}{\pi^2(a + b)}\right]^{0.2}$$

Notes: 0.2 represents the 5th root of data in brackets. Formulae assume identical coefficient of friction occurs between circular and rectangular ducts, i.e. same material used.

E.g. circular duct of 400 mm diameter to be converted to rectangular having an aspect ratio of 3:1. Therefore, a = 3b.

Substituting in formula 1:

$$400 = 1.265 \times \left[\frac{(3b \times b)^3}{3b + b}\right]^{0.2}$$

$$400 = 1.265 \times \left[\frac{(3b^2)^3}{4b}\right]^{0.2}$$
From this, b = 216 mm

a = 3b = 648 mm

Substituting in formula 2:

$$400 = \left[\frac{32(3 \times b^2)^3}{\pi^2(3b + b)}\right]^{0.2}$$

$$400 = \left[\frac{3.242(27b^5)}{4}\right]^{0.2}$$
From this, b = 216 mm

a = 3b = 648 mm

There are many scientific applications to frictional or pressure losses created as air flows through ductwork. One of the most established is derived from Bernoulli's theorem of energy loss and gain as applied to fluid and air flow physics. Interpretation by formula:

$$h = k\left(\frac{V^2}{2g} \times \frac{\text{Density of air}}{\text{Density of water}}\right)$$

Where: h = head or pressure loss (m)

k = velocity head loss factor

V = velocity of air flow (m/s)

g = gravity factor (9·81)

density of air = 1·2 kg/m^3 @ 20°C and 1013 mb

density of water = 1000 kg/m^3

'k' factors have been calculated by experimentation using different ductwork materials. They will also vary depending on the nature of fittings, i.e. tees, bends, etc., the profile, extent of direction change, effect of dampers and other restrictions to air flow. Lists of these factors are extensive and can be found in ventilation design manuals. The following is provided as a generalisation of some mid-range values for illustration purposes only:

Duct fitting	Typical 'k' factor
Radiused bend (90°)	0·30
Mitred bend (90°)	1·25
Branch (tee) piece (90°)	0·40–1·70*
Branch (tee) piece (45°)	0·12–0·80*
Reductions (abrupt)	0·25
Reductions (gradual)	0·04
Enlargements (abrupt)	0·35
Enlargements (gradual)	0·20
Obstructions (louvres/diffusers)	1·50
Obstructions (wire mesh)	0·40
Obstructions (dampers)	0·20–0·50†

Notes:

* Varies with area ratios of main duct to branch duct.

† Varies depending on extent of opening.

Resistances to Air Flow – Calculations

E.g. Calculate the pressure loss in a 10 m length of 400 mm diameter ductwork containing four 90° radiused bends. Velocity of air flow is 5 m/s.

$$k = \text{four No. bends @ } 0\cdot30 = 1\cdot20$$

Bernoulli's formula:

$$h = 1\cdot2\left(\frac{5^2}{2 \times 9\cdot81} \times \frac{1\cdot2}{1000}\right)$$

$$h = 0\cdot00183 \text{ m or } 1\cdot83 \text{ mm or approx. 18 Pa.}$$

From the duct sizing chart on page 153, the pressure loss for a 400 mm diameter duct at 5 m/s is approximately 0·8 Pa per metre.

For 10 m of ductwork = 10 x 0·8 = <u>8 Pa.</u>

Total pressure loss = 18 Pa + 8 Pa = 26 Pa.

An alternative to the duct sizing chart for finding air flow resistance is application of another established fluid and air flow theorem attributed to D'Arcy. This can be used for pipe sizing as well as for sizing small ducts.

D'Arcys formula:

$$h = \frac{4fLV^2}{2gD} \times \frac{\text{Density of air}}{\text{Density of water}}$$

where: f = friction coefficient, 0·005–0·007 depending on duct material

L = length of duct (m)

D = duct diameter (m).

Using the above example of a 10 m length of 400 mm (0·4 m) ductwork conveying air at 5 m/s:

$$h = \frac{4 \times 0\cdot0052 \times 10 \times 5^2}{2 \times 9\cdot81 \times 0\cdot4} \times \frac{1\cdot2}{1000}$$

h = 0·0008 m or 0·8 mm or approx. <u>8 Pa.</u>

6 AIR CONDITIONING

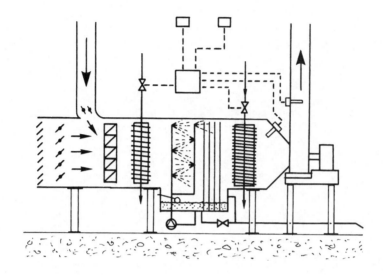

AIR CONDITIONING – PRINCIPLES

CENTRAL PLANT SYSTEM

AIR PROCESSING UNIT

HUMIDIFIERS

VARIABLE AIR VOLUME

INDUCTION (AIR/WATER) SYSTEM

FAN-COIL (AIR/WATER) UNIT AND INDUCTION DIFFUSER

DUAL DUCT SYSTEM

REFRIGERATION

COOLING SYSTEMS

PACKAGED AIR CONDITIONING SYSTEMS

PSYCHROMETRICS – PROCESSES AND APPLICATIONS

HEAT PUMPS

HEAT RECOVERY DEVICES

HEALTH CONSIDERATIONS

BUILDING RELATED ILLNESSES

Air conditioning is achieved by developing the principles of moving air in ducted ventilation systems to include a number of physical and scientific processes which enhance the air quality. The objective is to provide and maintain internal air conditions at a predetermined state, regardless of the time of year, the season and the external atmospheric environment. For buildings with human occupancy, the design specification is likely to include an internal air temperature of 19–23°C and relative humidity between 40 and 60%.

The following is a glossary of some of the terminology used in air conditioning design:

Dew point – temperature at which the air is saturated (100% RH) and further cooling manifests in condensation from water in the air.

Dry bulb temperature – temperature shown by a dry sensing element such as mercury in a glass tube thermometer (°C db).

Enthalpy – total heat energy, i.e. sensible heat + latent heat. Specific enthalpy (kJ/kg dry air).

Latent heat – heat energy added or removed as a substance changes state, whilst temperature remains constant, e.g. water changing to steam at 100°C and atmospheric pressure (W).

Moisture content – amount of moisture present in a unit mass of air (kg/kg dry air).

Percentage saturation – ratio of the amount of moisture in the air compared with the moisture content of saturated air at the same dry bulb temperature. Almost the same as RH and often used in place of it.

Relative humidity (RH) – ratio of water contained in air at a given dry bulb temperature, as a percentage of the maximum amount of water that could be held in air at that temperature.

Saturated air – air at 100% RH.

Sensible heat – heat energy which causes the temperature of a substance to change without changing its state (W).

Specific volume – quantity of air per unit mass (m³/kg).

Wet bulb temperature – depressed temperature measured on mercury in a glass thermometer with the sensing bulb kept wet by saturated muslin (°C wb).

This system is used where the air condition can be the same throughout the various parts of a building. It is also known as an all air system and may be categorised as low velocity for use in buildings with large open spaces, e.g. supermarkets, theatres, factories, assembly halls, etc. A variation could incorporate a heating and cooling element in sub-branch ductwork to smaller rooms such as offices. Very large and high rise buildings will require a high velocity and high pressure to overcome the resistances to air flow in long lengths of ductwork. Noise from the air velocity and pressure can be reduced just before the point of discharge, by incorporating an acoustic plenum chamber with low velocity sub-ducts conveying air to room diffusers.

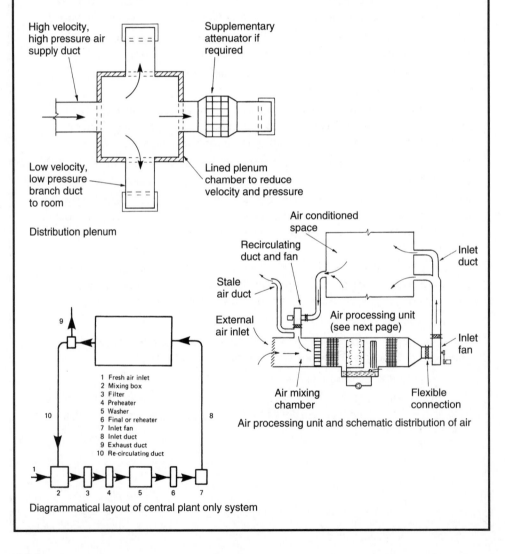

High velocity, high pressure air supply duct

Supplementary attenuator if required

Low velocity, low pressure branch duct to room

Lined plenum chamber to reduce velocity and pressure

Distribution plenum

Air conditioned space

Recirculating duct and fan

Inlet duct

Stale air duct

External air inlet

Air processing unit (see next page)

Inlet fan

1 Fresh air inlet
2 Mixing box
3 Filter
4 Preheater
5 Washer
6 Final or reheater
7 Inlet fan
8 Inlet duct
9 Exhaust duct
10 Re-circulating duct

Air mixing chamber

Flexible connection

Air processing unit and schematic distribution of air

Diagrammatical layout of central plant only system

Operation of the main air processing or air handling unit:

- Fresh air enters through a louvred inlet and mixes with the recirculated air. Maximum 75% recirculated to minimum 25% fresh air.
- The air is filtered to remove any suspended dust and dirt particles.
- In winter the air is pre-heated before passing through a humidifier. A spray wash humidifier may be used to cool the air up to dew point temperature. If a steam humidifier is used the air will gain slightly in temperature.
- In summer the air can be cooled by a chilled water coil or a direct expansion coil. The latter is the evaporator coil in a refrigeration cycle. Condensation of the air will begin, until at saturation level the air dehumidifies and reduces in temperature. Spray washing will also dehumidify the air.
- Air washers have zig-zag eliminator plates which remove drops of water and any dirt that may have escaped the filter.
- The final heater or reheater is used to adjust the supply air temperature and relative humidity before delivery through a system of insulated ductwork.

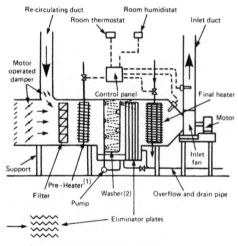

Section of main unit for the central plant system

Notes: (1) Pre-heater coil may be used with chilled water as a cooler in the summer months, but two separate coils are usually fitted.

(2) Steam humidifiers are the preferred replacement for spray wash humidifiers. The high temperature steam kills any bacteria.

Depending on the state of the air on entering a spray washer, it can be humidified or dehumidified. Humidification in the presence of moisture is understandable, but dehumidification is less easy to comprehend. It occurs when the spray is at a lower temperature than the air and the dewpoint of the air. In this condition the vapour pressure of the spray will be less than that of moisture in the air and some moisture from the air will transfer into the spray water. Hence, dehumidification.

Washers also remove some of the suspended dirt. Spray water pressure is usually between 200 and 300 kPa. Air velocity through the washer is between 2 and 2·5 m/s. Spray washers must be cleaned periodically and treated to neutralise any bacteria which could be living in the water. Water quality must also be monitored and findings documented. With numerous outbreaks of Legionnaires' disease originating from air conditioning systems, the Health and Safety Executive have identified these spray washers as a possible health risk.

Contemporary air processing units may incorporate steam injection humidifiers, but unlike washers, these should not be located immediately after the cooler coil. Here, the air will be close to saturation or even saturated (100% RH) and unable to accept further moisture. Therefore dry saturated steam at over 200°C is better injected into the air close to its final discharge.

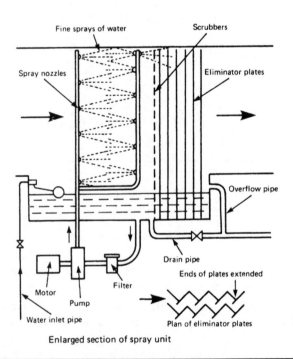

Enlarged section of spray unit

The VAV system has a central air processing unit to produce air at a specified temperature and relative humidity. The conditioned air from the main unit is conveyed in ductwork to ceiling diffusers which incorporate thermostatically controlled actuators. These can change the air volume to suit each room load. In a large room, several of these VAV ceiling units may be controlled by one room thermostat.

Several rooms/zones may have separate thermostats to control the air flow to each room. The inlet fan may have variable pitched impellers operated by compressed air. A pressure switch controls the pitch angle. Air distribution is usually medium to high velocity. The air temperature in each zone can be varied with the heat energy in the delivery air volume, but the system is only suitable for buildings having a fairly evenly distributed cooling load.

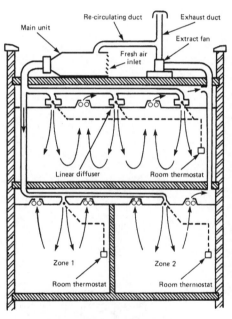

Layout of a typical variable air volume system

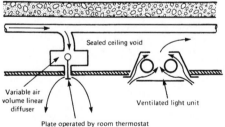

Note: The lighting fittings may require a fire damper

Section through plenum ceiling

Induction (Air/Water) System

Perimeter induction units – usually located under windows – blend primary air from the air processing unit with secondary air from within the room. The high velocity processed air delivery is induced into the unit through restrictive nozzles. This creates a negative pressure in its wake, drawing in the room secondary air for mixing and discharge. A damper regulates the volume of room air passing through a thermostatically controlled heating coil.

These coils may be used with chilled water as cooling coils in the summer months. If heating only is used, the system is known as the 'two-pipe induction system'. With the additional two pipes for cooling water, the system is known as the 'four-pipe change over induction system'. The latter system gives excellent control of the air temperature in various zones but is very capital intensive, therefore expensive to install.

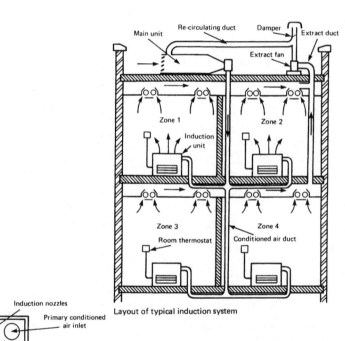

Layout of typical induction system

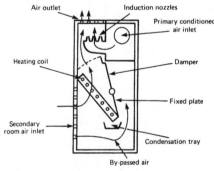

Section through an induction room unit

Fan-coil unit – an alternative discharge unit for application to the induction system shown on the previous page. Instead of nozzle injection of air, a low powered fan is used to disperse a mixture of primary and secondary air after reheating or cooling from an energy exchanger within the unit.

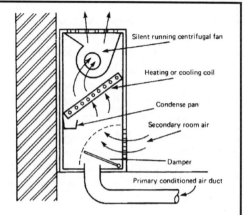

Section through a fan-coil room unit

Induction diffuser – another alternative which also uses a blend of recirculated room air with primary air. These locate at the end of branch ductwork and combine a diffuser with a simple primary and secondary air mixing chamber. The high velocity primary air mixes with low velocity secondary air drawn into a plenum ceiling from the room below. Light fitting extract grilles may be used to some advantage in this situation.

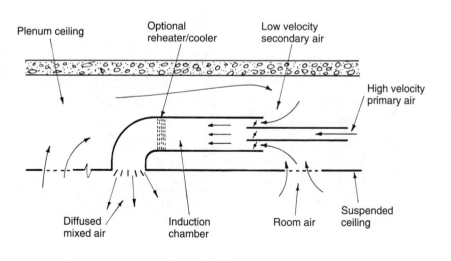

Section through an induction diffuser unit

Dual Duct System

The dual duct system is another means of providing varying air temperatures to different rooms in the same building. There is no water circulation to peripheral discharge units with terminal reheaters or coolers. This simplifies the plumbing installation as heating and cooling elements for each duct are located in the plant room. However, the system is space consuming and adequate provision must be made in suspended ceilings or raised flooring to accommodate both distribution ducts. The system is most energy economic when heating and cooling elements operate individually. For some of the year this will not be practical and simultaneous delivery of cold and hot air is provided for blending at the point of discharge.

Delivery is at high velocity with hot and cold air regulated by a damper connected to a room thermostat. A control plate in the mixing unit maintains constant air volume. As with all systems of air conditioning, fire dampers are required where the ductwork passes through compartment walls and floors.

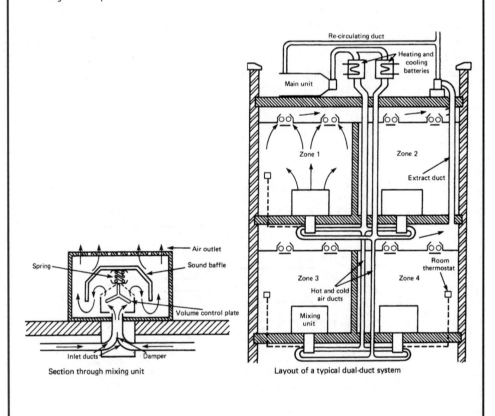

Section through mixing unit

Layout of a typical dual-duct system

Refrigeration systems are used to:

- Cool water for circulation through chiller coils. Brine may be used as a more efficient alternative to water.
- Directly chill air by suspending the cold evaporator coil in the air stream. When used in this way, the energy exchanger is known as a direct expansion (DX) coil.

The system most suited to air conditioning is the vapour compression cycle. It is a sealed pipe system containing refrigerant, compressor, condenser coil, expansion valve and evaporator coil, i.e. all the basic components of a domestic fridge.

Refrigerants are very volatile and boil at extremely low temperatures of –30 to –40°C. They are also capable of contributing to depletion of the ozone layer when released into the atmosphere. Dichlorodifluoromethane (R12), known as CFC, is used in many existing systems, but banned for new products. Monochlorodifluoromethane (R22), known as HCFC, is less ozone depleting. It is still used, whilst manufacturers research more environmentally friendly alternatives.

The refrigeration compression and evaporation cycle effects a change of temperature and state in the refrigerant, from liquid to gas and vice versa. Saturation pressure and temperature increase to emit heat at the condenser as heat energy is absorbed by the evaporator. As the liquid refrigerant changes to a gas through the expansion valve, it absorbs considerably more heat than during simple temperature change. This is known as the latent heat of vaporisation.

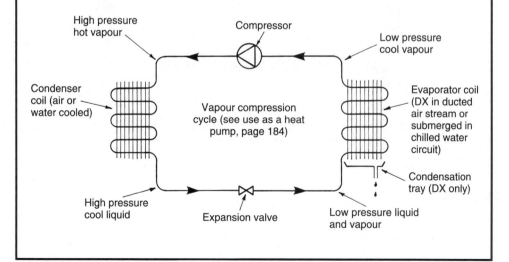

171

Efficient operation of refrigeration systems depends to a large extent on maintaining condenser temperature at an optimum level. This is necessary for correct reaction of the refrigerant. The cooling medium can be water or air. Water is more effective, but for practical purposes and health issues (see page 174), air cooling is becoming more widely used.

The condenser coil on a domestic fridge is suspended at the back of the unit and exposed to ambient air to cool. This same principle can be applied to small packaged and portable air conditioning units, possibly with the addition of a fan to enhance the cooling effect. Larger-scale air conditioning installations have several high powered fans to cool the condensers. These fans can be mounted horizontally or vertically to draw high velocity air through the condenser coils.

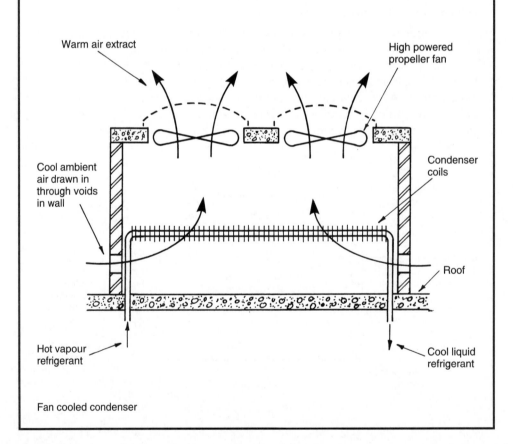

Fan cooled condenser

Natural draught water cooling can take many forms. The simplest and most inexpensive is a pond. Cooled water is drawn from one end and warm return water pumped into the other. Spray ponds are more efficient and may incorporate ornamental fountains as part of the process. Both have a tendency to accumulate debris and will require regular attention.

More common are evaporative atmospheric cooling towers. These are usually located on the building roof or within the roof structure plant room. Wall construction is louvred to permit crossflow of air. Internally the tower is either hollow or plastic baffled to increase the wetted contact area. Warm water from cooling the condenser is discharged through a bank of high level sprays to cool as it descends through the air draught. It is then recirculated to the condenser.

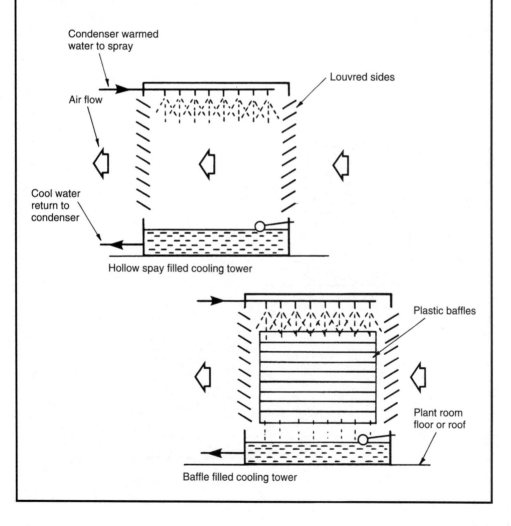

Hollow spay filled cooling tower

Baffle filled cooling tower

Mechanical fan draught cooling provides absolute control over the air supply, operating independently of fickle weather and wind direction. Fan draught cooling towers are of two types:

1. Forced draught – similar in construction and operating principle to the natural draught tower, but with one or more low level fans to force air through the tower.
2. Induced draught – a large high level fan draws or induces air flow through the tower. The relatively large single fan is more economic in use and less likely to generate system noise and vibration.

Note: All water cooling towers have become notorious as potential breeding areas for bacteria such as that associated with Legionnaires' disease. Therefore, towers must be maintained regularly and the water treated with a biocide, with regard to Workplace (Health, Safety and Welfare) Regulations 1992.

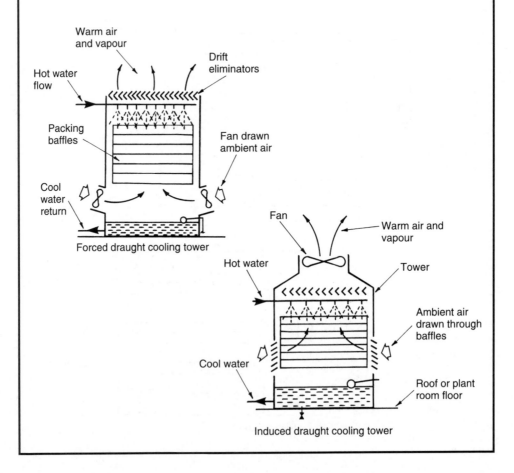

Forced draught cooling tower

Induced draught cooling tower

Packaged air conditioning systems are factory manufactured units, delivered to site for direct installation. They contain a vapour compression cycle refrigeration system, using the evaporator for cooling and the condenser for heating, with fan delivery of the processed air. They are available in a wide range of power capacity, fan output, refrigeration and heating load for adaptation to various building types and situations.

Small- to medium-sized buildings are best suited to these systems as it would be too costly and impractical to provide numerous units for use in multi-roomed large buildings. The smallest units (1–3 kW) are portable and free standing, simply plugging into an electrical wall socket. Larger, fixed units (generally 10–60 kW, but available up to 300 kW) can be unsightly and difficult to accommodate. These may be located in a store room and have short ductwork extensions to adjacent rooms.

Packages contain all the processes of conventional air handling units, with the exception of a steam or water humidifier. Humidification is achieved with condensation from the direct expansion (DX) refrigeration coil suspended in the air intake.

For summer use, the cold (DX) coil cools incoming and recirculated air. The hot condenser coil is fan cooled externally. For winter use, the refrigeration cycle is reversed by a changeover valve to become a heat pump – see page 184. Now the cold incoming air is warmed or pre-heated through the hot condenser coil and may be further heated by an electric element or hot water coil at the point of discharge.

System types:

• Self-contained (single) package.

• Split (double) package.

Self-contained (single) package – suitable for relatively small rooms, e.g. shops, restaurants and classrooms. May be free standing or attached to the structure.

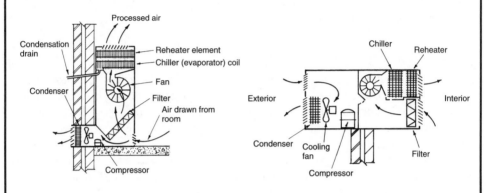

Single duct packaged unit   Single package wall opening unit

Split (double) package – two separate units. One contains fan, filter, evaporator and expansion valve for interior location. The other contains condenser, fan and compressor for external location. The two link by refrigeration pipework. This has the advantage that one external unit can serve several interior units.

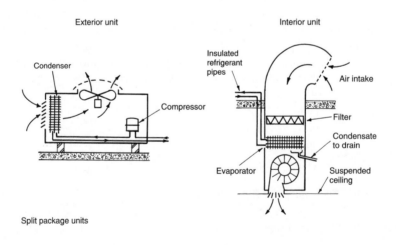

Split package units

Psychrometry – the science of moist air conditions, i.e. the characteristics of mixed air and water vapour. This can be calculated or design manuals consulted for tabulated information. Graphical psychrometric details are also available for simplified presentation of data. The chart outlined below is based on the calculated interrelationship of air properties at varying temperatures and conditions. In more detailed format, reasonably accurate design calculations can be applied. These are based on the processes shown plotted on the next page.

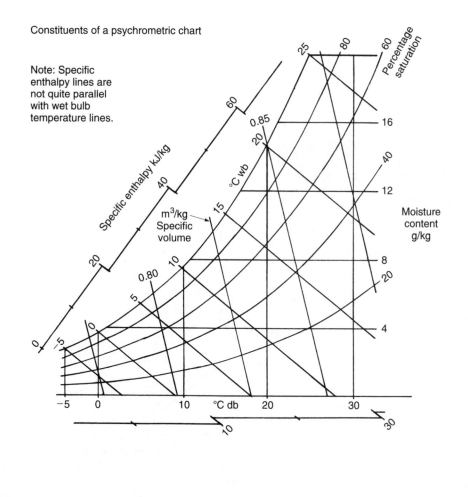

Constituents of a psychrometric chart

Note: Specific enthalpy lines are not quite parallel with wet bulb temperature lines.

To locate a representative air condition on the psychrometric chart, two properties of the air must be known. The easiest coordinates to obtain are the dry and wet bulb temperatures. These can be measured from a sling psychrometer, also known as a whirling or sling hygrometer. Two mercury-in-glass thermometers are mounted in a frame for rotation about the handle axis. One thermometer bulb has a wetted muslin wick. After rotation, the wet bulb temperature will be lower than the dry bulb due to the evaporation effect of moisture from the muslin. The extent of evaporation will depend on the moisture content of the air.

For example, a sling psychrometer indicates 10°C db and 5°C wb temperatures. From the chart the following can be determined:

Percentage saturation = 42%
Moisture content = 3·3 g/kg dry air
Specific volume = 0·805 m³/kg
Specific enthalpy = 18·5 kJ/kg

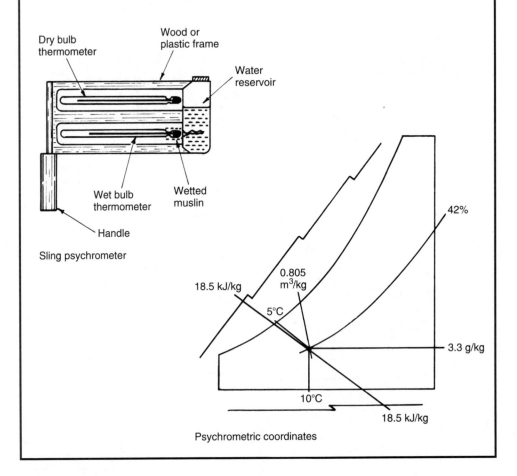

Dry bulb thermometer
Wood or plastic frame
Water reservoir
Wet bulb thermometer
Wetted muslin
Handle
Sling psychrometer

42%
0.805 m³/kg
18.5 kJ/kg
5°C
3.3 g/kg
10°C
18.5 kJ/kg

Psychrometric coordinates

Treatment of air is based on heating, cooling, humidification and dehumidification. These processes can be represented by lines drawn on the psychrometric chart.

- Heating (sensible) is depicted by a horizontal line drawn left to right. Dry bulb temperature increases with no change in moisture content, but there is a reduction in percentage saturation.
- Heating (latent) is the effect of steam humidification and is represented by a rising vertical line. Dry bulb temperature remains the same, moisture content and percentage saturation increase.
- Cooling (sensible) is depicted by a horizontal line drawn right to left. Dry bulb temperature decreases with no change in moisture content. Cooling by water spray humidifier is represented by an incline following the wet bulb temperature line. This is known as adiabatic humidification. Both cooling processes show an increase in percentage saturation.
- Dehumidification is shown with a descending vertical line. Moisture content and percentage saturation decrease.

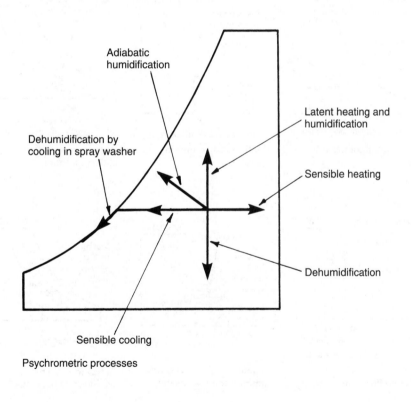

Psychrometric processes

Sensible heating of air may reduce its percentage saturation or relative humidity to an unacceptable level, i.e. <30%. Conversely, sensible cooling may increase the percentage saturation or humidity to an unacceptable level, i.e. >70%.

Applications:

1. Air enters the air handling unit at 5°C db with an RH of 60%. Conditioned air is required at 20°C db with an RH of 50%. The air is preheated to 18·5°C db, cooled to 9°C dew point temperature (dry and wet bulb temperatures identical) and reheated to 20°C db (see lower diagram, centre).
2. Air enters the a.h.u. at 30°C db with an RH of 70%. Conditioned air is required at 20°C db with an RH of 50%. The air is cooled to 9°C dew point temperature and reheated to 20°C db (see lower diagram, right).

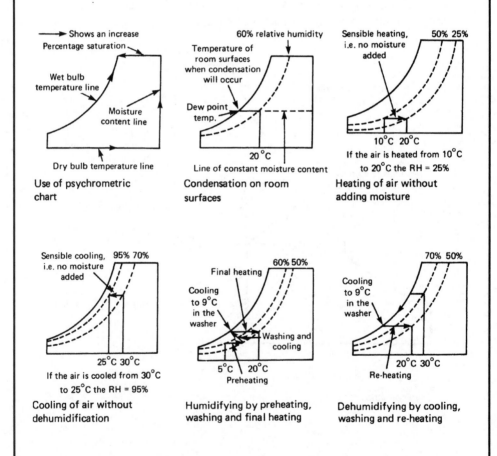

Use of psychrometric chart	Condensation on room surfaces	Heating of air without adding moisture
Cooling of air without dehumidification	Humidifying by preheating, washing and final heating	Dehumidifying by cooling, washing and re-heating

Mixing of two airstreams frequently occurs when combining fresh air with recirculated air from within the building. The process can be represented on a psychrometric chart by drawing a straight line between the two conditions and calculating a point relative to the proportions of mass flow rates.

Example 1:

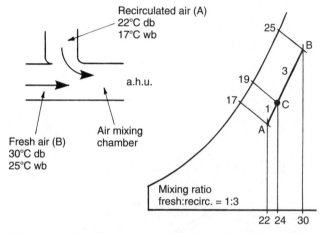

Mixed air condition (C) is found on straight line linking (A) and (B) proportioned 1:3, i.e. 24°C db, 19°C wb, 63% RH and 12g/kg m.c.

Example 2:

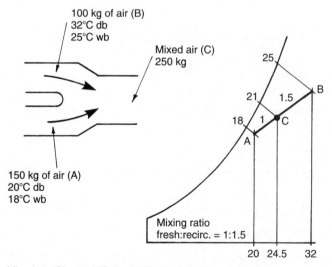

Mixed air (C) = 24.5°C db, 21°C wb, 72% RH and 14 g/kg m.c.

The calculation below relates to the example on page 180, where cool intake air at 5°C db, 60% RH is conditioned to 20°C db, 50% RH.

Applied to an office of 2400 m³ volume, requiring three air changes per hour, the quantity of air (Q) delivered will be:

$$Q = \frac{\text{Volume} \times \text{Air changes per hour}}{3600} = \frac{2400 \times 3}{3600} = 2\text{m}^3/\text{s}$$

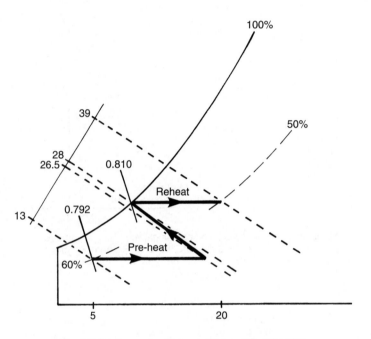

Humidifying by pre-heating, washing and final heating

Pre-heater enthalpy = 26·5 – 13 = 13·5 kJ/kg. Specific volume = 0·792 m³/kg

Reheater enthalpy = 39 – 28 = 11 kJ/kg. Specific volume = 0.810 m³/kg

Pre-heater

Specific volume converted to kg/s: 2·0 m³/s ÷ 0·792 m³/kg = 2·53 kg/s

Pre-heater rating: 2·53 kg/s × 13·5 kJ/kg = 34·2 kW

Reheater

Specific volume converted to kg/s: 2·0 m³/s ÷ 0·810 m³/kg = 2·47 kg/s

Reheater rating: 2·47 kg/s x 11 kJ/kg = 27·2 kW

The calculation below relates to the example on page 180, where warm intake air at 30°C db, 70% RH is conditioned to 20°C db, 50% RH.

With reference to the situation given on the previous page, the quantity of air delivered will be taken as 2 m³/s.

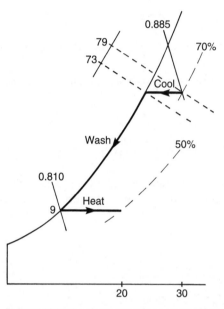

Dehumidifying by cooling, washing and reheating

Chiller enthalpy = 79 – 73 = 6 kJ/kg. Specific volume = 0·885 m³/kg

Specific volume converted to kg/s: 2·0 m³/s ÷ 0·885 m³/kg = 2·26 kg/s

Chiller rating: 2·26 kg/s × 6 kJ/kg = 13·6 kW

Note: Calculations on this and the preceding page assume 100% efficiency of plant. This is unrealistic, therefore energy exchangers should be overrated to accommodate this.

A heat pump is in principle a refrigeration cycle. It differs in application, extracting heat from a low temperature source and upgrading it to a higher temperature for heat emission or water heating. The low temperature heat source may be from water, air or soil which surrounds the evaporator.

A heat pump must be energy efficient; it must generate more power than that used to operate it. A measure of theoretical coefficient of performance (COP) can be expressed as:

$$COP = T_c/T_c - T_e$$

where: T_c = condenser temperature based on degrees Kelvin (0°C = 273 K)

T_e = evaporator temperature based on degrees Kelvin

E.g. T_c = 60°C, Te = 2°C.

$$COP = \frac{60 + 273}{(60 + 273) - (2 + 273)} = 5 \cdot 74$$

i.e. 5·74 kW of energy produced for every 1 kW absorbed. Allowing for efficiency of equipment and installation, a COP of 2 to 3 is more likely.

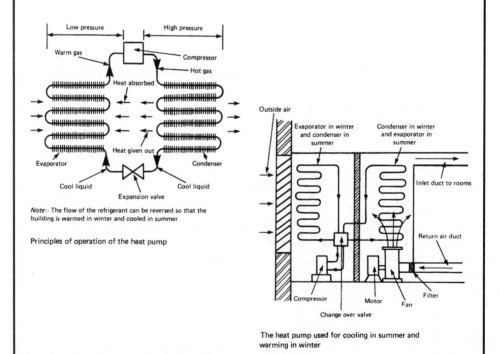

Note:- The flow of the refrigerant can be reversed so that the building is warmed in winter and cooled in summer

Principles of operation of the heat pump

The heat pump used for cooling in summer and warming in winter

Heat pump units are available as large items of plant that can be used to warm a whole building. However, small self-contained units are more common. These are usually located under window openings for warm and cool air distribution in winter and summer respectively.

To transfer the warmth in stale extract duct air, water may be circulated through coils or energy exchangers in both the extract and cool air intake ducts. This is known as a run-around coil. Using water as an energy transfer medium is inexpensive but limited in efficiency. Use of a refrigerant is more effective, with an evaporator coil in the warm extract duct and a condenser coil in the cold air inlet duct.

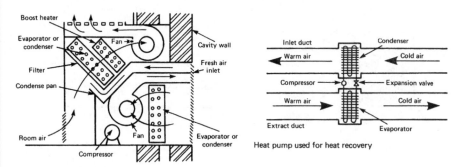

Unit heat pump fixed below window

Heat pump used for heat recovery

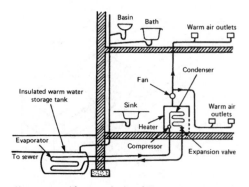

Heat pump used for extracting heat from
warm waste water

Heat energy in warm waste water from sanitary fittings may be retrieved and used to supplement space heating by using a heat pump. An insulated tank buried below ground receives the waste water before it flows to the sewer. Heat energy is extracted through an evaporator inside the tank.

The concept of a thermal or heat wheel was devised about 50 years ago by Carl Munter, a Swedish engineer. Wheels range from 600 mm to 4 m in diameter, therefore sufficient space must be allowed for their accommodation. They have an extended surface of wire mesh or fibrous paper impregnated with lithium chloride. Lithium chloride is an effective absorbent of latent heat energy in the moisture contained in stale air. A low power (700 W) electric motor rotates the wheel at an angular velocity of 10–20 rpm. Heat from the exhaust air transfers to the inlet air and the purging section extracts the contaminants. Efficiency can be up to 90%.

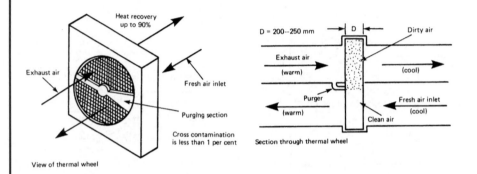

View of thermal wheel

Section through thermal wheel

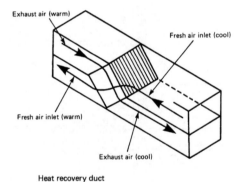

Heat recovery duct

The heat recovery duct or plate heat exchanger has warm exhaust air separated from the cool inlet air by metal or glass vanes. Heat from the exhaust vanes is transferred to the inlet vanes to warm the incoming air. Ducts must be well insulated to conserve energy and to reduce condensation. Condensation should be drained at the base of the unit. Efficiency is unlikely to exceed 50%.

Buildings are designed with the intention of providing a comfortable internal environment. To achieve this efficiently, many incorporate air conditioning and ventilation systems. Misuse of some of the system equipment may cause the following health hazards:

- Legionnaires' disease.
- Humidifier fever (see next page).
- Sick building syndrome (see next page).

Legionnaires' disease – obtained its name from the first significant outbreak that occurred during an American Legionnaires' convention in Philadelphia, USA, in 1976. The bacterial infection was contracted by 182 people; it has similar symptons to pneumonia. Of these, 29 died. Subsequently, numerous outbreaks have been identified worldwide. They are generally associated with hot water systems (see page 58) and air conditioning water cooling towers.

The organisms responsible occur naturally in swamps and similar humid conditions. In limited numbers they are harmless, but when concentrated they contaminate the water in which they live. If this water is suspended in the air as an aerosol spray, it can be inhaled to establish lung disease in susceptible persons.

Areas for concern – water systems with a temperature between 20°C and 60°C, as the optimum breeding temperature of the bacteria is about 40°C; water cooling towers, particularly the older type with coarse timber packing in dirty/dusty atmospheres, e.g. city centres and adjacency with building sites; contaminated spray dispersing in the atmosphere can be inhaled by people in the locality or it may be drawn into a ventilation inlet and distributed through the ductwork; spray humidifiers in air handling units are also possible breeding areas – the water in these should be treated with a biocide or they should be replaced with steam humidifiers.

People at risk – the elderly, those with existing respiratory problems, heavy smokers and those in a generally poor state of health. Nevertheless, there have been cases of fit, healthy, young people being infected.

Solution – abolition of wet cooling towers and replacement with air cooled condensers. Use of packaged air conditioning with air cooling. Documented maintenance of existing wet cooling towers, i.e. regular draining and replacement of water, cleaning of towers and treatment of new water with a biocide.

Ref: Workplace (Health, Safety and Welfare) Regulations 1992.

Humidifier fever – this is not an infection, but an allergic reaction producing flu-like symptons such as headaches, aches, pains and shivering. It is caused by micro-organisms which breed in the water reservoirs of humidifiers whilst they are shut down, i.e. weekends or holidays. When the plant restarts, concentrations of the micro-organisms and their dead husks are drawn into the air stream and inhaled. After a few days' use of the plant, the reaction diminishes and recommences again after the next shutdown. Water treatment with a biocide is a possible treatment or replacement with a steam humidifier.

Sick building syndrome – this is something of a mystery as no particular cause has been identified for the discomfort generally attributed to this disorder. The symptoms vary and can include headaches, throat irritations, dry or running nose, aches, pains and loss of concentration. All or some may be responsible for personnel inefficiency and absenteeism from work. Whilst symptoms are apparent, the causes are the subject of continued research. Some may be attributed to physical factors such as:

- Noise from computers, machinery, lighting or ducted air movement.
- Strobing from fluorescent strip lights.
- Static electricity from computer screens, copiers, etc.
- Fumes from cleaning agents.
- Glare from lighting and monitors.
- Unsympathetic internal colour schemes.
- Carpet mites.

Other factors are psychological:

- Lack of personal control over an air conditioned environment.
- No direct link with the outside world, i.e. no openable windows.
- Disorientation caused by tinted windows.
- Working in rooms with no windows.
- Dissatisfaction with air conditioning does not provide the ideal environment.

More apparent may be lack of maintenance and misuse of air conditioning plant. Energy economising by continually recirculating the same air is known to cause discomfort for building occupants. The research continues and as a result of sick building syndrome, new building designs often favour more individual control of the workplace environment or application of traditional air movement principles such as stack effect.

7 DRAINAGE SYSTEMS, SEWAGE TREATMENT AND REFUSE DISPOSAL

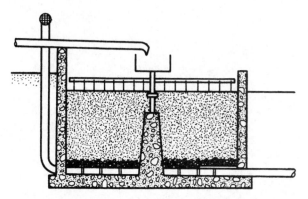

COMBINED AND SEPARATE SYSTEMS

PARTIALLY SEPARATE SYSTEM

RODDING POINT SYSTEM

SEWER CONNECTION

DRAINAGE VENTILATION

UNVENTILATED SPACES

DRAIN LAYING

MEANS OF ACCESS

BEDDING OF DRAINS

DRAINS UNDER OR NEAR BUILDINGS

JOINTS USED ON DRAIN PIPES

ANTI-FLOOD DEVICES

GARAGE DRAINAGE

DRAINAGE PUMPING

SUBSOIL DRAINAGE

TESTS ON DRAINS

SOAKAWAYS

CESSPOOLS AND SEPTIC TANKS

DRAINAGE FIELDS AND MOUNDS

DRAINAGE DESIGN

WASTE AND REFUSE PROCESSING

The type of drainage system selected for a building will be determined by the local water authority's established sewer arrangements. These will be installed with regard to foul water processing and the possibility of disposing surface water via a sewer into a local water course or directly into a soakaway.

Combined system – this uses a single drain to convey both foul water from sanitary appliances and rainwater from roofs and other surfaces to a shared sewer. The system is economical to install, but the processing costs at the sewage treatment plant are high.

Separate system – this has foul water from the sanitary appliances conveyed in a foul water drain to a foul water sewer. The rainwater from roofs and other surfaces is conveyed in a surface water drain into a surface water sewer or a soakaway. This system is relatively expensive to install, particularly if the ground has poor drainage qualities and soakaways cannot be used. However, the benefit is reduced volume and treatment costs at the processing plant.

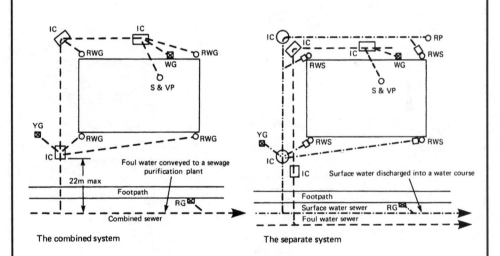

The combined system

The separate system

Key:
IC = Inspection chamber   RWG = Rainwater gully
WG = Waste gully   RG = Road gully
YG = Yard gully   RWS = Rainwater shoe
RP = Rodding point   S & VP = Soil and vent pipe (discharge
                                                  stack)

Partially separate system – most of the rainwater is conveyed by the surface water drain into the surface water sewer. For convenience and to reduce site costs, the local water authority may permit an isolated rainwater inlet to be connected to the foul water drain. This is shown with the rainwater inlet at A connected to the foul water inspection chamber. Also, a rodding point is shown at B. These are often used at the head of a drain, as an alternative to a more costly inspection chamber.

A back inlet gully can be used for connecting a rainwater down pipe or a waste pipe to a drain. The bend or trap provides a useful reservoir to trap leaves. When used with a foul water drain, the seal prevents air contamination. A yard gully is solely for collecting surface water and connecting this with a drain. It is similar to a road gully, but smaller. A rainwater shoe is only for connecting a rainwater pipe to a surface water drain. The soil and vent pipe or discharge stack is connected to the foul water drain with a rest bend at its base. This can be purpose made or produced with two 135° bends. It must have a centre-line radius of at least 200 mm.

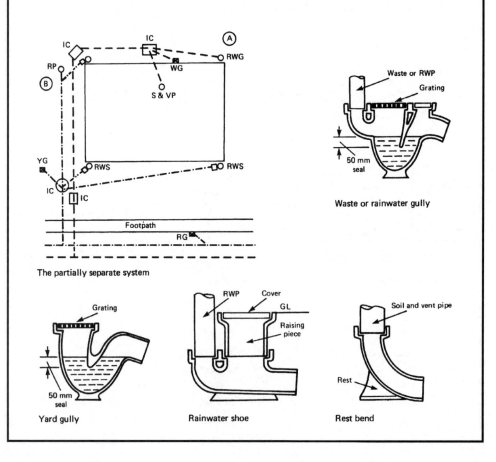

The partially separate system

Waste or rainwater gully

Yard gully

Rainwater shoe

Rest bend

Rodding points or rodding eyes provide a simple and inexpensive means of access at the head of a drain or on shallow drain runs for rodding in the direction of flow. They eliminate isolated loads that manholes and inspection chambers can impose on the ground, thus reducing the possibility of uneven settlement. The system is also neater, with less surface interruptions. Prior to installation, it is essential to consult with the local authority to determine whether the system is acceptable and, if so, to determine the maximum depth of application and any other limitations on use. As rodding is only practical in one direction, an inspection chamber or manhole is usually required before connection to a sewer.

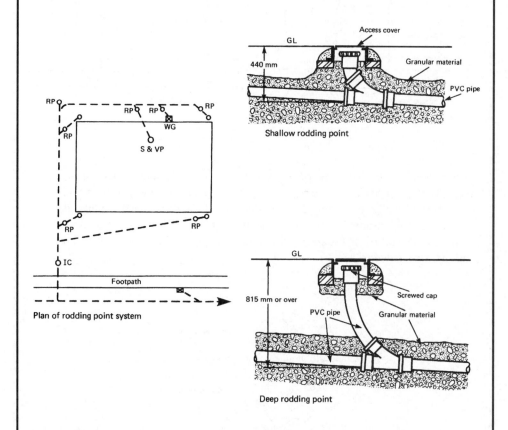

Plan of rodding point system

Shallow rodding point

Deep rodding point

Refs: Building Regulations, Approved Documents H1: Foul water drainage and H3: Rainwater drainage.
BS EN 752: Drain and sewer systems outside buildings.

Sewer Connection

Connections between drains and sewers must be obliquely in the direction of flow. Drains may be connected independently to the public sewer so that each building owner is responsible for the maintenance of the drainage system for that building. In situations where there would be long drain runs, it may be more economical to connect each drain to a private sewer. This requires only one sewer connection for several buildings. Maintenance of the private sewer is shared between the separate users.

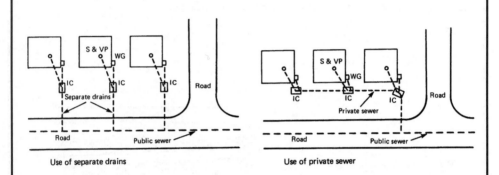

Use of separate drains   Use of private sewer

Connection of a drain or private sewer to the public sewer can be made with a manhole. If one of these is used at every connection, the road surface is unnecessarily disrupted. Therefore a saddle is preferred, but manhole access is still required at no more than 90 m intervals. Saddles are bedded in cement mortar in a hole made in the top of the sewer.

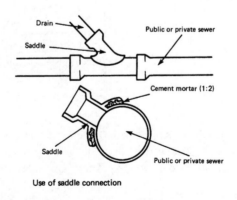

Use of saddle connection

Venting of foul water drains is necessary to prevent a concentration of gases and to retain the air inside the drain at atmospheric pressure. This is essential to prevent the loss of trap water seals by siphonage or compression. The current practice of direct connection of the discharge stack and drain to the public sewer provides a simple means of ventilation through every stack. In older systems, generally pre-1950s, an interceptor trap with a 65 mm water seal separates the drain from the sewer. The sewer is independently vented by infrequently spaced high level vent stacks. Through ventilation of the drain is by fresh air inlet at the lowest means of access and the discharge stack. It may still be necessary to use this system where new buildings are constructed where it exists. It is also a useful means of controlling rodent penetration from the sewer.

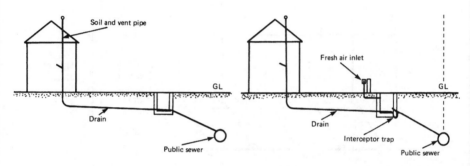

Without the use of an interceptor trap

With the use of an interceptor trap

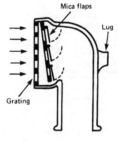

Fresh air inlet

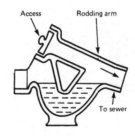

Interceptor trap

To reduce installation costs and to eliminate roof penetration of ventilating stacks, discharge stacks can terminate inside a building. This is normally within the roof space, i.e. above the highest water level of an appliance connected to the stack, provided the top of the stack is fitted with an air admittance valve (AAV). An AAV prevents the emission of foul air, but admits air into the stack under conditions of reduced atmospheric pressure. AAVs are limited in use to dwellings of no more than three storeys, in up to four adjacent buildings. The fifth building must have a conventional vent stack to ventilate the sewer.

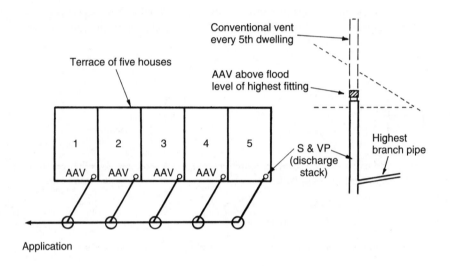

Application

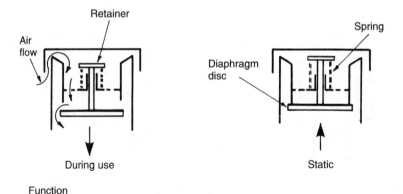

Function

Direct connection – a WC may discharge directly into a drain, without connection to a soil and ventilating stack. Application is limited to a maximum distance between the centre line of the WC trap outlet and the drain invert of 1·5 m.

Stub stack – this is an extension of the above requirement and may apply to a group of sanitary fittings. In addition to the WC requirement, no branch pipes to other fittings may be higher than 2 m above a connection to a ventilated stack or the drain invert.

The maximum length of branch drain from a single appliance to a means of drain access is 6 m. For a group of appliances, it is 12 m.

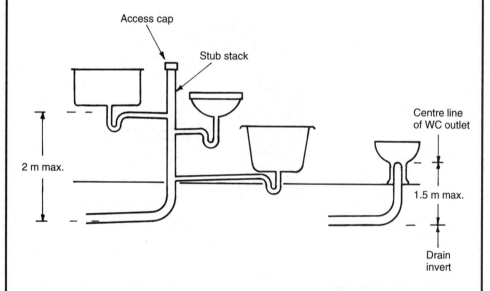

Stub stack for a group of fittings   WC direct connection

Ref: Building Regulations, Approved Document H1, Section 1: Sanitary pipework.
BS EN 12056-2: Gravity drainage systems inside buildings.

Drain Laying

The bottom of a drain trench must be excavated to a gradient. This is achieved by setting up sight rails, suitably marked to show the centre of the drain. These are located above the trench and aligned to the gradient required. At least three sight rails should be used. A boning rod (rather like a long 'T' square) is sighted between the rails to establish the level and gradient of the trench bottom. Wooden pegs are driven into the trench bottom at about 1 m intervals. The required level is achieved by placing the bottom of the boning rod on each peg and checking top alignment with the sight rails. Pegs are adjusted accordingly and removed before laying the drains. For safe working in a trench, it is essential to provide temporary support to the excavation.

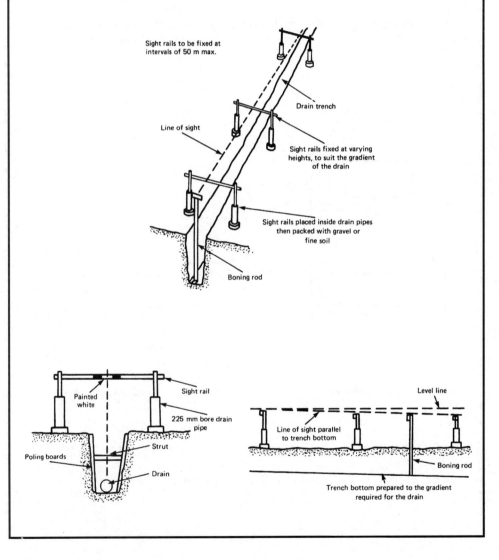

Sight rails to be fixed at intervals of 50 m max.

Line of sight

Drain trench

Sight rails fixed at varying heights, to suit the gradient of the drain

Sight rails placed inside drain pipes then packed with gravel or fine soil

Boning rod

Painted white

Sight rail

225 mm bore drain pipe

Poling boards

Strut

Drain

Level line

Line of sight parallel to trench bottom

Boning rod

Trench bottom prepared to the gradient required for the drain

Drain access may be obtained through rodding points (page 193), shallow access chambers, inspection chambers and manholes. Pipe runs should be straight and access provided only where needed, i.e.:

- at significant changes in direction
- at significant changes in gradient
- near to, or at the head of a drain
- where the drain changes in size
- at junctions
- on long straight runs.

Maximum spacing (m) of access points based on Table 10 of Approved Document H1 to the Building Regulations:

To \ From	Access fitting Small	Large	Junction	Inspection Chamber	Manhole
Start of drain	12	12	–	22	45
Rodding eye	22	22	22	45	45
Access fitting:					
150 diam	–	–	12	22	22
150 × 100	–	–	12	22	22
225 × 100	–	–	22	45	45
Inspection chamber	22	45	22	45	45
Manhole	22	45	45	45	90

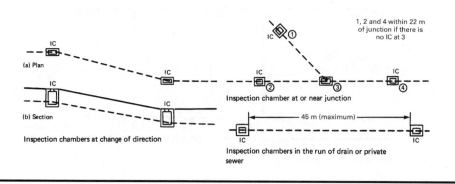

(a) Plan

(b) Section

Inspection chambers at change of direction

1, 2 and 4 within 22 m of junction if there is no IC at 3

Inspection chamber at or near junction

45 m (maximum)

Inspection chambers in the run of drain or private sewer

Shallow access chambers or access fittings are small compartments similar in size and concept to rodding points, but providing drain access in both directions and possibly into a branch. They are an inexpensive application for accessing shallow depths up to 600 mm to invert. Within this classification manufacturers have created a variety of fittings to suit their drain products. The uPVC bowl variation shown combines the facility of an inspection chamber and a rodding point.

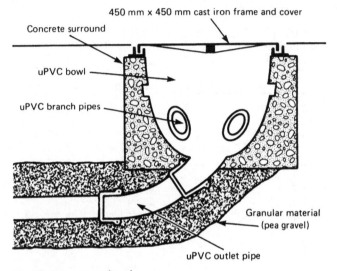

The Marscar access bowl

Note: Small lightweight cover plates should be secured with screws, to prevent unauthorised access, e.g. children.

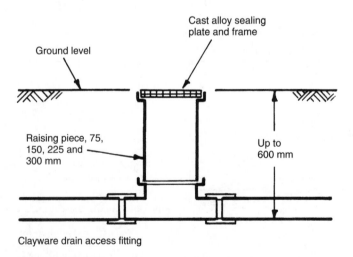

Clayware drain access fitting

Inspection chambers are larger than access chambers, having an open channel and space for several branches. They may be circular or rectangular on plan and preformed from uPVC, precast in concrete sections or traditionally constructed with dense bricks from a concrete base. The purpose of an inspection chamber is to provide surface access only, therefore the depth to invert level does not exceed 1 m.

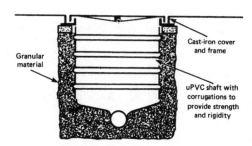

Granular material

Cast-iron cover and frame

uPVC shaft with corrugations to provide strength and rigidity

uPVC inspection chamber

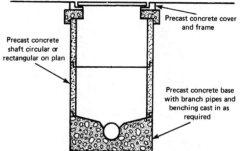

Precast concrete cover and frame

Precast concrete shaft circular or rectangular on plan

Precast concrete base with branch pipes and benching cast in as required

Precast concrete inspection chamber

Size of chamber

Depth	Length	Width
Up to 600 mm	750 mm	700 mm
600 to 1000 mm	1·2 m	750 mm

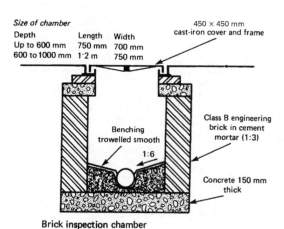

450 × 450 mm cast-iron cover and frame

Benching trowelled smooth

1:6

Class B engineering brick in cement mortar (1:3)

Concrete 150 mm thick

Brick inspection chamber

201

The term manhole is used generally to describe drain and sewer access. By comparison, manholes are large chambers with sufficient space for a person to gain access at drain level. Where the depth to invert exceeds 1 m, step irons should be provided at 300 mm vertical and horizontal spacing. A built-in ladder may be used for very deep chambers. Chambers in excess of 2·7 m may have a reduced area of access known as a shaft (min. 900 × 840 mm or 900 mm diameter), otherwise the following applies:

Depth (m)	Internal dimensions (mm) l × b	Cover size
<1·5	1200 × 750 or 1050 diam.	Min. dimension 600 mm
1·5–2.7	1200 × 750 or 1200 diam.	Min. dimension 600 mm
>2·7	1200 × 840 or 1200 diam.	Min. dimension 600 mm

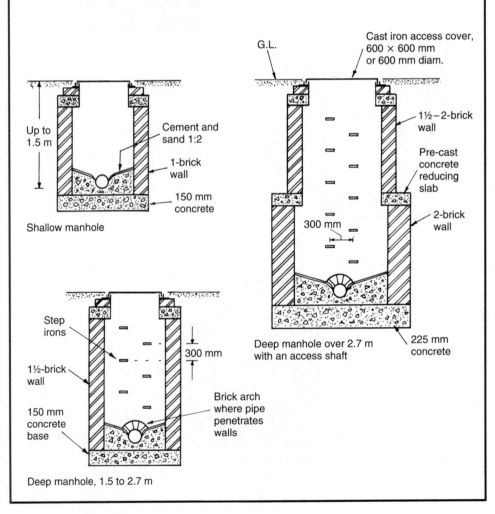

Shallow manhole

Deep manhole, 1.5 to 2.7 m

Deep manhole over 2.7 m with an access shaft

Where there is a significant difference in level between a drain and a private or public sewer, a back-drop may be used to reduce excavation costs. Back-drops have also been used on sloping sites to limit the drain gradient, as at one time it was thought necessary to regulate the velocity of flow. This is now considered unnecessary and the drain may be laid to the same slope as the ground surface. For use with cast-iron and uPVC pipes up to 150 mm bore, the back-drop may be secured inside the manhole. For other situations, the back-drop is located outside the manhole and surrounded with concrete.

The access shaft should be 900 × 840 mm minimum and the working area in the shaft at least 1·2 m × 840 mm.

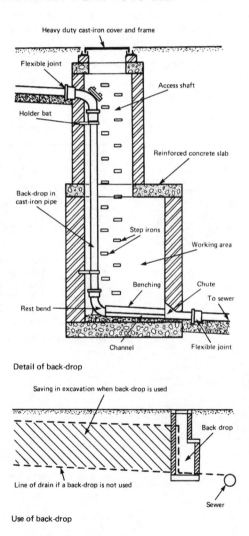

Detail of back-drop

Use of back-drop

Drains must be laid with due regard for the sub-soil condition and the imposed loading. The term bedding factor is applied to laying rigid drain pipes. This describes the ratio of the pipe strength when bedded to the pipe test strength as given in the relevant British Standard.

Class A bedding gives a bedding factor of 2·6, which means that a rigid drain pipe layed in this manner could support up to 2·6 times the quoted BS strength. This is due to the cradling effect of concrete, with a facility for movement at every pipe joint. This method may be used where extra pipe strength is required or great accuracy in pipe gradient is necessary. Class B bedding is more practical, considerably less expensive and quicker to use. This has a more than adequate bedding factor of 1·9. If used with plastic pipes, it is essential to bed and completely surround the pipe with granular material to prevent the pipe from distortion.

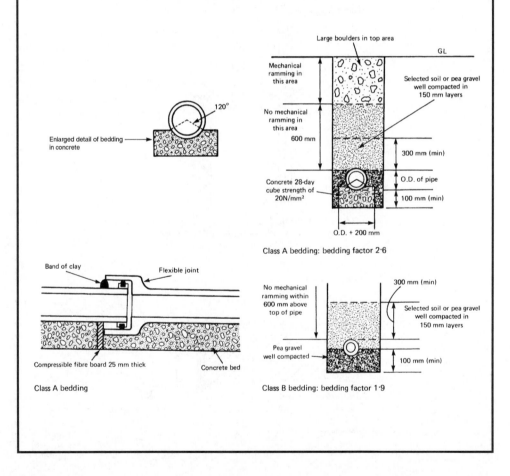

Enlarged detail of bedding in concrete

120°

Large boulders in top area

GL

Mechanical ramming in this area

Selected soil or pea gravel well compacted in 150 mm layers

No mechanical ramming in this area
600 mm

300 mm (min)

Concrete 28-day cube strength of 20N/mm²

O.D. of pipe

100 mm (min)

O.D. + 200 mm

Class A bedding: bedding factor 2·6

Band of clay

Flexible joint

Compressible fibre board 25 mm thick

Concrete bed

Class A bedding

No mechanical ramming within 600 mm above top of pipe

300 mm (min)

Selected soil or pea gravel well compacted in 150 mm layers

Pea gravel well compacted

100 mm (min)

Class B bedding: bedding factor 1·9

Approved Document H to the Building Regulations provides many methods which will support, protect and allow limited angular and lineal movement to flexibly jointed clay drain pipes. Those shown below include three further classifications and corresponding bedding factors. Also shown is a suitable method of bedding flexible plastic pipes. In water-logged trenches it may be necessary to temporarily fill plastic pipes with water to prevent them floating upwards whilst laying. In all examples shown, space to the sides of pipes should be at least 150 mm.

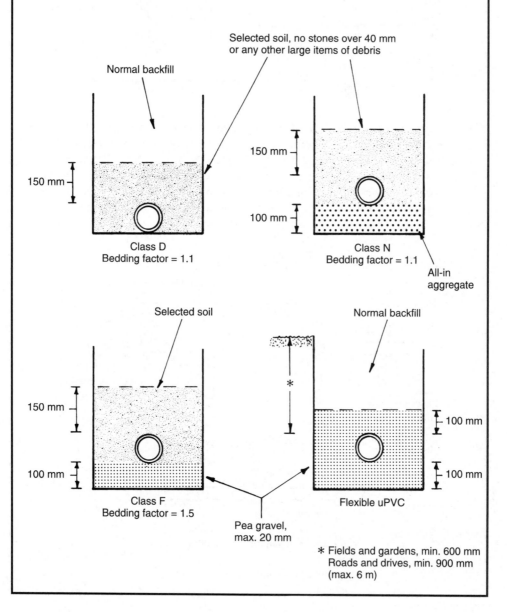

Drains Under or Near Buildings

Drain trenches should be avoided near to and lower than building foundations. If it is unavoidable and the trench is within 1 m of the building, the trench is filled with concrete to the lowest level of the building. If the trench distance exceeds 1 m, concrete is filled to a point below the lowest level of the building equal to the trench distance less 150 mm.

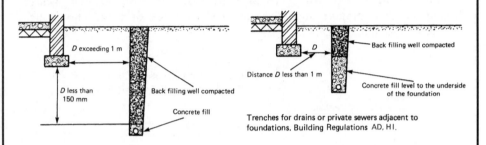

Trenches for drains or private sewers adjacent to foundations. Building Regulations AD, H1.

Drains under buildings should be avoided. Where it is impossible to do so, the pipe should be completely protected by concrete and integrated with the floor slab. If the pipe is more than 300 mm below the floor slab, it is provided with a granular surround. Pipes penetrating a wall below ground should be installed with regard for building settlement. Access through a void or with flexible pipe joints each side of the wall are both acceptable.

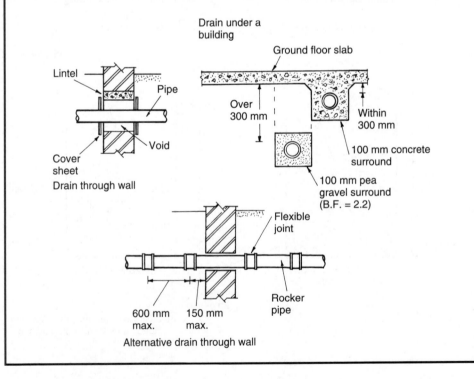

Rigid jointing of clay drain pipes is now rarely specified as flexible joints have significant advantages:

- They are quicker and simpler to make.
- The pipeline can be tested immediately.
- There is no delay in joint setting due to the weather.
- They absorb ground movement and vibration without fracturing the pipe.

Existing clay drains will be found with cement and sand mortar joints between spigot and socket. Modern pipe manufacturers have produced their own variations on flexible jointing, most using plain ended pipes with a polypropylene sleeve coupling containing a sealing ring. Cast iron pipes can have spigot and sockets cold caulked with lead wool. Alternatively, the pipe can be produced with plain ends and jointed by rubber sleeve and two bolted couplings. Spigot and socket uPVC pipes may be jointed by solvent cement or with a push-fit rubber 'O' ring seal. They may also have plain ends jointed with a uPVC sleeve coupling containing a sealing ring.

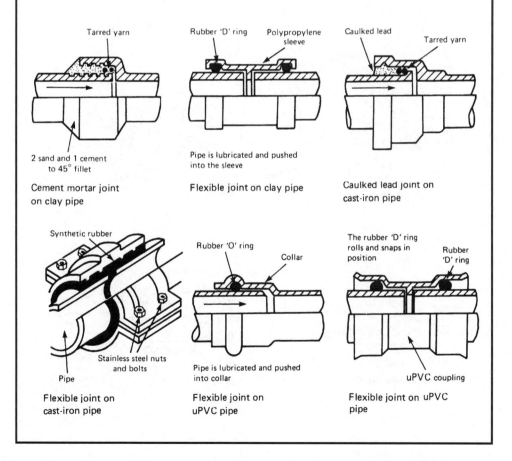

Tarred yarn

2 sand and 1 cement to 45° fillet

Cement mortar joint on clay pipe

Rubber 'D' ring   Polypropylene sleeve

Pipe is lubricated and pushed into the sleeve

Flexible joint on clay pipe

Caulked lead   Tarred yarn

Caulked lead joint on cast-iron pipe

Synthetic rubber

Stainless steel nuts and bolts

Pipe

Flexible joint on cast-iron pipe

Rubber 'O' ring   Collar

Pipe is lubricated and pushed into collar

Flexible joint on uPVC pipe

The rubber 'D' ring rolls and snaps in position   Rubber 'D' ring

uPVC coupling

Flexible joint on uPVC pipe

Where there is a possibility of a sewer surcharging and back flooding a drain, an anti-flooding facility must be fitted. For conventional drainage systems without an interceptor trap, an anti-flooding trunk valve may be fitted within the access chamber nearest the sewer. If an interceptor trap is required, an anti-flooding type can be used in place of a conventional interceptor. An anti-flooding gully may be used in place of a conventional fitting, where back flooding may occur in a drain.

Waste water from canteen sinks or dishwashers contains a considerable amount of grease. If not removed it could build up and block the drain. Using a grease trap allows the grease to be cooled by a large volume of water. The grease solidifies and floats to the surface. At regular intervals a tray may be lifted out of the trap and cleaned to remove the grease.

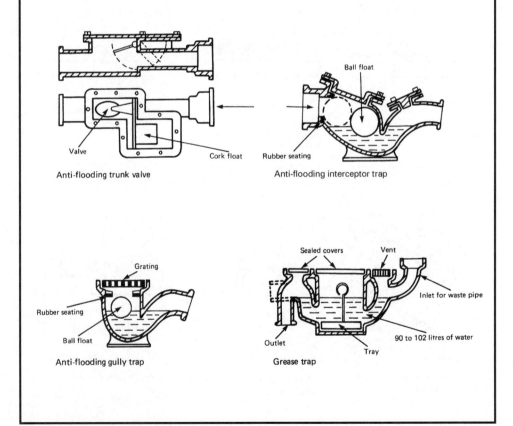

Anti-flooding trunk valve

Anti-flooding interceptor trap

Anti-flooding gully trap

Grease trap

The Public Health Act prohibits discharge of petroleum and oil into a sewer. Garage floor washings will contain petrochemicals and these must be prevented from entering a sewer. The floor layout should be arranged so that one garage gully serves up to 50 m² of floor area. The gully will retain some oil and other debris, which can be removed by emptying the inner bucket. A petrol interceptor will remove both petrol and oil. Both rise to the surface with some evaporation through the vent pipes. The remaining oil is removed when the tanks are emptied and cleaned. The first chamber will also intercept debris and this compartment will require more regular cleaning. Contemporary petrol interceptors are manufactured from reinforced plastics for simple installation in a prepared excavation.

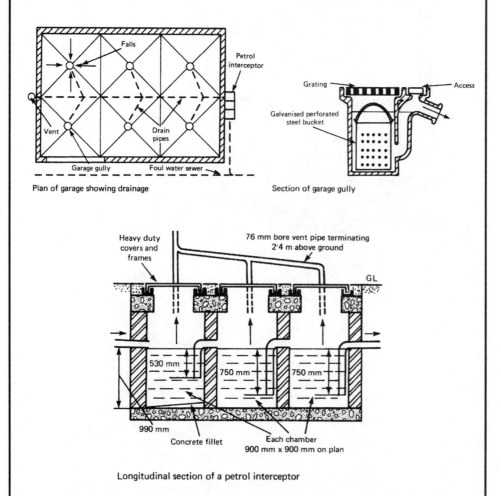

Plan of garage showing drainage

Section of garage gully

Longitudinal section of a petrol interceptor

The contents of drainage pipe lines should gravitate to the sewer and sewage processing plant. In some situations site levels or basement sanitary facilities will be lower than adjacent sewers and it becomes necessary to pump the drainage flows. A pumping station or plant room can be arranged with a motor room above or below surface level. Fluid movement is by centrifugal pump, usually immersed and therefore fully primed. For large schemes, two pumps should be installed with one on standby in the event of the duty pump failing. The pump impeller is curved on plan to complement movement of sewage and to reduce the possibility of blockage. The high level discharge should pass through a manhole before connecting to the sewer.

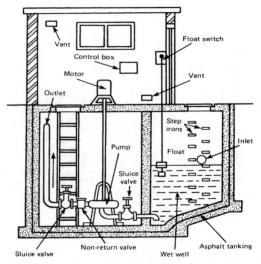

Section through pumping station

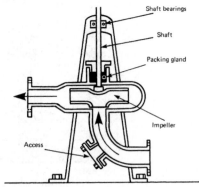

Section through centrifugal pump

Refs: BS EN 12056-4: Gravity drainage systems inside buildings. Wastewater lifting plants. Layout and calculation.
BS EN 12050: Waste water lifting plants . . . .

A sewage ejector may be used as an alternative to a centrifugal pump for lifting foul water. The advantages of an ejector are:

- Less risk of blockage.
- Fewer moving parts and less maintenance.
- A wet well is not required.
- One compressor unit can supply air to several ejectors.

Operation:

- Incoming sewage flows through inlet pipe A into ejector body B.
- Float rises to the top collar.
- Rod is forced upwards opening an air inlet valve and closing an exhaust valve.
- Compressed air enters the ejector body forcing sewage out through pipe C.
- The float falls to the bottom collar and its weight plus the rocking weight closes the air inlet valve and opens the exhaust valve.

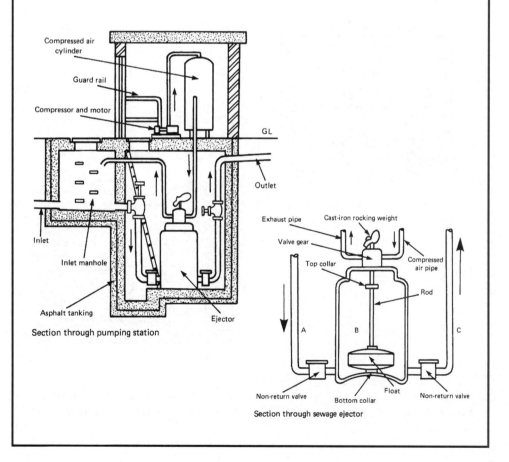

Section through pumping station

Section through sewage ejector

When considering methods of drainage pumping, equipment manufacturers should be consulted with the following details:

- Drainage medium – foul or surface water, or both.
- Maximum quantity – anticipated flow in m^3/h.
- Height to which the sewage has to be elevated.
- Length of delivery pipe.
- Availability of electricity – mains or generated.
- Planning constraints, regarding appearance and siting of pump station.

In the interests of visual impact, it is preferable to construct the motor room below ground. This will also absorb some of the operating noise.

In basements there may be some infiltration of ground water. This can be drained to a sump and pumped out as the level rises. In plant rooms a sump pump may be installed to collect and remove water from any leakage that may occur. It is also useful for water extraction when draining down boilers for servicing.

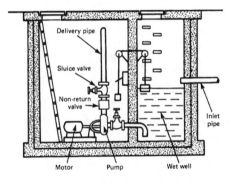

Pumping station with motor room below ground level

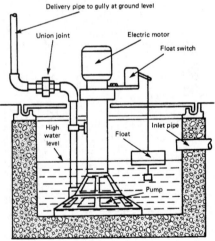

Sump pump

Design guidance for external pumped installations may be found in BS EN 752-6: Drain and sewer systems outside buildings. Pumping installations.

Ideally, buildings should be constructed with foundations above the subsoil water table. Where this is unavoidable or it is considered necessary to generally control the ground water, a subsoil drainage system is installed to permanently lower the natural water table. Various ground drainage systems are available, the type selected will depend on site conditions. The simplest is a French drain. It comprises a series of strategically located rubble-filled trenches excavated to a fall and to a depth below high water table. This is best undertaken after the summer, when the water table is at its lowest. Flow can be directed to a ditch, stream or other convenient outfall. In time the rubble will become silted up and need replacing.

An improvement uses a polyethylene/polypropylene filament fabric membrane to line the trench. This is permeable in one direction only and will also function as a silt filter. This type of drain is often used at the side of highways with an open rubble surface.

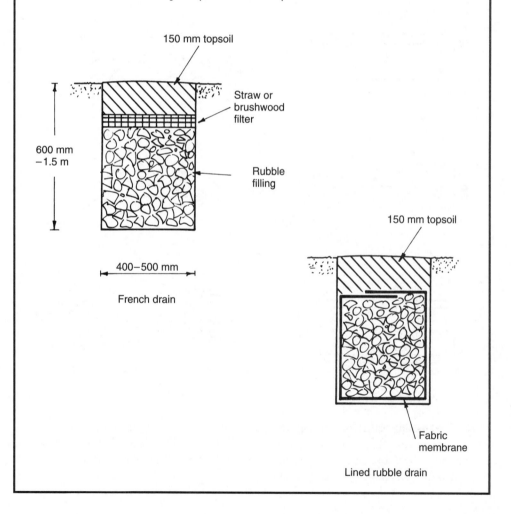

150 mm topsoil

Straw or brushwood filter

600 mm –1.5 m

Rubble filling

400–500 mm

French drain

150 mm topsoil

Fabric membrane

Lined rubble drain

The layout and spacing of subsoil drainage systems depends on the composition and drainage qualities of the subsoil and the disposition of buildings. For construction sites the depth of drainage trench will be between 600 mm and 1·5 m. Shallower depths may be used in agricultural situations and for draining surface water from playing fields. Installation of pipes within the rubble drainage medium has the advantage of creating a permanent void to assist water flow. Suitable pipes are produced in a variety of materials including clay (open jointed, porous or perforated), concrete (porous (no-fine aggregate) or perforated) and uPVC (perforated). The pipe void can be accessed for cleaning and the system may incorporate silt traps at appropriate intervals. Piped outlets may connect to a surface water sewer with a reverse acting interceptor trap at the junction.

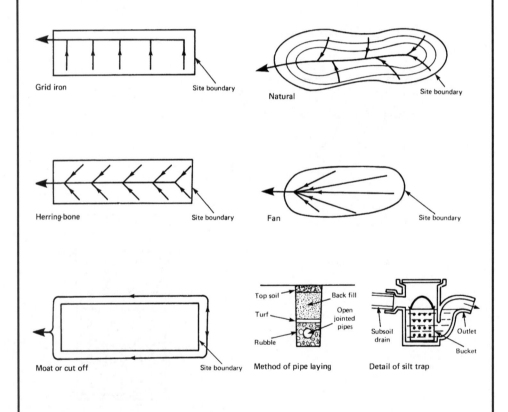

Grid iron Site boundary

Natural Site boundary

Herring-bone Site boundary

Fan Site boundary

Moat or cut off Site boundary

Top soil Back fill
Turf Open jointed pipes
Rubble
Method of pipe laying

Subsoil drain Outlet
Bucket
Detail of silt trap

Note: The installation of subsoil drainage may be necessary under the requirements of Building Regulation C3. The purpose of this is to prevent the passage of ground moisture into a building and the possibility of damage to a building.

British Standard pipes commonly used for subsoil drainage:

- Perforated clay, BSEN 295–5.
- Porous clay, BS 1196.
- Porous concrete, BS 5911–114.
- Profiled and slotted plastics, BS 4962.
- Perforated uPVC, BS 4660.

Silt and other suspended particles will eventually block the drain unless purpose-made traps are strategically located for regular cleaning. The example shown on the previous page is adequate for short drain runs, but complete systems will require a pit which can be physically accessed. This is an essential requirement if the drain is to connect to a public surface water sewer. In order to protect flow conditions in the sewer, the local water authority may only permit connection via a reverse acting interceptor trap. This item does not have the capacity to function as a silt trap.

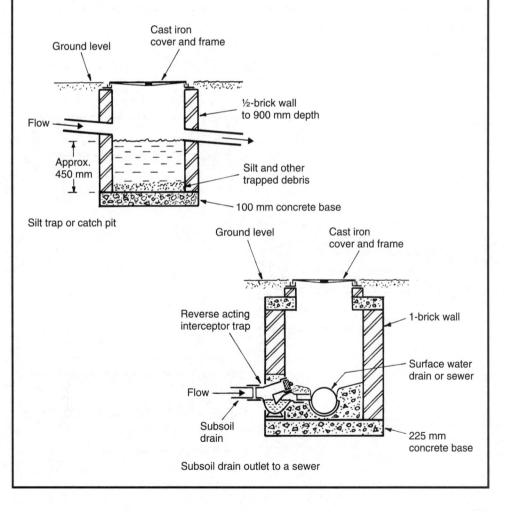

Silt trap or catch pit

Subsoil drain outlet to a sewer

Drains must be tested before and after backfilling trenches.

Air test – the drain is sealed between access chambers and pressure tested to 100 mm water gauge with hand bellows and a 'U' gauge (manometer). The pressure must not fall below 75 mm during the first 5 minutes.

Smoke test – may be used to detect leakage. The length of drain to be tested is sealed and smoke pumped into the pipes from the lower end. The pipes should then be inspected for any trace of smoke. Smoke pellets may be used in the smoke machine or with clay and concrete pipes they may be applied directly to the pipe line.

Water test – effected by stopping the lower part of the drain and filling the pipe run with water from the upper end. This requires a purpose made test bend with an extension pipe to produce a 1·5 m head of water. This should stand for 2 hours and if necessary topped up to allow for limited porosity. For the next 30 minutes, maximum leakage for 100 mm and 150 mm pipes is 0·05 and 0·08 litres per metre run respectively.

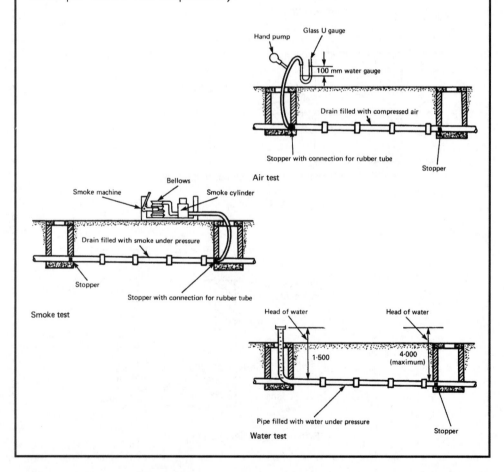

Air test

Smoke test

Water test

Where a surface water sewer is not available, it may be possible to dispose of rainwater into a soakaway. A soakaway will only be effective in porous soils and above the water table. Water must not be allowed to flow under a building and soakaways should be positioned at least 3 m away (most local authorities require 5 m). A filled soakaway is inexpensive to construct, but it will have limited capacity. Unfilled or hollow soakaways can be built of precast concrete or masonry.

Soakaway capacity can be determined by applying a rainfall intensity of at least 50 mm per hour to the following formula:

$$C = A \times R \div 3$$

where C = capacity in m^3
     A = area to be drained in m^2
     R = rainfall in metres per hour.

E.g. a drained area of 150 m^2
     $C = 150 \times 0.050 \div 3 = 2.5\ m^3$

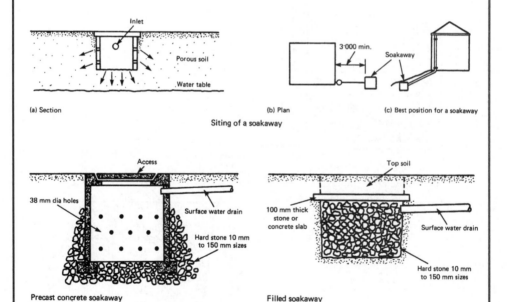

(a) Section

Inlet

Porous soil

Water table

(b) Plan

3·000 min.

Soakaway

(c) Best position for a soakaway

Siting of a soakaway

Precast concrete soakaway

Access

38 mm dia holes

Surface water drain

Hard stone 10 mm to 150 mm sizes

Filled soakaway

Top soil

100 mm thick stone or concrete slab

Surface water drain

Hard stone 10 mm to 150 mm sizes

Note: BRE Digest 365: Soakaway Design, provides a more detailed approach to capacity calculation.

Cesspools

A cesspool is an acceptable method of foul water containment where main drainage is not available. It is an impervious chamber requiring periodic emptying, sited below ground level. Traditional cesspools were constructed of brickwork rendered inside with waterproof cement mortar. Precast concrete rings supported on a concrete base have also been used, but factory manufactured glass reinforced plastic units are now preferred. The Building Regulations require a minimum capacity below inlet level of 18 000 litres. A cesspool must be impervious to rainwater, well ventilated and have no outlets or overflows. It should be sited at least 15 m from a dwelling.

Capacity is based on 150 litres per person per day at 45 day emptying cycles, e.g. a four-person house:

$$= 4 \times 150 \times 45 = 27\,000 \text{ litres } (27 \text{ m}^3)$$

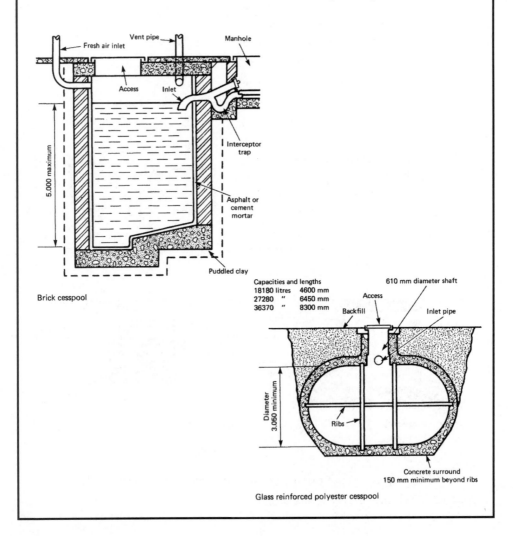

Brick cesspool

Capacities and lengths
18180 litres   4600 mm
27280   "   6450 mm
36370   "   8300 mm

Glass reinforced polyester cesspool

218

Where main drainage is not available a septic tank is preferable to a cesspool. A septic tank is self-cleansing and will only require annual desludging. It is in effect a private sewage disposal plant, which is quite common for buildings in rural areas. The tank is a watertight chamber in which the sewage is liquefied by anaerobic bacterial activity. This type of bacteria lives in the absence of oxygen which is ensured by a sealed cover and the natural occurrence of a surface scum or crust. Traditionally built tanks are divided into two compartments with an overall length of three times the breadth. Final processing of sewage is achieved by conveying it through subsoil drainage pipes or a biological filter. Capacity is determined from the simple formula:

$$C = (180 \times P) + 2000$$

where: C = capacity in litres
      P = no. of persons served

E.g. 10 persons; C = (180 × 10) + 2000 = 3800 litres (3·8 m³).

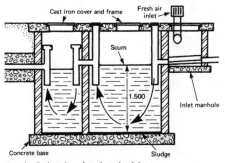

Longitudinal section of septic tank minimum volume under Building Regulations = 2.7 m³

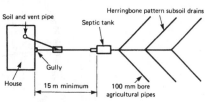

Site plan of installation

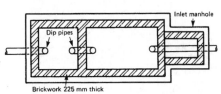

Plan of septic tank

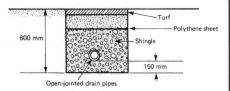

Subsoil irrigation pipe trench

The Klargester settlement tank is a simple, reliable and cost-effective sewage disposal system manufactured from glass reinforced plastics for location in a site prepared excavation. The tanks are produced in capacities ranging from 2700 to 100 000 litres, to suit a variety of applications from individual houses to modest developments including factories and commercial premises. The sewage flows through three compartments (1,2,3) on illustration where it is liquefied by anaerobic bacterial activity. In similarity with traditionally built tanks, sludge settlement at the base of the unit must be removed annually. This is achieved by pushing away the floating ball to give extraction tube access into the lowest chamber. Processed sewage may be dispersed by subsoil irrigation or a biological filter.

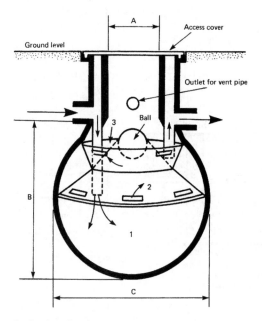

Section through tank

Capacity of tank in litres	Number of users with flow rate per head per day		Nominal dimensions in mm.		
	180 litres	250 litres	A	B	C
2700	4	3	610	1850	1800
3750	9	7	610	2060	2000
4500	14	10	610	2150	2100
6000	22	16	610	2400	2300
7500	30	22	610	2630	2500
10000	44	32	610	2800	2740

Ref: Building Regulations, Approved Document H2: Waste water treatment and cesspools.

The biological disc has many successful applications to modest size buildings such as schools, prisons, country clubs, etc. It is capable of treating relatively large volumes of sewage by an accelerated process. Crude sewage enters the biozone chamber via a deflector box which slows down the flow. The heavier solids sink to the bottom of the compartment and disperse into the main sludge zone. Lighter solids remain suspended in the biozone chamber. Within this chamber, micro-organisms present in the sewage adhere to the partially immersed slowly rotating discs to form a biologically active film feeding on impurities and rendering them inoffensive. Baffles separate the series of rotating fluted discs to direct sewage through each disc in turn. The sludge from the primary settlement zone must be removed every 6 months.

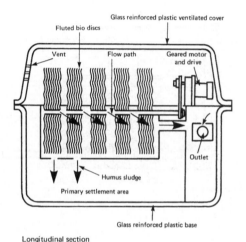

Longitudinal section

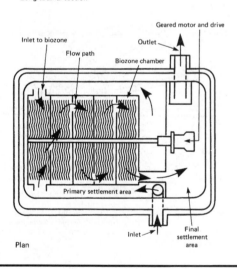

Plan

Biological Filter

Treatment of septic tank effluent – liquid effluent from a septic tank is dispersed from a rotating sprinkler pipe over a filter of broken stone, clinker, coke or polythene shingle. The filter surfaces become coated with an organic film which assimilates and oxidises the pollutants by aerobic bacterial activity. This type of bacteria lives in the presence of oxygen, encouraged by ventilation through underdrains leading to a vertical vent pipe. An alternative process is conveyance and dispersal of septic tank effluent through a system of subsoil drains or a drainage field. To succeed, the subsoil must be porous and the pipes laid above the highest water table level. Alternatively, the primary treated effluent can be naturally processed in constructed wetland phragmite or reed beds (see page 225). Whatever method of sewage containment and processing is preferred, the local water authority will have to be consulted for approval.

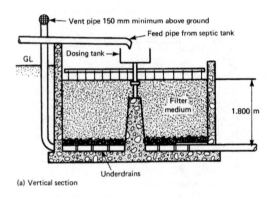

(a) Vertical section

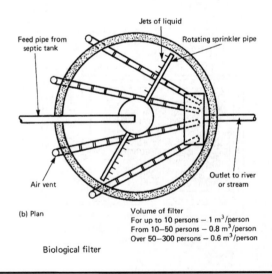

(b) Plan

Volume of filter
For up to 10 persons – 1 m³/person
From 10–50 persons – 0.8 m³/person
Over 50–300 persons – 0.6 m³/person

Biological filter

Drainage fields and mounds are a less conspicuous alternative to use of a biological filter for secondary processing of sewage. Disposal and dispersal is through a system of perforated pipes laid in a suitable drainage medium.

Location:
- Min. 10 m from any watercourse or permeable drain.
- Min. 50 m from any underground water supply.
- Min. distance from a building:

< 5 people	15 m
6–30 people	25 m
31–100 people,	40 m
> 100 people	70 m

- Downslope of any water source.
- Unencroached by any other services.
- Unencroached by access roads or paved areas.

Ground quality:
- Preferably granular, with good percolation qualities. Subsoils of clay composition are unlikely to be suited.
- Natural water table should not rise to within 1 m of distribution pipes invert level.
- Ground percolation test:

  1. Dig several holes 300 x 300 mm, 300 mm below the expected distribution pipe location.

  2. Fill holes to a 300 mm depth of water and allow to seep away overnight.

  3. Next day refill holes to 300 mm depth and observe time in seconds for the water to fall from 225 mm depth to 75 mm. Divide time by 150 mm to ascertain average time (Vp) for water to drop 1 mm.

  4. Apply floor area formula for drainage field:

  $$At = p \times Vp \times 0.25$$

  where, At = floor area (m^2)

  p = no. of persons served

  e.g. 40 min (2400 secs) soil percolation test time in a system serving 6 persons.

  $$Vp = 2400 \div 150 = 16$$

  $$At = 6 \times 16 \times 0.25 = 24 \text{ m}^2$$

Note: Vp should be between 12 and 100. Less than 12 indicates that untreated effluent would percolate into the ground too rapidly. A figure greater than 100 suggests that the field may become saturated.

Typical drainage field

Foul water inlet →

Septic tank or biological processor

Inspection chamber

100 mm diameter perforated pipes

Selected backfill

Geotextile filter membrane

50 mm

500 mm cover

Perforated distribution pipe, max. gradient 1 in 200

300 mm

Clean shingle or 20–50 mm granular material

300 mm minimum

A — A

2 m min. separation

Plan

Section A-A

Typical drainage field

Typical constructed drainage mound

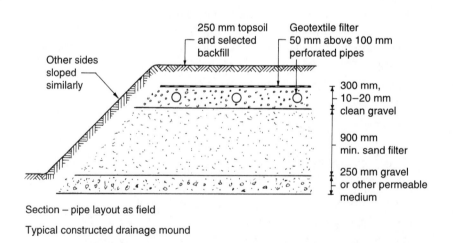

250 mm topsoil and selected backfill

Geotextile filter 50 mm above 100 mm perforated pipes

Other sides sloped similarly

300 mm, 10–20 mm clean gravel

900 mm min. sand filter

250 mm gravel or other permeable medium

Section – pipe layout as field

Typical constructed drainage mound

These provide a natural method for secondary treatment of sewage from septic tanks or biological processing equipment.

Common reeds (Phragmites australis) are located in prepared beds of selected soil or fine gravel. A minimum bed area of 20 m² is considered adequate for up to four users. 5 m² should be added for each additional person. Reeds should be spaced about every 600 mm and planted between May and September. For practical purposes application is limited to about 30 people, due to the large area of land occupied. Regular maintenance is necessary to reduce unwanted weed growth which could restrict fluid percolation and natural processing. The site owners have a legal responsibility to ensure that the beds are not a source of pollution, a danger to health or a nuisance.

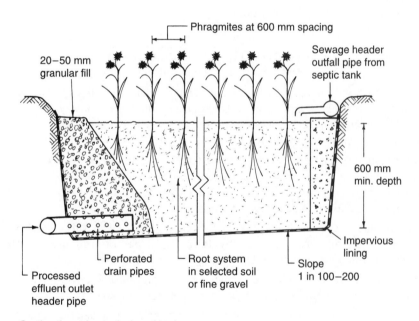

Section through a typical reed bed

Ref. Building Regulations, Approved Document H2: Waste treatment systems and cesspools.

Sustainable Urban Drainage Systems (SUDS)

Extreme weather situations in the UK have led to serious property damage from flooding, as drains, rivers and other watercourses are unable to cope with the unexpected volumes of surface water. A possible means of alleviating this and moderating the flow of surface water is construction of SUDS between the drainage system and its outfall.

Objectives are to:
- decrease the volume of water discharging or running-off from a site or building
- reduce the run-off rate
- filter and cleanse the debris from the water flow.

Formats:
- soakaways
- swales
- infiltration basins and permeable surfaces
- filter drains
- retention or detention ponds
- reed beds.

Soakaways – See page 217. For application to larger areas, see BS EN 752-4: Drain and sewer systems outside buildings.

Swales – Channels lined with grass. These slow the flow of water, allowing some to disperse into the ground as they convey water to an infiltration device or watercourse. They are best suited to housing, car parks and roads.

Infiltration basins and permeable surfaces – Purposely located depressions lined with grass and positioned to concentrate surface water into the ground. Permeable surfaces such as porous asphalt or paving can also be used to the same effect.

Filter drains – Otherwise known as French drains, see page 213. Note that drainage may be assisted by locating a perforated pipe in the centre of the gravel or rubble filling.

Retention or detention ponds – These are man-made catchments to contain water temporarily, for controlled release later.

Reed beds – These are not restricted to processing septic tank effluent, as shown on page 225. They are also a useful filter mechanism for surface water, breaking down pollutants and settlement of solids.

Ref: Sustainable Urban Drainage Systems – A design manual for England and Wales – CIRIA.

The size of gutters and downpipes will depend on the effective surface area to be drained. For flat roofs this is the plan area, whilst pitched roof effective area (A_e) can be calculated from:

Roof plan area ÷ Cosine pitch angle

Roofs over 70° pitch are treated as walls, with the effective area taken as:

Elevational area × 0·5.

Actual rainfall varies throughout the world. For UK purposes, a rate of 75 mm/h (R) is suitable for all but the most extreme conditions. Rainfall run-off (Q) can be calculated from:

$$Q = (A_e \times R) \div 3600 = l/s$$

E.g. a 45° pitched roof of 40 m² plan area.

$$Q = ([40 \div \text{Cos } 45°] \times 75) \div 3600$$
$$Q = ([40 \div 0·707] \times 75) \div 3600$$
$$Q = 1·18 \text{ l/s}$$

Size of gutter and downpipe will depend on profile selected, i.e. half round, ogee, box, etc. Manufacturers' catalogues should be consulted to determine a suitable size. For guidance only, the following is generally appropriate for half round eaves gutters with one end outlet:

Half round gutter (mm)	Outlet dia. (mm)	Flow capacity (l/s)
75	50	0·38
100	63	0·78
115	63	1·11
125	75	1·37
150	90	2·16

Therefore the example of a roof with a flow rate of 1·18 l/s would be adequately served by a 125 mm gutter and 75 mm downpipe.

- Where an outlet is not at the end, the gutter should be sized to the larger area draining into it.
- The distance between a stopped end and an outlet should not exceed 50 times the flow depth.
- The distance between two or more outlets should not exceed 100 times the flow depth (see example below).
- For design purposes, gutter slope is taken as less than 1 in 350.

E.g. a 100 mm half round gutter has a 50 mm depth of flow, therefore:

100 × 50 = 5000 mm or 5m spacing of downpipes.

Refs: Building Regulations, Approved Document H3: Rainwater Drainage.
BS EN 12056-3: Roof drainage, layout and calculations.

When designing rainfall run-off calculations for car parks, playgrounds, roads and other made up areas, a rainfall intensity of 50 mm/h is considered adequate. An allowance for surface permeability (P) should be included, to slightly modify the formula from the preceding page:

$$Q = (A \times R \times P)\ 3600 = l/s$$

Permeability factors:

Asphalt	0·85–0·95
Concrete	0·85–0·95
Concrete blocks (open joint)	0·40–0·50
Gravel drives	0·15–0·30
Grass	0·05–0·25
Paving (sealed joints)	0·75–0·85
Paving (open joints)	0·50–0·70

E.g. a paved area (P = 0·75) 50 m × 24 m (1200 m²).

$$Q = (1200 \times 50 \times 0·75) \div 3600$$
$$Q = 12.5\ l/s\ \text{or}\ 0·0125\ m^3/s$$

The paved area will be served by several gullies (at 1 per 300 m² = 4) with subdrains flowing into a main surface water drain. Each drain can be sized according to the area served, but for illustration purposes, only the main drain is considered here. The pipe sizing formula is:

$$Q = V \times A$$

where: Q = quantity of water (m³/s)
   V = velocity of flow (min. 0·75 m/s) – see next page
   A = area of water flowing (m²)

Drains should not be designed to flow full bore as this leaves no spare capacity for future additions. Also, fluid flow is eased by the presence of air space. Assuming half full bore, using the above figure of 0.0125 m³/s, and the minimum velocity of flow of 0·75 m/s:

$$Q = V \times A$$
$$0·0125 = 0·75 \times A$$

Transposing,

$$A = 0·0125 \div 0·75$$
$$A = 0·017\ m^2$$

This represents the area of half the pipe bore, so the total pipe area is double, i.e. 0.034 m².
Area of a circle (pipe) = πr^2 where r = radius of pipe (m).

Transposing,

$$r = \sqrt{Area \div \pi}$$
$$r = \sqrt{0·034 \div \pi}$$
$$r = 0·104\ m\ \text{or}\ 104\ mm$$

Therefore the pipe diameter = 2 × 104 = 208 mm.
The nearest commercial size is 225 mm nominal inside diameter.

Velocity of flow – 0·75 m/s – is the accepted minimum to achieve self-cleansing. It is recognised that an upper limit is required to prevent separation of liquids from solids. A reasonable limit is 1·8 m/s for both surface and foul water drainage, although figures up to 3 m/s can be used especially if grit is present. The selected flow rate will have a direct effect on drain gradient, therefore to moderate excavation costs a figure nearer the lower limit is preferred. Also, if there is a natural land slope and excavation is a constant depth, this will determine the gradient and velocity of flow.

Hydraulic mean depth (HMD) – otherwise known as hydraulic radius represents the proportion or depth of flow in a drain. It will have an effect on velocity and can be calculated by dividing the area of water flowing in a drain by the contact or wetted perimeter. Thus for half full bore:

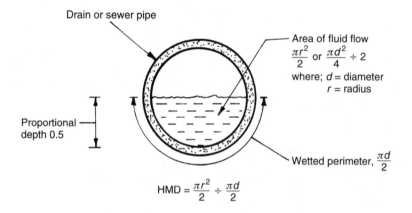

Drain or sewer pipe

Area of fluid flow
$\frac{\pi r^2}{2}$ or $\frac{\pi d^2}{4} \div 2$
where; d = diameter
r = radius

Proportional depth 0.5

Wetted perimeter, $\frac{\pi d}{2}$

$$HMD = \frac{\pi r^2}{2} \div \frac{\pi d}{2}$$

This table summarises HMD for proportional flows:

Depth of flow	HMD
0·25	Pipe dia. (m) ÷ 6·67
0·33	Pipe dia. (m) ÷ 5·26
0·50	Pipe dia. (m) ÷ 4·00
0·66	Pipe dia. (m) ÷ 3·45
0·75	Pipe dia. (m) ÷ 3·33
Full	Pipe dia. (m) ÷ 4·00

E.g. a 225 mm (0·225 m) drain flowing half bore:

$$HMD = 0·225 \div 4 = 0·05625$$

The fall, slope or inclination of a drain or sewer will relate to the velocity of flow and the pipe diameter. The minimum diameter for surface water and foul water drains is 75 mm and 100 mm respectively.

Maguire's rule of thumb is an established measure of adequate fall on drains and small sewers. Expressing the fall as 1 in x, where 1 is the vertical proportion to horizontal distance x, then:

$$x = \text{pipe diameter in mm} \div 2 \cdot 5$$

E.g. a 150 mm nominal bore drain pipe:

$$x = 150 \div 2 \cdot 5 = 60, \text{ i.e. 1 in 60 minimum gradient.}$$

Pipe dia. (mm)	Gradient
100	1 in 40
150	1 in 60
225	1 in 90
300	1 in 120

For full and half bore situations, these gradients produce a velocity of flow of about 1·4 m/s.

The Building Regulations, Approved Documents H1 and H3, provide guidance on discharge capacities for surface water drains running full and foul water drains running 0·75 proportional depth. The chart below is derived from this data:

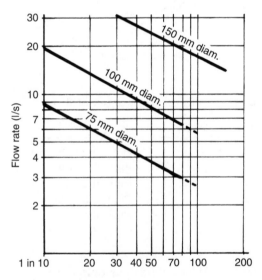

Notes: (1) – – = Rainwater only.
(2) 75 mm is for rainwater or waste water (no soil) only.
(3) To convert l/s to m^3/s, divide by 1000.

An alternative approach to drainage design is attributed to the established fluid flow research of Antoine Chezy and Robert Manning. This can provide lower gradients:

Chezy's formula: $V = C\sqrt{m} \times i$

where, V = velocity of flow (min. 0·75 m/s)

C = Chezy coefficient

m = HMD (see page 229)

i = inclination or gradient as 1/X or 1÷X.

Manning's formula: $C = (1 \div n) \times (m)^{\frac{1}{6}}$

where:
C = Chezy coefficient

n = coefficient for pipe roughness 0·010*

m = HMD

$\frac{1}{6}$ = sixth root

*A figure of 0·010 is appropriate for modern high quality uPVC and clay drainware – for comparison purposes it could increase to 0·015 for a cast concrete surface.

E.g. A 300 mm (0·3 m) nominal bore drain pipe flowing 0·5 proportional depth half full bore). The Chezy coefficient can be calculated from Manning's formula:

$$HMD = 0·3 \div 4 = 0·075 \text{ (see page 229)}$$
$$C = (1 \div n) \times (m)^{\frac{1}{6}}$$
$$C = (1 \div 0.010) \times (0·075)^{\frac{1}{6}} = 65$$

Using a velocity of flow shown on the previous page of 1·4 m/s, the minimum gradient can be calculated from Chezy's formula:

$$V = C\sqrt{m} \times i$$
$$1·4 = 65\sqrt{0·075 \times i}$$
$$(1·4 \div 65)^2 = 0·075 \times i$$
$$0·00046 \div 0·075 = i$$
$$i = 0·00617$$
$$i = 1 \div X$$
$$\text{So, } X = 1 \div 0·00617 = 162, \text{ i.e. 1 in 162}$$

Drainage Design – Foul Water (1)

Small drainage schemes:

<20 dwellings, 100 mm nom. bore pipe, min. gradient 1 in 80.

20–150 dwellings, 150 mm nom. bore pipe, min. gradient 1 in 150.

Minimum size for a public sewer is 150 mm. Most water authorities will require a pipe of at least 225 mm to allow for future developments and additions to the system.

For other situations, estimates of foul water flow may be based on water consumption of 225 litres per person per day. A suitable formula for average flow would be:

$$l/s = \frac{\text{Half consumption per person per day}}{6 \text{ hours} \times 3600 \text{ seconds}}$$

Note: 6 hours is assumed for half daily flow.

E.g. A sewer for an estate of 500, four-person dwellings:

$$l/s = \frac{112 \times 4 \times 500}{6 \times 3600} = 10 \cdot 4$$

Assuming maximum of 5 times average flow = 52 l/s or 0·052 m³/s.

Using the formula Q = V × A (see 228) with a velocity of flow of, say, 0·8 m/s flowing half full bore (0·5 proportional depth):

$$Q = 0 \cdot 052 \text{ m}^3/\text{s}$$
$$V = 0 \cdot 8 \text{ m/s}$$
$$A = \text{half bore (m}^2)$$

Transposing the formula:

$$A = Q \div V$$
$$A = 0 \cdot 052 \div 0 \cdot 8 = 0 \cdot 065 \text{ m}^2$$

A represents half the bore, therefore the full bore area = 0·130 m².

Area of a circle (pipe) = πr^2, therefore $\pi r^2 = 0 \cdot 130$
Transposing: $r = \sqrt{0 \cdot 130 \div \pi}$
$r = 0 \cdot 203$ m radius
Therefore diameter = 0·406 m or 406 mm
Nearest commercial size is 450 mm nominal bore.

An alternative approach to estimating drain and sewer flows is by summation of discharge units and converting these to a suitable pipe size. Discharge units represent frequency of use and load producing properties of sanitary appliances. They are derived from data in BS EN 12056-2 and BS EN 752, standards for drainage systems inside and outside buildings, respectively. Although intended primarily for sizing discharge stacks, they are equally well applied to drains and sewers.

Appliance	Situation	No. of units
WC	Domestic	7
	Commercial	14
	Public	28
Basin	Domestic	1
	Commercial	3
	Public	6
Bath	Domestic	7
	Commercial	18
Sink	Domestic	6
	Commercial	14
	Public	27
Shower	Domestic	1
	Commercial	2
Urinal		0·3
Washing machine		4–7
Dishwasher		4–7
Waste disposal unit		7
Group of WC, bath and 1 or 2 basins		14
Other fittings with an outlet of:		
	50 mm nom. i.d.	7
	65 mm nom. i.d.	7
	75 mm nom. i.d.	10
	90 mm nom. i.d.	10
	100 mm nom. i.d.	14

Note:

Domestic = houses and flats.

Commercial = offices, factories, hotels, schools, hospitals, etc.

Public or peak = cinemas, theatres, stadia, sports centres, etc.

Using the example from page 232, i.e. 500, four-person dwellings. Assuming 1 WC, 1 shower, 2 basins, 2 sinks, 1 group of appliances, washing machine and dishwasher per dwelling.

WC	7 discharge units
Shower	1 discharge unit
Basins	2 discharge units
Sinks	12 discharge units
Group	14 discharge units
Washing machine	4 discharge units
Dishwasher	4 discharge units

Total = 44 discharge units × 500 dwellings = 22000 discharge units.

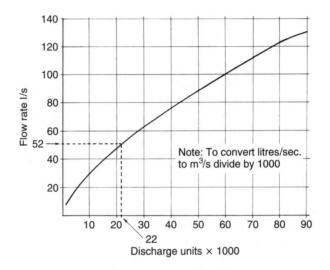

Sewer size can be calculated for a 0·052 m³/s flow at half full bore using the formula, Q = V × A as shown page 232. Gradient can be calculated using the Chezy and Manning formulas as shown on page 231.

Combined surface and foul water drains will require separate calculations for both flow conditions. Drain size can be based on the greater flow, not the total flow as the chance of the peak flows of both coinciding is remote.

The quantity and location of refuse chutes depends upon:

- layout of the building
- number of dwellings served – max. six per hopper
- type of material stored
- frequency of collection
- volume of refuse
- refuse vehicle access – within 25 m.

The chute should be sited away from habitable rooms, but not more than 30 m horizontal distance from each dwelling. It is more economical to provide space for additional storage beneath the chute, than to provide additional chutes. Chute linings are prefabricated from refractory or Portland cement concrete with a smooth and impervious internal surface. The structure containing the chute void should have a fire resistance of 1 hour. The refuse chamber should also have a 1 hour fire resistance and be constructed with a dense impervious surface for ease of cleaning.

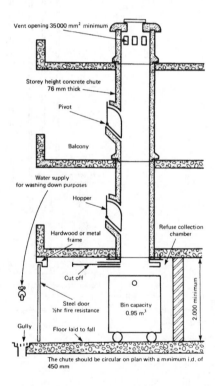

Vent opening 35000 mm² minimum

Storey height concrete chute 76 mm thick

Pivot

Balcony

Water supply for washing down purposes

Hopper

Refuse collection chamber

Hardwood or metal frame

Cut off

Steel door ½hr fire resistance

Bin capacity 0.95 m³

2 000 minimum

Gully

Floor laid to fall

The chute should be circular on plan with a minimum i.d. of 450 mm

Ref: BS 5906: Code of practice for storage and on-site treatment of solid waste from buildings.

235

On-site Incineration of Refuse

This system has a flue to discharge the incinerated gaseous products of combustion above roof level. A fan ensures negative pressure in the discharge chute to prevent smoke and fumes being misdirected. A large combustion chamber receives and stores the refuse until it is ignited by an automatic burner. Duration of burning is thermostatically and time controlled. Waste gases are washed and cleaned before discharging into the flue. There is no restriction on wet or dry materials, and glass, metal or plastics may be processed.

Health risks associated with storing putrefying rubbish are entirely eliminated as the residue from combustion is odourless and sterile. Refuse removal costs are reduced because the residual waste is only about 10% of the initial volume.

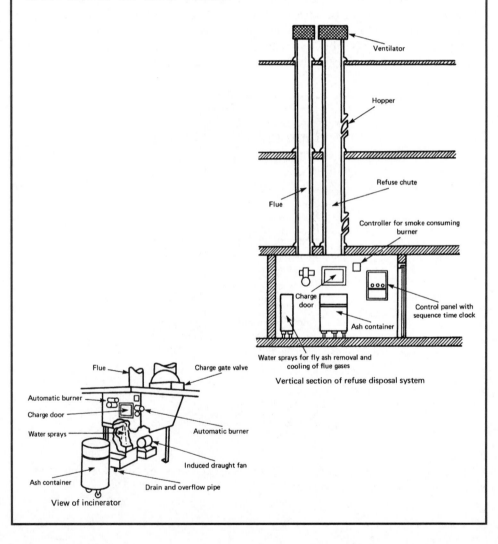

Vertical section of refuse disposal system

View of incinerator

Incinerators are the quickest, easiest and most hygienic method for disposing of dressings, swabs and sanitary towels. They are usually installed in office lavatories, hospitals and hotels. When the incinerator door is opened, gas burners automatically ignite and burn the contents. After a predetermined time, the gas supply is cut off by a time switch. Each time the door is opened, the time switch reverts to its original position to commence another burning cycle. Incinerators have a removable ash pan and a fan assisted flue to ensure efficient extraction of the gaseous combustion products. In event of fan failure, a sensor ensures that gas burners cannot function. The gas pilot light has a thermocoupled flame failure device.

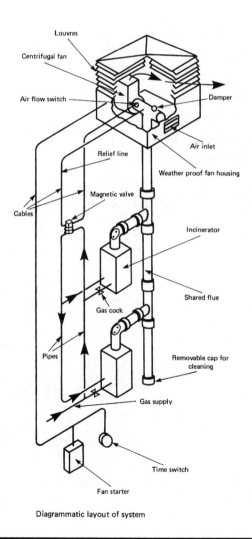

Diagrammatic layout of system

The Matthew-Hall Garchey System

Food waste, bottles, cans and cartons are disposed of at source, without the need to grind or crush the refuse. A bowl beneath the sink retains the normal waste water. Refuse is placed inside a central tube in the sink. When the tube is raised the waste water and the refuse are carried away down a stack or discharge pipe to a chamber at the base of the building. Refuse from the chamber is collected at weekly intervals by a specially equipped tanker in which the refuse is compacted into a damp, semi-solid mass that is easy to tip. One tanker has sufficient capacity to contain the refuse from up to 200 dwellings. Waste water from the tanker is discharged into a foul water sewer.

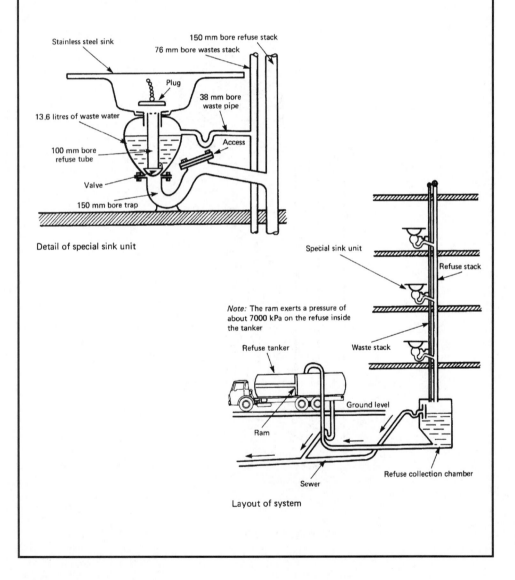

Detail of special sink unit

Layout of system

Note: The ram exerts a pressure of about 7000 kPa on the refuse inside the tanker

Refuse from conventional chutes is collected in a pulveriser and disintegrated by grinder into pieces of about 10 mm across. The refuse is then blown a short distance down a 75 mm bore pipe in which it is retained, until at predetermined intervals a flat disc valve opens. This allows the small pieces of refuse to be conveyed by vacuum or air stream at 75 to 90 km/h through a common underground service pipe of 150–300 mm bore. The refuse collection silo may be up to 2·5 km from the source of refuse. At the collection point the refuse is transferred by a positive pressure pneumatic system to a treatment plant where dust and other suspended debris is separated from bulk rubbish. The process can be adapted to segregate salvagable materials such as metals, glass and paper.

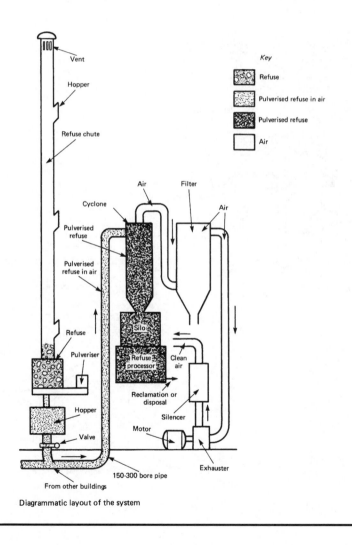

Diagrammatic layout of the system

Food Waste Disposal Units

Food waste disposal units are designed for application to domestic and commercial kitchen sinks. They are specifically for processing organic food waste and do not have the facility to dispose of glass, metals, rags or plastics. Where a chute or Garchey system is not installed, these units may be used to reduce the volume otherwise deposited in dustbins or refuse bags.

Food waste is fed through the sink waste outlet to the unit. A grinder powered by a small electric motor cuts the food into fine particles which is then washed away with the waste water from the sink. The partially liquefied food particles discharge through a standard 40 mm nominal bore waste pipe into a back inlet gully. As with all electrical appliances and extraneous metalwork, it is essential that the unit and the sink are earthed.

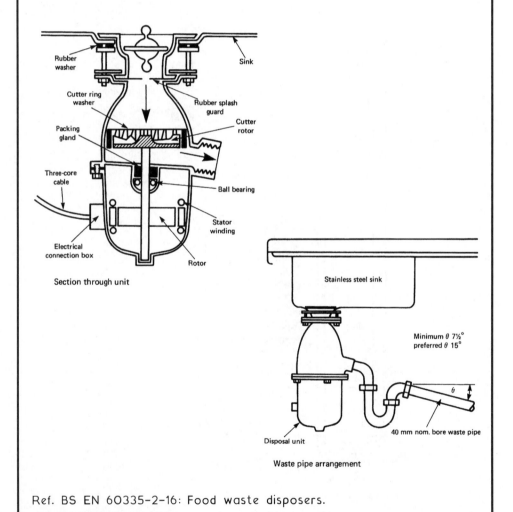

Section through unit

Waste pipe arrangement

Ref. BS EN 60335-2-16: Food waste disposers.

240

8 SANITARY FITMENTS AND APPLIANCES: DISCHARGE AND WASTE SYSTEMS

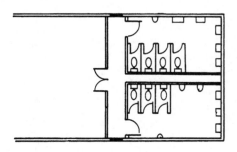

FLUSHING CISTERNS, TROUGHS AND VALVES

WATER CLOSETS

BIDETS

SHOWERS

BATHS

SINKS

WASH BASINS AND TROUGHS

URINALS

HOSPITAL SANITARY APPLIANCES

SANITARY CONVENIENCES

TRAPS AND WASTE VALVE

SINGLE STACK SYSTEM AND VARIATIONS

ONE- AND TWO-PIPE SYSTEMS

PUMPED WASTE SYSTEM

WASH BASINS – WASTE ARRANGEMENTS

WASHING MACHINE AND DISHWASHER WASTES

AIR TEST

SANITATION – DATA

OFFSETS

GROUND FLOUR APPLIANCES – HIGH RISE BUILDINGS

FIRE STOPS AND SEALS

FLOW RATES AND DISCHARGE UNITS

SANITATION DESIGN – DISCHARGE STACK SIZING

Bell type – this form of flushing cistern is now virtually obsolete, although some reproductions are available for use in keeping with refurbishment of historic premises. Cast iron originals may still be found in use in old factories, schools and similar established buildings. It is activated by the chain being pulled which also lifts the bell. As the chain is released the bell falls to displace water down the stand pipe, effecting a siphon which empties the cistern. The whole process is relatively noisy.

Disc type – manufactured in a variety of materials including plastics and ceramics for application to all categories of building. Depressing the lever raises the piston and water is displaced over the siphon. A siphonic action is created to empty the cistern. Some cisterns incorporate an economy or dual flush siphon. When the lever is depressed and released promptly, air passing through the vent pipe breaks the siphonic action to give a 4·5 litre flush. When the lever is held down a 7·5 litre flush is obtained. From 2001 the maximum permitted single flush to a WC pan is 6 litres.

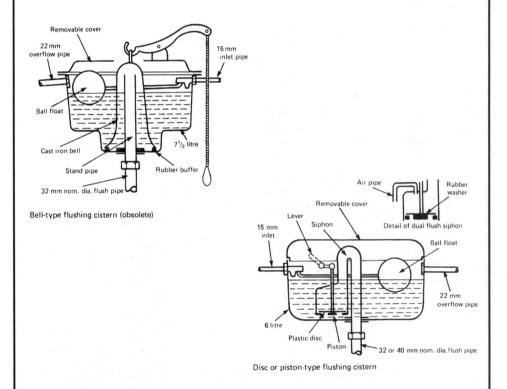

Bell-type flushing cistern (obsolete)

Disc or piston-type flushing cistern

Refs: BS 1125 and 7357: Specifications for WC flushing cisterns.
The Water Supply (Water Fittings) Regulations 1999.

Flushing Trough

A flushing trough may be used as an alternative to several separate flushing cisterns where a range of WCs are installed. They are particularly applicable to school, factory and office sanitary accommodation. Trough installation is economic in equipment and time. It is also more efficient in use as there is no waiting between consecutive flushes. The disadvantage is that if it needs maintenance or repair, the whole range of WCs are unusable. The trough may be bracketed from the rear wall and hidden from view by a false wall or ceiling.

The siphon operates in the same manner as in a separate cistern, except that as water flows through the siphon, air is drawn out of the air pipe. Water is therefore siphoned out of the anti-siphon device, the flush terminated and the device refilled through the small hole.

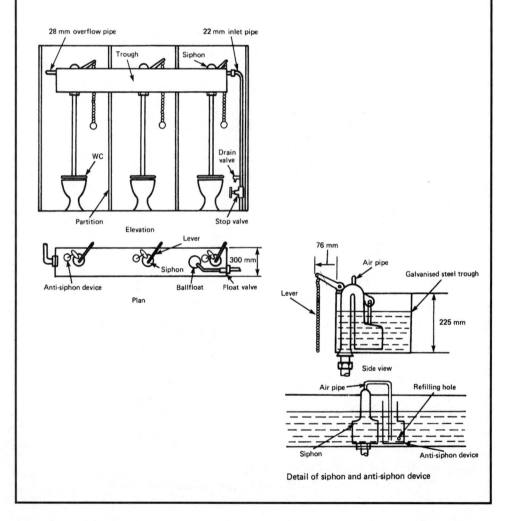

Detail of siphon and anti-siphon device

Roger Field's flushing cistern is used for automatically flushing WCs. It has application to children's lavatories and other situations where the users are unable to operate a manual flush device. As the cistern fills, air in the stand pipe is gradually compressed. When the head of water 'H' is slightly above the head of water 'h', water in the trap is forced out. Siphonic action is established and the cistern flushes the WC until air enters under the dome to break the siphon.

With the smaller urinal flush cistern, water rises inside the cistern until it reaches an air hole. Air inside the dome is trapped and compressed as the water rises. When water rises above the dome, compressed air forces water out of the U tube. This lowers the air pressure in the stand pipe creating a siphon to empty the cistern. Water in the reserve chamber is siphoned through the siphon tube to the lower well.

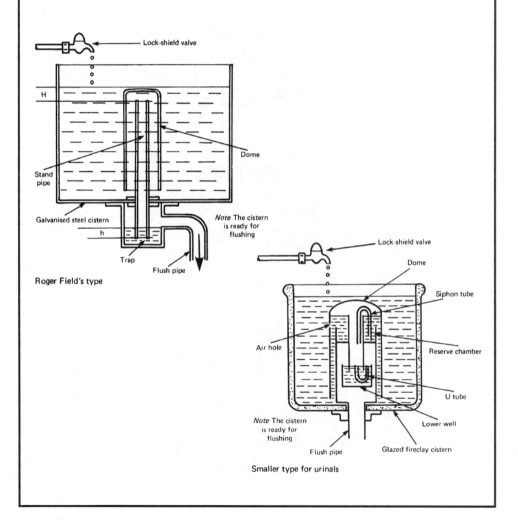

Lock-shield valve

H

Dome

Stand pipe

Galvanised steel cistern

Note The cistern is ready for flushing

h

Trap

Flush pipe

Roger Field's type

Lock-shield valve

Dome

Siphon tube

Air hole

Reserve chamber

U tube

Note The cistern is ready for flushing

Flush pipe

Lower well

Glazed fireclay cistern

Smaller type for urinals

Flushing Valves

Flushing valves are a more compact alternative to flushing cisterns, often used in marine applications, but may only be used in buildings with approval of the local water authority. The device is a large equilibrium valve that can be flushed at any time without delay, provided there is a constant source of water from a storage cistern. The minimum and maximum head of water above valves is 2·2 m and 36 m respectively. When the flushing handle is operated, the release valve is tilted and water displaced from the upper chamber. The greater force of water under piston 'A' lifts valve 'B' from its seating and water flows through the outlet. Water flows through the by-pass and refills the upper chamber to cancel out the upward force acting under piston 'A'. Valve 'B' closes under its own weight.

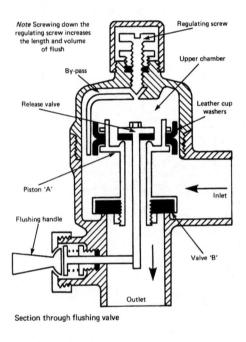

Section through flushing valve

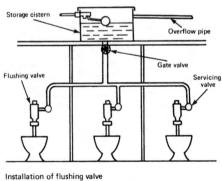

Installation of flushing valve

- The minimum flow rate at an appliance is 1·2 litres per second. By domestic standards this is unrealistically high, therefore pressure flushing valves are not permitted in houses.
- Where connected to a mains supply pipe or a cistern distributing pipe, a flushing valve must include a backflow prevention device having a permanently vented pipe interrupter situated at least 300 mm above the spillover level of the served WC.
- If a permanently vented pipe interrupter is not fitted, the water supply to a flushing valve must be from a dedicated cistern with a type B air gap (see page 16) below its float valve delivery.
- The maximum flush in a single operation is 6 litres.
- Flushing valves may be used to flush urinals. In this situation they should deliver no more than 1·5 litres of water to each bowl or position per operation. See page 261.

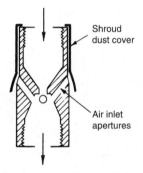

Shroud dust cover

Air inlet apertures

Pipe interrupter with a permanent atmospheric vent

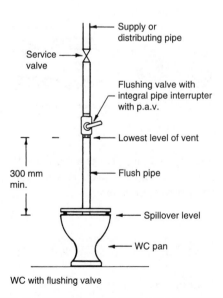

Supply or distributing pipe

Service valve

Flushing valve with integral pipe interrupter with p.a.v.

Lowest level of vent

300 mm min.

Flush pipe

Spillover level

WC pan

WC with flushing valve

Washdown Water Closet and Joints

The washdown WC pan is economic, simple and efficient. It rarely becomes blocked and can be used in all types of buildings with colour variations to suit internal decor. Manufacture is primarily from vitreous china, although glazed fireclay and stoneware have been used. Stainless steel WCs can be specified for use in certain public areas and prisons. Pan outlet may be horizontal, P, S, left or right handed. Horizontal outlet pans are now standard, with push-fit adaptors to convert the pan to whatever configuration is required. Plastic connectors are commonly used for joining the outlet to the soil branch pipe. The flush pipe joint is usually made with a rubber cone connector which fits tightly between WC and pipe.

WC pan outlet < 80 mm, trap diameter = 75 mm

WC pan outlet > 80 mm, trap diameter = 100 mm

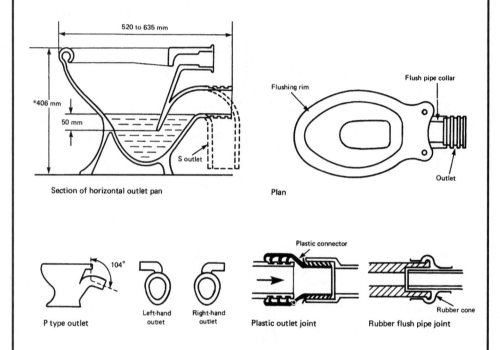

Section of horizontal outlet pan

Plan

P type outlet

Left-hand outlet

Right-hand outlet

Plastic outlet joint

Rubber flush pipe joint

*Note: Add approximately 25 mm to the top of the WC to allow for seat height. Overall height for disabled is 450 mm, junior school children 355 mm and infants 305 mm.

Refs: BS 5503: Vitreous china washdown WC pans with horizontal outlet.
BS 5504: Wall hung WC pan.

Siphonic WCs are much quieter in operation than washdown WCs and they require less flush action to effect an efficient discharge. They are not suitable for schools, factories and public buildings as they are more readily blocked if not used carefully.

The double trap type may be found in house and hotel bathrooms. When flushed, water flows through the pressure reducing fitting 'A'. This reduces the air pressure in chamber 'B'. Siphonic action is established and the contents of the first trap are removed. This is replenished from reserve chamber 'C'.

The single trap variant is simpler and has limited application to domestic bathrooms. When the cistern is flushed, the content is discharged through the trap and restricted in flow by the specially shaped pan outlet pipe. The pipe fills with water which causes a siphonic effect. Sufficient water remains in the reserve chamber to replenish the seal.

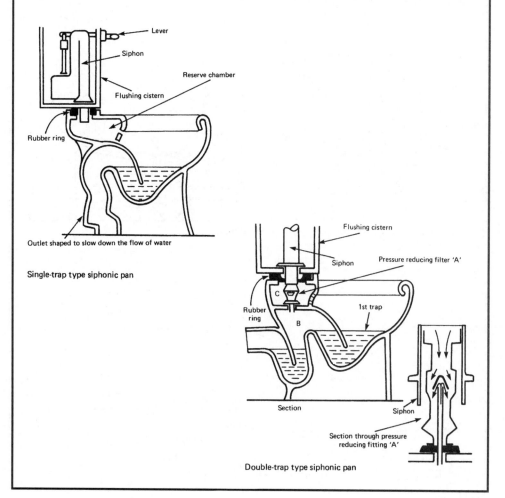

Single-trap type siphonic pan

Double-trap type siphonic pan

Bidets

A bidet is classified as a waste fitting. The requirements for a discharge pipe from a bidet may therefore be treated in the same manner as a basin waste of the same diameter – nominally 32 mm. It is an ablutionary fitting used for washing the excretory organs, but may also be used as a foot bath. Hot and cold water supplies are mixed to the required temperature for the ascending spray. For greater comfort the rim of the fitting may be heated from the warm water. Ascending spray type bidets are not favoured by the water authorities because the spray nozzle is below the spill level, risking water being back-siphoned into other draw off points. This is prevented by having independent supply pipes to the bidet which are not connected to any other fittings. A further precaution would be installation of check valves on the bidet supply pipes or a thermostatic regulator with integral check valves. Over the rim hot and cold supplies are preferred with a type 'A' air gap (see page 16) between rim and tap outlets.

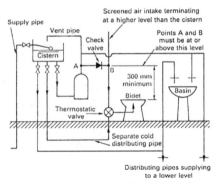

Installation pipework for bidet

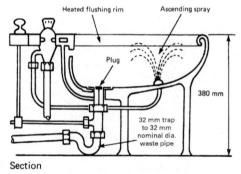

Section

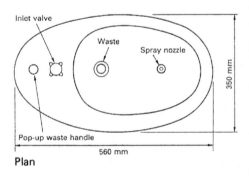

Plan

Ref: BS 5505: Specification for bidets.

A shower is more economic to use than a bath as it takes less hot water (about one-third), it is arguably more hygienic and it takes up less space. The mixing valve may be non-thermostatic or thermostatic: the latter is preferred to avoid the risk of scalding. A minimum 1 m head of water should be allowed above the shower outlet. If this is impractical, a pumped delivery could be considered (see next page). The shower outlet (rose) should also be at least 2 m above the floor of the shower tray. Supply pipes to individual showers are normally 15 mm o.d. copper or equivalent. These should incorporate double check valves if there is a flexible hose to the rose, as this could be left in dirty tray water which could back-siphon. An exception to check valves is where the shower head is fixed and therefore well above the air gap requirements and spill over level of the tray.

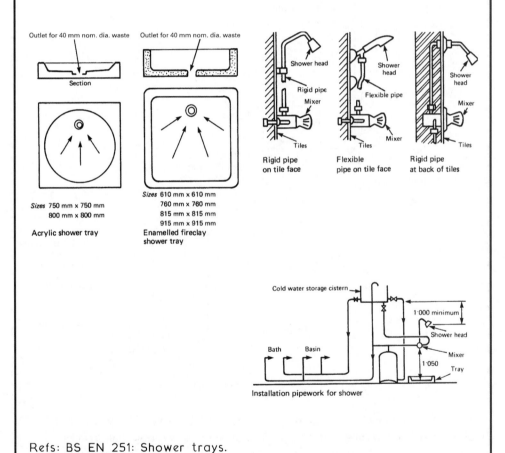

Outlet for 40 mm nom. dia. waste

Section

Sizes 750 mm x 750 mm
800 mm x 800 mm

Acrylic shower tray

Outlet for 40 mm nom. dia. waste

Sizes 610 mm x 610 mm
760 mm x 760 mm
815 mm x 815 mm
915 mm x 915 mm

Enamelled fireclay
shower tray

Shower head
Rigid pipe
Mixer
Tiles

Rigid pipe
on tile face

Shower head
Flexible pipe
Mixer
Tiles

Flexible
pipe on tile face

Shower head
Mixer
Tiles

Rigid pipe
at back of tiles

Cold water storage cistern

1·000 minimum
Shower head
Mixer
1·050
Tray

Bath   Basin

Installation pipework for shower

Refs: BS EN 251: Shower trays.
       BS 7015: Specification for cast acrylic shower trays.
       BS 6340: Shower units (various specifications).

Pumped Showers

Where the 1 m minimum head of water above the shower outlet is not available and it is impractical to raise the level of the supply cistern, a pump can be fitted to the mixer outlet pipe or on the supply pipes to the mixer. The pump is relatively compact and small enough to be installed on the floor of an airing cupboard or under the bath. It must be accessible for maintenance, as the pump should be installed with filters or strainers which will require periodic attention, particularly in hard water areas. The pump will operate automatically in response to the shower mixer being opened. A pressure sensor and flow switch detect water movement to activate the pump and vice versa. Electricity supply can be from an isolating switch or pull cord switch with a 3 amp fuse overload protection spurred off the power socket ring main.

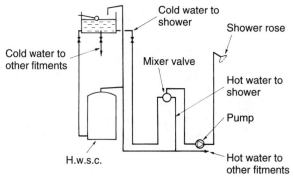

Pump on mixed water outlet

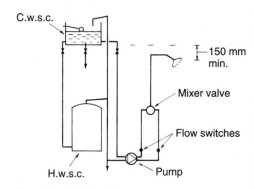

Pumped supply to mixer

Note: Double check valves may be required on the supply pipes as described on the previous page. The mixing valve and pump may incorporate check valves – refer to manufacturer's information.

Instantaneous electric water heating for showers is an economic, simple to install alternative to a pumped shower. This is particularly apparent where there would otherwise be a long secondary flowpipe from the hot water storage cylinder to the shower outlet, possibly requiring additional secondary return pipework to avoid a long 'dead leg'. Cold water supply is taken from the rising main in 15 mm o.d. copper tube. This will provide a regulated delivery through the shower unit of up to 3 litres/min. The unit contains an electric element, usually of 7·2 or 8·4 kW rating. It also has a number of built in safety features:

- Automatic low pressure switch to isolate the element if water pressure falls significantly or the supply is cut off.
- Thermal cut off. This is set by the manufacturer at approximately 50°C to prevent the water overheating and scalding the user.
- Non-return or check valve on the outlet to prevent back-siphonage.

Electricity supply is an independent radial circuit, originating at the consumer unit with a miniature circuit breaker (MCB) appropriately rated. Alternatively a suitable rated fuse way may be used in the consumer unit and added protection provided with an in-line residual current device (RCD) trip switch. All this, of course, is dependent on there being a spare way in the consumer unit. If there is not, there will be additional expenditure in providing a new consumer unit or a supplementary fuse box. A double pole cord operated pull switch is located in the shower room to isolate supply.

Shower rating (kW)	Cable length (m)	Fuse of MCB rating (amps)	Cable size (mm²)
7·2	<13	30 or 32	4
7·2	13–20	30 or 32	6
7·2	20–35	30 or 32	10
8·4	<17	40 or 45	6
8·4	17–28	40 or 45	10

Ref: BS 6340: Shower units.

Unit detail and installation:

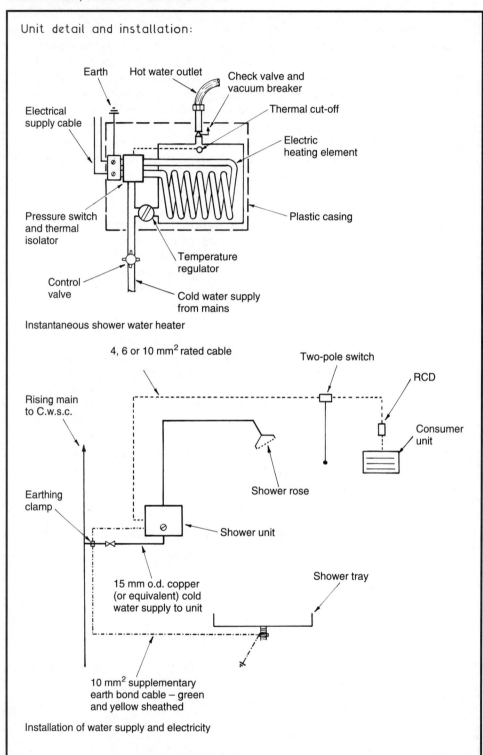

Earth

Hot water outlet

Check valve and vacuum breaker

Electrical supply cable

Thermal cut-off

Electric heating element

Pressure switch and thermal isolator

Temperature regulator

Control valve

Cold water supply from mains

Plastic casing

Instantaneous shower water heater

4, 6 or 10 mm^2 rated cable

Two-pole switch

RCD

Rising main to C.w.s.c.

Consumer unit

Shower rose

Earthing clamp

Shower unit

Shower tray

15 mm o.d. copper (or equivalent) cold water supply to unit

10 mm^2 supplementary earth bond cable – green and yellow sheathed

Installation of water supply and electricity

Baths are manufactured in acrylic sheet, reinforced glass fibre, enamelled pressed steel and enamelled cast iron. The acrylic sheet bath has the advantage of light weight to ease installation, it is comparatively inexpensive and is available in a wide range of colours. However, special cleaning agents must be used otherwise the surface can become scratched. It will require a timber support framework, normally laid across metal cradles. Traditional cast iron baths are produced with ornate feet and other features. Less elaborate, standard baths in all materials can be panelled in a variety of materials including plastic, veneered chipboard and plywood.

The corner bath is something of a luxury. It may have taps located to one side to ease accessibility. A Sitz bath is stepped to form a seat. It has particular application to nursing homes and hospitals for use with the elderly and infirm.

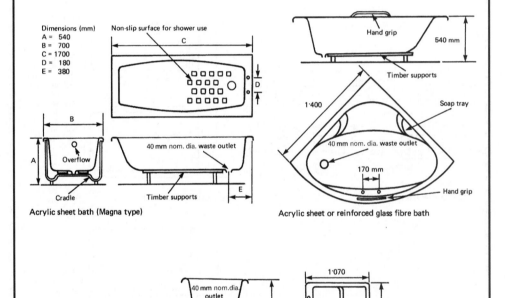

Dimensions (mm)
A = 540
B = 700
C = 1700
D = 180
E = 380

Non-slip surface for shower use

Hand grip

540 mm

Timber supports

1·400

Soap tray

40 mm nom. dia. waste outlet

170 mm

Overflow

Cradle

40 mm nom. dia. waste outlet

Timber supports

Hand grip

Acrylic sheet bath (Magna type)

Acrylic sheet or reinforced glass fibre bath

1·070

40 mm nom.dia. outlet

Section

760 mm

Plan

Enamelled cast iron Sitz bath

685 mm

Refs: BS 1189 and 1390: Specifications for baths made from porcelain enamelled cast iron and vitreous enamelled sheet steel, respectively.
BS 4305: Baths for domestic purposes made of acrylic material.
BS EN 232: Baths - connecting dimensions.

Sinks

Sinks are designed for culinary and other domestic uses. They may be made from glazed fireclay, enamelled cast iron or steel, stainless steel or from glass fibre reinforced polyester.

The Belfast sink has an integral weir overflow and water may pass through this to the waste pipe via a slotted waste fitting. It may have a hardwood or dense plastic draining board fitted at one end only or a draining board fitted on each end. Alternatively, the sink may be provided with a fluted drainer of fireclay. The London sink has similar features, but it does not have an integral overflow. In recent years sinks of this type have lost favour to surface built-in metal and plastic materials, but there is now something of a resurgence of interest in these traditional fittings. Stainless steel sinks may have single or double bowls, with left- or right-hand drainers or double drainers. These can be built into a work surface or be provided as a sink unit with cupboards under. The waste outlet is a standard 40 mm nominal diameter.

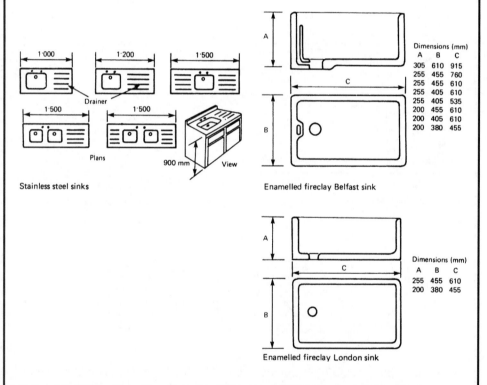

Stainless steel sinks

Enamelled fireclay Belfast sink

Dimensions (mm)

A	B	C
305	610	915
255	455	760
255	455	610
255	405	610
255	405	535
200	455	610
200	405	610
200	380	455

Enamelled fireclay London sink

Dimensions (mm)

A	B	C
255	455	610
200	380	455

Refs: BS 1206: Specification for fireclay sinks, dimensions and workmanship.
BS 1244: Metal sinks for domestic purposes.

These are rarely necessary in domestic situations, but have an application to commercial premises, schools, hospitals and similar public buildings. They are usually located inside the cleaning contractor's cubicle and are fitted at quite a low level to facilitate ease of use with a bucket. They are normally supported by built-in cantilever brackets and are additionally screwed direct to the wall to prevent forward movement. 13 mm bore (half inch) hot and cold water draw off bib-taps may be fitted over the sink, at sufficient height for a bucket to clear below them. 19 mm bore (three-quarter inch) taps may be used for more rapid flow. A hinged stainless steel grating is fitted to the sink as a support for the bucket. The grating rests on a hardwood pad fitted to the front edge of the sink to protect the glazed finish. A 40 mm nominal diameter waste pipe is adequate for this type of sink.

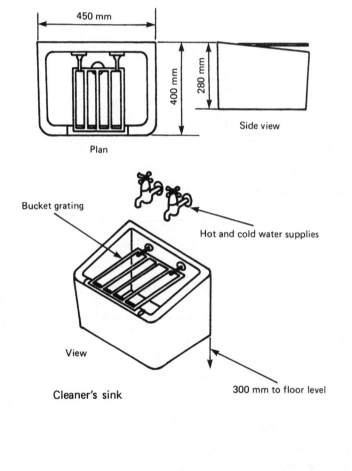

Plan

Side view

Bucket grating

Hot and cold water supplies

View

Cleaner's sink

300 mm to floor level

Wash Basins

There are various types of basin, ranging in size and function from hand rinsing to surgical use. A standard basin for domestic application to bathrooms and cloakrooms consists of a bowl, soap tray, weir overflow and holes for taps and outlet. It may be supported by cast iron brackets screwed to the wall, a corbel which is an integral part of the basin built into the wall or a floor pedestal which conceals the pipework. Water supply is through 13 mm (half inch) pillar taps for both hot and cold. A standard 32 mm nominal diameter waste outlet with a slot to receive the integral overflow connects to a trap and waste pipe of the same diameter. A plug and chain normally controls outflow, but some fittings have a pop-up waste facility.

Most basins are made from coloured ceramic ware or glazed fireclay. There are also some metal basins: stainless steel to BS 1329 and porcelain enamelled sheet steel or cast iron.

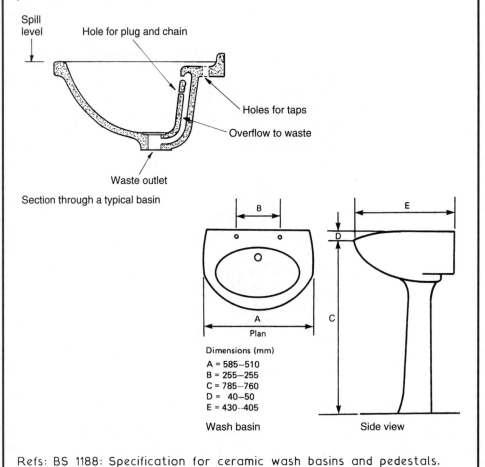

Spill level

Hole for plug and chain

Holes for taps

Overflow to waste

Waste outlet

Section through a typical basin

B

A
Plan

Wash basin

Dimensions (mm)
A = 585–510
B = 255–255
C = 785–760
D = 40–50
E = 430–405

E

D

C

Side view

Refs: BS 1188: Specification for ceramic wash basins and pedestals.
BS 5506: Specification for wash basins.

Washing troughs are manufactured circular or rectangular on plan in ceramic materials or stainless steel. They are an economic and space saving alternative to a range of basins, for use in factory, school and public lavatories. Some variations have an overall umbrella spray or fountain effect operated by a foot pedal. These are no longer favoured by the water supply undertakings as a trough must have a separate draw-off tap for every placement. In common with other sanitary fitments, there must be provision to prevent the possibility of back-siphonage, i.e. an adequate air gap between tap outlet and spill level of the trough. Hot and cold water supply to the taps is thermostatically blended to about 45°C.

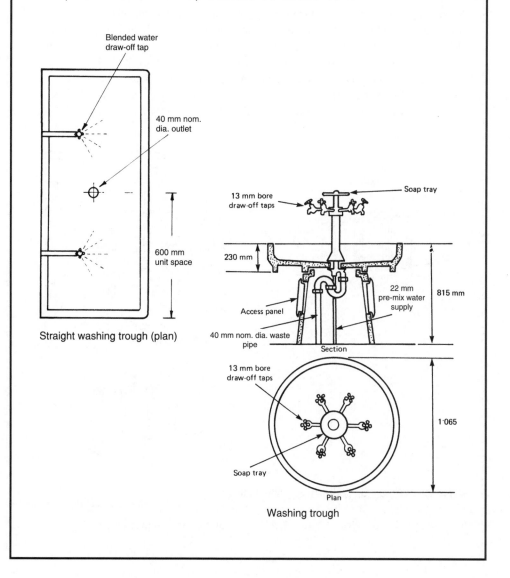

Straight washing trough (plan)

Washing trough

Urinals

These are used in virtually all buildings and public lavatories containing common facilities for male conveniences. They reduce the need for a large number of WCs. Three formats are available in ceramic ware or stainless steel:

- Bowl – secured to the wall and provided with division pieces where more than one is installed.
- Flat slab – fixed against the wall with projecting return end slabs and a low level channel.
- Stall – contains curved stalls, dividing pieces and low level channel.

Urinals are washed at intervals of 20 minutes by means of an automatic flushing cistern discharging 4·5 litres of water per bowl of 610 mm of slab/stall width. The water supply to the cistern should be isolated by a motorised valve on a time control, to shut off when the building is not occupied. A hydraulically operated inlet valve to the automatic flushing cistern can be fitted. This closes when the building is unoccupied and other fittings not used.

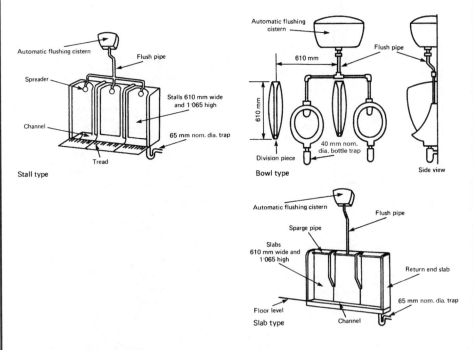

Refs: BS 4880: Specification for urinals.
      BS 5520: Specification for vitreous china bowl urinals.
      BS EN 80: Wall hung urinals – connecting dimensions.

See page 245 and preceding page for automatic devices.

Urinals usually have automatically operated flushing mechanisms. However, manual operation is also acceptable by use of:

- Flushing cistern.
- Flushing valve.
- Wash basin tap and hydraulic valve (combination of manual and automatic).

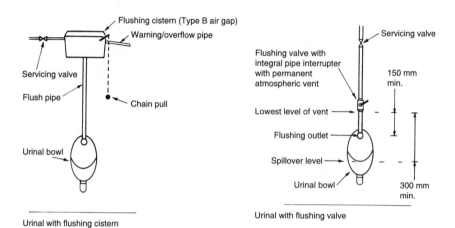

Urinal with flushing cistern

Urinal with flushing valve

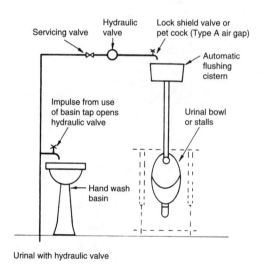

Urinal with hydraulic valve

Special types of sanitary appliances are required for hospital sluice rooms. The slop hopper is used for the efficient disposal of bed pan excrement and general waste. It is similar in design to the washdown WC pan, but has a hinged stainless steel grating for a bucket rest. Another grating inside the pan prevents the entry of large objects which could cause a blockage.

The bed pan washer has a spray nozzle for cleaning bed pans and urine bottles. To prevent possible contamination of water supply, it is essential that the water supplying the nozzle is taken from an interposed cold water storage cistern used solely to supply the bed pan washer. Alternatively, the design of the bed pan washer must allow for a type 'A' air gap (min. 20 mm) between spray outlet nozzle and water spill level. A 90 mm nominal diameter outlet is provided for the pan.

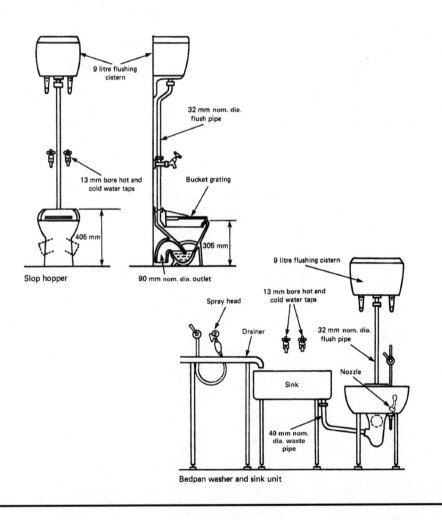

Slop hopper

Bedpan washer and sink unit

Approved Document G provides for minimum quantity, use and disposition of sanitary conveniences. These should contain sufficient appliances relative to a building's purpose and be separated from places where food is stored or prepared. Layout of appliances and installation should allow for access and cleaning. The diagrams illustrate various locations for sanitary conveniences, with an intermediate lobby or ventilated space as required. En-suite facilities are acceptable direct from a bedroom, provided another sanitary convenience is available in the building. All dwellings must have at least one WC and one wash basin. The wash basin should be located in the room containing the WC or in a room or space giving direct access to the WC room (provided that it is not used for the preparation of food). A dwelling occupying more than one family should have the sanitary facilities available to all occupants.

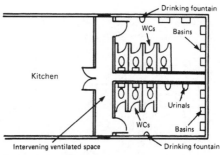

Sanitary accommodation from a kitchen

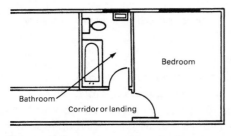

Entry to a bathroom via a corridor or landing

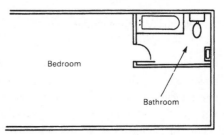

Entry to a bathroom directly from a bedroom

Refs: Building Regulations, Approved Document G – Hygiene.
Building Regulations, Approved Document F – Ventilation. (See Chapter 5.)

The British Standard recommends that every new dwelling is fitted with at least one WC, one bath or shower, one wash basin and one sink. In dwellings accommodating five or more people there should be two WCs, one of which may be in a bathroom. Any WC compartment not adjoining a bathroom shall also contain a wash basin. Where two or more WCs are provided, it is preferable to site them on different floors.

The number of appliances recommended for non-domestic premises such as offices, factories, shops, etc. varies considerably. BS 6465-1 should be consulted for specific situations.

Bathroom arrangements are detailed in BS 6465-2. Some simple domestic layouts are shown below, with minimum dimensions to suit standard appliances and activity space.

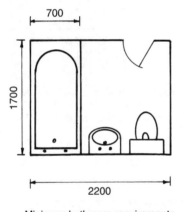

Minimum bathroom requirements

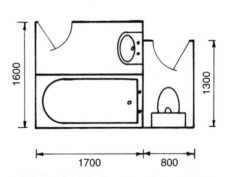

Alternative with adjacent WC compartment

Design of appliances should be such that they are smooth, impervious and manufactured from non-corrosive materials. They should be self-cleansing in operation and easily accessible for manual cleaning. Simplicity in design and a regard to satisfactory appearance are also important criteria.

Refs: BS 6465-1: Sanitary installations – scale of provision.
BS 6465-2: Sanitary installations – space requirements.

Foul air from the drain and sewer is prevented from penetrating buildings by applying a water trap to all sanitary appliances. A water seal trap is an integral part of gullies and WCs, being moulded in during manufacture. Smaller fittings, i.e. sinks, basins, etc., must be fitted with a trap. The format of a traditional tubular trap follows the outline of the letter 'P' or 'S'. The outlet on a 'P' trap is slightly less than horizontal ($2\frac{1}{2}°$) and on an 'S' trap it is vertical. A 'Q' trap has an outlet inclined at an angle of 45°, i.e. half way between 'P' and 'S'. These are no longer used for sanitation but have an application to gullies.

Depth of water seal:

- WCs and gullies – 50 mm (less than smaller fittings as these are unlikely to lose their seal due to the volume of water retained).
- Sanitary appliances other than WCs with waste pipes of 50 mm nominal diameter or less – 75 mm, where the branch pipe connects directly to a discharge stack. However, because of the slow run-off, seal depth may be reduced to 50 mm for baths and shower trays.
- Sinks, baths and showers – 38 mm, where appliance waste pipes discharge over a trapped gully.

Note: Under working and test conditions, the depth of water seal must be retained at not less than 25 mm.

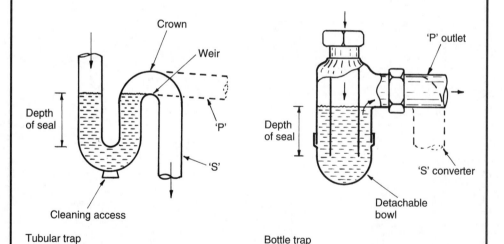

Tubular trap

Bottle trap

Refs: BS 1184: Copper and copper alloy traps (obsolescent).
BS 3943: Specification for plastic waste traps.
BS EN 274: Sanitary tapware. Waste fittings for basins, bidets and baths. General technical specifications.

Loss of Trap Water Seal

The most obvious cause of water seal loss is leakage due to defective fittings or poor workmanship. Otherwise, it may be caused by poor system design and/or installation:

- Self siphonage – as an appliance discharges, the water fills the waste pipe and creates a vacuum to draw out the seal. Causes are a waste pipe that is too long, too steep or too small in diameter.
- Induced siphonage – the discharge from one appliance draws out the seal in the trap of an adjacent appliance by creating a vacuum in that appliance's branch pipe. Causes are the same as for self-siphonage, but most commonly a shared waste pipe that is undersized. Discharge into inadequately sized stacks can have the same effect on waste branch appliances.
- Back pressure – compression occurs due to resistance to flow at the base of a stack. The positive pressure displaces water in the lowest trap. Causes are a too small radius bottom bend, an undersized stack or the lowest branch fitting too close to the base of the stack.
- Capillary action – a piece of rag, string or hair caught on the trap outlet.
- Wavering out – gusts of wind blowing over the top of the stack can cause a partial vacuum to disturb water seals.

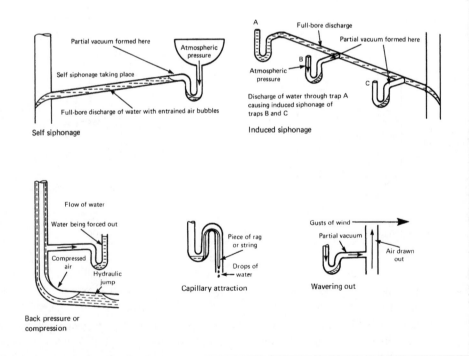

Self siphonage

Induced siphonage

Back pressure or compression

Capillary attraction

Wavering out

Where trap water seal loss is apparent, the problem may be relieved by fitting either a resealing or an anti-siphon trap. A number of proprietory trap variations are available, some of which include:

- McAlpine trap – this has a reserve chamber into which water is retained as siphonage occurs. After siphonage, the retained water descends to reseal the trap.
- Grevak trap – contains an anti-siphonage pipe through which air flows to break any siphonic action.
- Econa trap – contains a cylinder on the outlet into which water flows during siphonic action. After siphonage the water in the cylinder replenishes the trap.
- Anti-siphon trap – as siphonage commences, a valve on the outlet crown opens allowing air to enter. This maintains normal pressure during water discharge, preventing loss of water seal.

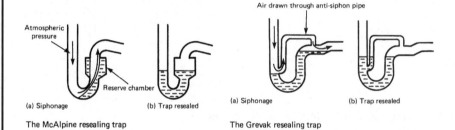

The McAlpine resealing trap

The Grevak resealing trap

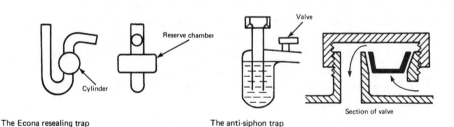

The Econa resealing trap

The anti-siphon trap

Note: Resealing and anti-siphon traps will require regular maintenance to ensure they are functioning correctly. They can be noisy in use.

This compact device has been developed by Hepworth Building Products for use on all sanitary appliances with a 32 or 40 mm nominal diameter outlet. Unlike conventional water seal traps it is a straight section of pipe containing a flexible tubular sealed membrane. This opens with the delivery of waste water and fresh air into the sanitary pipework, resealing or closing after discharge. System design is less constrained, as entry of fresh air into the waste pipework equalises pressures, eliminating the need for traps with air admittance/anti-siphon valves on long waste pipe lengths.

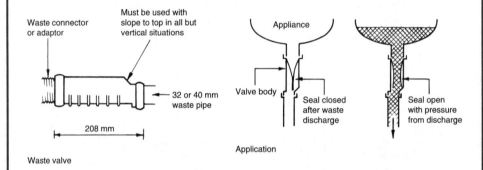

Waste connector or adaptor

Must be used with slope to top in all but vertical situations

32 or 40 mm waste pipe

208 mm

Waste valve

Appliance

Valve body

Seal closed after waste discharge

Seal open with pressure from discharge

Application

- No siphonage with full-bore discharge.
- Full-bore discharge provides better cleansing of pipework.
- Smaller diameter waste pipes possible as there is no water seal to siphon.
- Anti-siphon and ventilating pipes are not required.
- Ranges of appliances do not need auxiliary venting to stacks.
- No maximum waste pipe lengths or gradients (min. 18 mm/m).
- Space saving, i.e. fits unobtrusively within a basin pedestal.
- Tight radius bends will not affect performance.
- In many situations will provide a saving in materials and installation time.

Note: Manufacturers state compliance with British Standard Codes of Practice and Building Regulations, Approved Documents for drainage and waste disposal.

The single stack system was developed by the Building Research Establishment during the 1960s, as a means of simplifying the extensive pipework previously associated with above ground drainage. The concept is to group appliances around the stack with a separate branch pipe serving each. Branch pipe lengths and falls are constrained. Initially the system was limited to five storeys, but applications have proved successful in high rise buildings of over 20 storeys. Branch vent pipes are not required unless the system is modified. Lengths and falls of waste pipes are carefully selected to prevent loss of trap water seals. Water seals on the waste traps must be 75 mm (50 mm bath and shower).

Branch pipe slope or fall:

Sink and bath – 18 to 90 mm/m

Basin and bidet – 20 to 120 mm/m

WC – 9 mm/m.

The stack should be vertical below the highest sanitary appliance branch. If an offset is unavoidable, there should be no connection within 750 mm of the offset.

The branch bath waste connection must be at least 200 mm below the centre of the WC branch to avoid crossflow. This may require a 50 mm nom. dia. parallel pipe to offset the bath waste pipe, or an 'S' trap WC to offset its connection.

The vent part of the stack may reduce to 75 mm nom. dia. when it is above the highest branch.

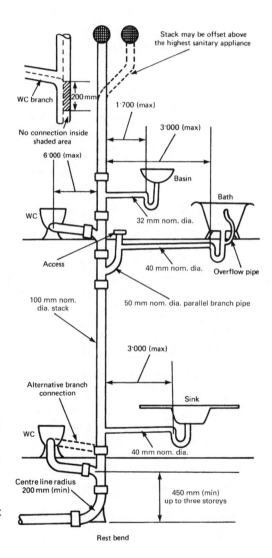

WC branch | 200 mm

No connection inside shaded area

6·000 (max)

Stack may be offset above the highest sanitary appliance

1·700 (max)

3·000 (max)

Basin

WC

32 mm nom. dia.

Bath

Access

40 mm nom. dia.

Overflow pipe

100 mm nom. dia. stack

50 mm nom. dia. parallel branch pipe

3·000 (max)

Alternative branch connection

Sink

WC

40 mm nom. dia.

Centre line radius 200 mm (min)

450 mm (min) up to three storeys

Rest bend

Single Stack System – Modified

If it is impractical to satisfy all the requirements for waste pipe branches in a standard single stack system, some modification is permitted in order to maintain an acceptable system performance:

- Appliances may be fitted with resealing or anti-siphon traps (see page 267).
- Branch waste pipes can be ventilated (see pages 272 and 273).
- Larger than standard diameter waste pipes may be fitted.

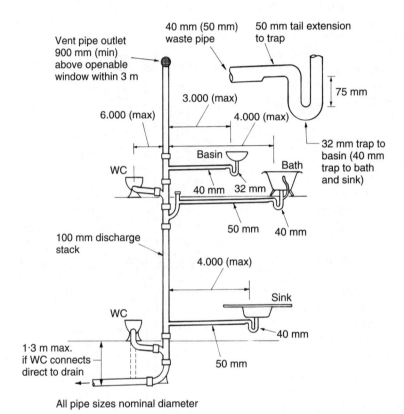

Vent pipe outlet 900 mm (min) above openable window within 3 m

40 mm (50 mm) waste pipe

50 mm tail extension to trap

75 mm

3.000 (max)

6.000 (max)

4.000 (max)

32 mm trap to basin (40 mm trap to bath and sink)

Basin

WC

Bath

40 mm   32 mm

50 mm   40 mm

100 mm discharge stack

4.000 (max)

Sink

WC

40 mm

1·3 m max. if WC connects direct to drain

50 mm

All pipe sizes nominal diameter

Note: Where larger than standard branch pipes are used, the trap size remains as standard. Each trap is fitted with a 50 mm tail extension before connecting to a larger waste pipe.

Refs: Building Regulations, Approved Document H1, Section 1: Sanitary pipework.
BS EN 12056: Gravity drainage systems inside buildings (in 6 parts).

The collar boss system is another modification to the standard single stack system. It was developed by the Marley company for use with their uPVC pipe products. The collar is in effect a gallery with purpose-made bosses for connection of waste pipes to the discharge stack without the problem of crossflow interference. This simplifies the bath waste connection and is less structurally disruptive.

Small diameter loop vent pipes on (or close to) the basin and sink traps also connect to the collar. These allow the use of 'S' traps and vertical waste pipes without the possibility of siphonage, even when the bath waste discharges and flows into the combined bath and basin waste pipe. Vertical outlets are also likely to be less obtrusive and less exposed than higher level 'P' trap waste pipes.

If the branch waste pipes are kept to minimal lengths, the loop vents may not be required. However, the system must be shown to perform adequately under test without the loss of trap water seals.

All pipe sizes shown are nominal inside diameter. There may be some slight variation between different product manufacturers, particularly those using outside diameter specifications. Note that there is not always compatibility between different manufacturers' components.

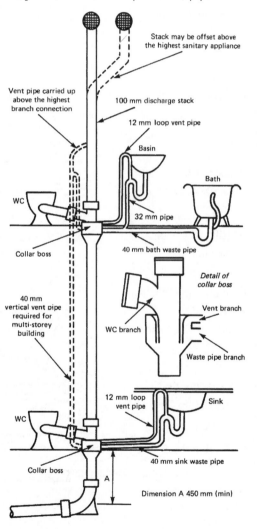

Stack may be offset above the highest sanitary appliance

Vent pipe carried up above the highest branch connection

100 mm discharge stack

12 mm loop vent pipe

Basin

Bath

WC

32 mm pipe

Collar boss

40 mm bath waste pipe

Detail of collar boss

40 mm vertical vent pipe required for multi-storey building

Vent branch

WC branch

Waste pipe branch

12 mm loop vent pipe

Sink

WC

40 mm sink waste pipe

Collar boss

A

Dimension A 450 mm (min)

The ventilated stack system is used in buildings where close grouping of sanitary appliances occurs – typical of lavatories in commercial premises. The appliances need to be sufficiently close together and limited in number not to be individually vented.

Requirements:

WCs:

8 maximum

100 mm branch pipe

15 m maximum length

Gradient between 9 and 90 mm/m ($\theta = 90\frac{1}{2}°$–95°).

Basins:

4 maximum

50 mm pipe

4 m maximum length

Gradient between 18 and 45 mm/m ($\theta = 91°$–$92\frac{1}{2}°$).

Urinals (bowls):

5 maximum

50 mm pipe

Branch pipe as short as possible

Gradient between 18 and 90 mm/m.

Urinals (stalls):

7 maximum

65 mm pipe

Branch pipe as for bowls.

All pipe sizes are nominal inside diameter.

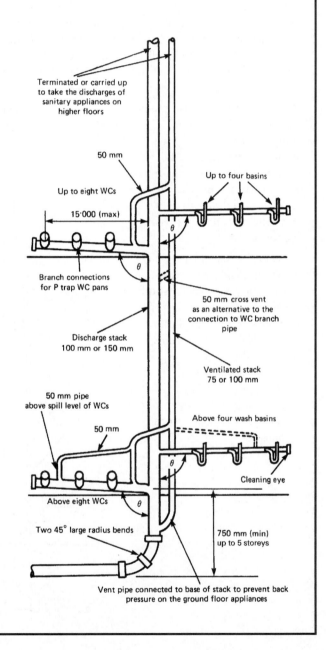

Terminated or carried up to take the discharges of sanitary appliances on higher floors

50 mm

Up to four basins

Up to eight WCs

15·000 (max)

θ

Branch connections for P trap WC pans

θ

50 mm cross vent as an alternative to the connection to WC branch pipe

Discharge stack 100 mm or 150 mm

Ventilated stack 75 or 100 mm

50 mm pipe above spill level of WCs

Above four wash basins

50 mm

θ

Cleaning eye

Above eight WCs

θ

Two 45° large radius bends

750 mm (min) up to 5 storeys

Vent pipe connected to base of stack to prevent back pressure on the ground floor appliances

The fully vented one-pipe system is used in buildings where there are a large number of sanitary appliances in ranges, e.g. factories, schools, offices and hospitals.

The trap on each appliance is fitted with an anti-siphon or vent pipe. This must be connected within 300 mm of the crown of the trap.

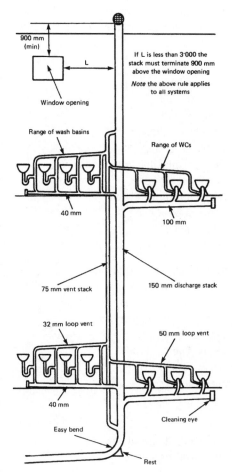

900 mm (min)

L

Window opening

If L is less than 3·000 the stack must terminate 900 mm above the window opening

Note the above rule applies to all systems

Range of wash basins

Range of WCs

40 mm

100 mm

75 mm vent stack

150 mm discharge stack

32 mm loop vent

50 mm loop vent

40 mm

Cleaning eye

Easy bend

Rest

Individual vent pipes combine in a common vent for the range, which is inclined until it meets the vertical vent stack. This vent stack may be carried to outside air or it may connect to the discharge stack at a point above the spillover level of the highest appliance.

The base of the vent stack should be connected to the discharge stack close to the bottom rest bend to relieve any compression at this point.

Size of branch and stack vents:

Discharge pipe or stack (D) (mm)	Vent pipe (mm)
<75	0·67D
75–100	50
>100	0·50D

All pipe sizes are nominal inside diameter.

This system was devised to comply with the old London County Council requirements for connection of soil (WC and urinal) and waste (basin, bath, bidet, sink) appliances to separate stacks. For modern systems the terms soil and waste pipes are generally replaced by the preferred terminology, discharge pipes and discharge stacks.

There many examples of the two-pipe system in use. Although relatively expensive to install, it is still permissible and may be retained in existing buildings that are the subject of refurbishment.

It may also be used where the sanitary appliances are widely spaced or remote and a separate waste stack is the only viable method for connecting these to the drain.

A variation typical of 1930s dwellings has first floor bath and basin wastes discharging through the wall into a hopper. The waste stack from this and the ground floor sink waste discharge over a gully.

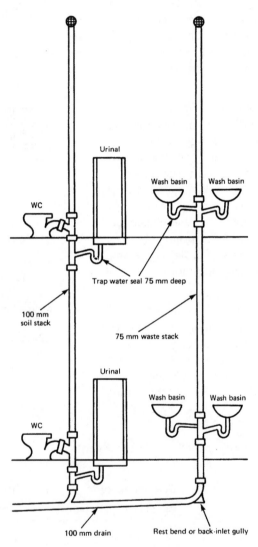

A gully may be used as an alternative to a rest bend before the drain.

These systems are particularly useful where sanitary appliance location is impractical, relative to the existing discharge pipework and stack, e.g. loft conversions and basements. The macerator, pump and small diameter discharge pipe are fairly compact, and unlikely to cause structural disruption on the scale of modifications to a conventional gravity flow system.

There are a variety of proprietary pumping systems, most capable of delivering WC and basin discharge over 20 m horizontally and 4 m vertically. Only products that have been tested and approved by the European Organisation for Technical Approvals (EOTA) or their recognised members, e.g. British Board of Agrément (BBA), are acceptable for installation under the Building Regulations.

Installation is at the discretion of the local water and building control authorities. They will not accept the permanent connection of a WC to a macerator and pump, unless there is another WC connected to a gravity discharge system within the same building.

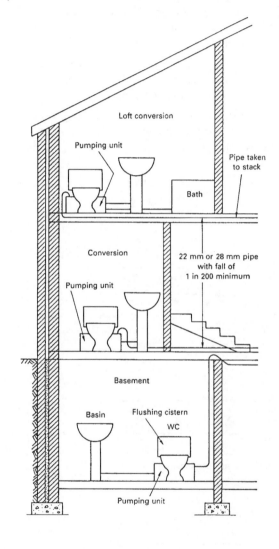

Pipework may be in 22 or 28 mm outside diameter copper tube or equivalent in stainless steel or polypropylene. Easy bends, not elbow fittings must be deployed at changes in direction.

Wash Basins – Waste Arrangements

The arrangement of waste and vent pipes for ranges of basins depends upon the type of building and the number of basins in the range. See BS 6465: Pt.1: Sanitary installations – scale of provision, to determine exact requirements for different purpose groups of building.

For ranges of up to four basins, branch ventilating pipes are not necessary, providing that the inside diameter of the main waste pipe is at least 50 mm and its slope is between 1° and $2\frac{1}{2}°$ (18 mm to 45 mm/m).

For ranges above four basins, the inside diameter and slope is the same, but a 32 mm nominal inside diameter vent pipe is required. Alternatively, resealing or anti-siphon traps may be used.

In schools and factories a running trap may be used, providing that the length of main waste pipe does not exceed 5 m. Alternatively, the wastes may discharge into a glazed channel with a trapped gully outlet to the stack.

For best quality installation work, all traps may be provided with a vent or anti-siphon pipework.

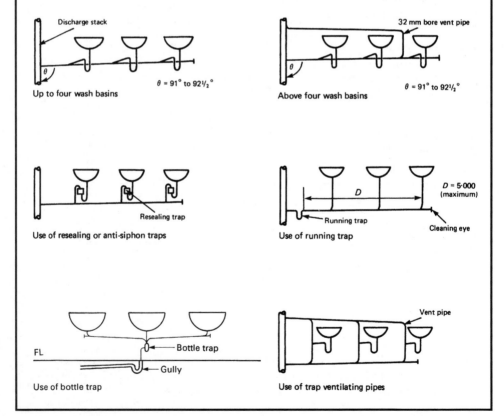

The simplest method for discharging the hose pipe from a washing machine or dishwasher is to bend the hose pipe over the rim of the sink. However, this is unattractive and may be inconvenient if the hose pipe creates an obstruction. A more discrete and less obtrusive arrangement is to couple the hose to a tee fitting or purpose-made adaptor located between the trap and waste outlet of the sink. If a horizontal waste pipe is required at low level behind kitchen fitments, it must be independently trapped and some provision must be made for the machine outlet to ventilate to atmosphere (a purpose-made vent must not be connected to a ventilating stack). Alternatively, the machine hose pipe may be inserted loosely into the vertical waste pipe leaving an air gap between the two pipes.

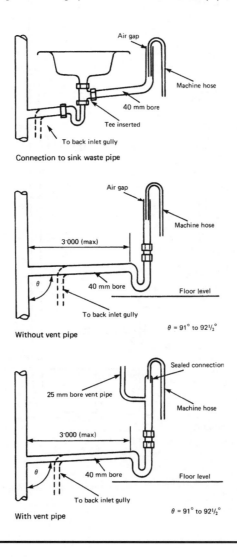

Connection to sink waste pipe

Without vent pipe

$\theta = 91°$ to $92\frac{1}{2}°$

With vent pipe

$\theta = 91°$ to $92\frac{1}{2}°$

Approved Document H1 to the Building Regulations provides guidance on an acceptable method for determining air tightness of sanitary pipework systems. Installations must be capable of withstanding an air or smoke test pressure at least equal to a 38 mm head of water for a minimum of 3 minutes. Smoke testing is not recommended for use with uPVC pipe work.

Equipment for the air test:

Manometer (U gauge), rubber tube, hand bellows and two drain plugs or stoppers.

Procedure:

Stoppers are inserted at the top and bottom of the discharge stack. Each stopper is sealed with water, the lower seal with a flush from a WC. Traps to each appliance are primed to normal depth of seal. The rubber tube connected to the manometer and bellows is passed through the water seal in a WC. Hand bellows are used to pump air into the stack until the manometer shows a 38 mm water displacement. After a few minutes for air temperature stabilisation, the water level in the manometer must remain stationary for 3 minutes. During this time, every trap must maintain at least 25 mm of water seal.

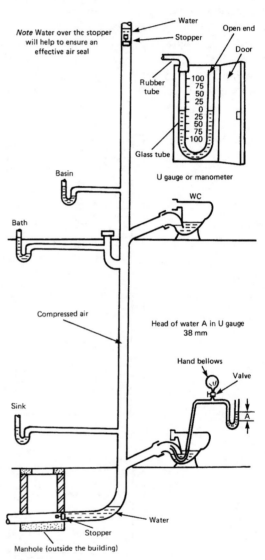

Note Water over the stopper will help to ensure an effective air seal

Water
Stopper
Open end
Door
Rubber tube

100
75
50
25
0
25
50
75
100

Glass tube

U gauge or manometer

Basin

Bath

WC

Compressed air

Head of water A in U gauge 38 mm

Hand bellows

Valve

A

Sink

Water

Stopper

Manhole (outside the building)

Appliances:

Fitment	Capacity (l)	Discharge flow rate (l/s)
Basin	6	0·6
Basin − spray tap	−	0·06
Bath	80	1·1
Shower	−	0·1
Sink	23	0·9
Urinal	4·5	0·15
Washing machine	180	0·7
Water closet	6	2·3

All appliances in a dwelling are unlikely to be used simultaneously, therefore the flow rate that stacks and drains have to accommodate is not the summation of their respective discharges. Allowing for normal usage, the anticipated flow rates from dwellings contained one WC, one bath, one or two wash basins and one sink are as follows:

Flow rates per dwelling:

No. of dwellings	Flow rate (l/s)
1	2·5
5	3·5
10	4·1
15	4·6
20	5·1
25	5·4
30	5·8

Discharge stack sizes:

Min. stack size (nom. i.d.)	Max. capacity (l/s)
50	1·2
65	2·1
75	3·4
90	5·3
100	7·2

Stacks serving urinals, not less than 50 mm.

Stack serving one or more washdown WCs, not less than 100 mm.

If one siphonic WC with a 75 mm outlet, stack size also 75 mm.

Discharge pipe and trap sizes:

Fitment	Trap and pipe nom. i.d. (mm)	Trap water seal (mm)
Basin	32	75
Bidet	32	75
Bath	40	75*
Shower	40	75*
Sink	40	75*
Dishwasher	40	75
Washing machine	40	75
Domestic food waste disposal unit	40	75
Commercial food waste disposal unit	50	75
Urinal bowl	40	75
Urinal bowls (2–5)	50	75
Urinal stalls (1–7)	65	50
WC pan – siphonic	75	50‡
WC pan – washdown	100†	50‡
Slop hopper	100†	50‡

* 38 mm if discharging to a gully.

† Nominally 100 mm but approx. 90 mm (min. 75 mm).

‡ Trap integral with fitment.

Bath and shower trays may be fitted with 50 mm seal traps.

The following materials are acceptable for sanitary pipework:

Application	Material	Standard
Discharge pipes and stacks	Cast iron	BS 415 and BS EN 877
	Copper	BS EN 1254 and 1057
	Galv. steel	BS 3868
	uPVC	BS EN 1329
	Polypropylene	BS EN 1451
	MuPVC	BS 5255
	ABS	BS EN 1455
Traps	Copper	BS 1184 (obsolescent)
	Polypropylene	BS 3943 and BS EN 274

Offsets have two interpretations:

1. Branch waste or discharge pipe connections to the discharge stack. Typically the 200 mm offset required for opposing bath and WC discharge pipes – see page 269. Additional requirements are shown below.
2. Stack offsets – to be avoided, but may be necessary due to the structural outline of the building to which the stack is attached. Large radius bends should be used and no branch connections are permitted within 750 mm of the offset in buildings up to three storeys. In buildings over three storeys a separate vent stack may be needed. This is cross-vented to the discharge stack above and below the offset to relieve pressure. Bends and offsets are acceptable above the highest spillover level of an appliance. They are usually necessary where external stacks avoid eaves projections.

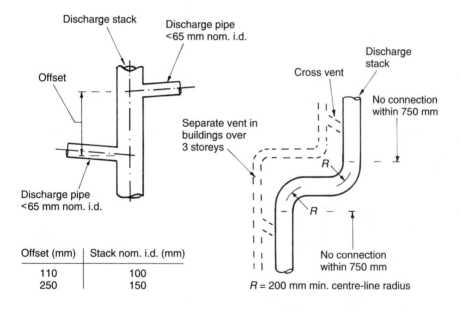

Offset (mm)	Stack nom. i.d. (mm)
110	100
250	150

Discharge pipes offset   Stack off-set

Note: Discharge stacks may be located internally or externally to buildings up to three storeys. Above three storeys, stacks should be located internally.

Lowest discharge pipe connection to stack:

Up to three storeys – 450 mm min. from stack base (page 269).
Up to five storeys – 750 mm min. from stack base (page 272).

Above five storeys, the ground floor appliances should not connect into the common stack, as pressure fluctuations at the stack base could disturb the lower appliance trap water seals. Above 20 storeys, both ground and first floor appliances should not connect into the common stack. Ground and first floor appliances so affected can connect directly to a drain or gully, or be provided with a stack specifically for lower level use.

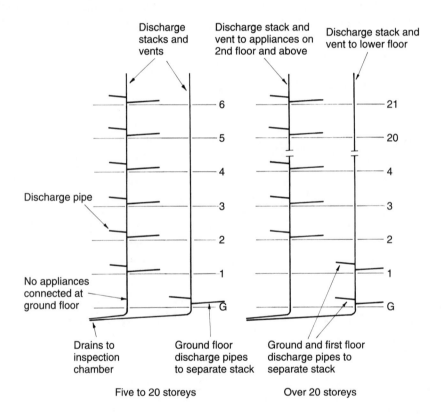

Discharge stacks and vents

Discharge stack and vent to appliances on 2nd floor and above

Discharge stack and vent to lower floor

Discharge pipe

No appliances connected at ground floor

Drains to inspection chamber

Ground floor discharge pipes to separate stack

Ground and first floor discharge pipes to separate stack

Five to 20 storeys

Over 20 storeys

Access – required for clearing blockages. Rodding points should be fitted at the end of discharge pipes, unless trap removal provides access to the full pipe length. Discharge stacks are accessed from the top and through access plates located midway between floors at a maximum spacing of three storeys apart.

For fire protection and containment purposes, the Building Regulations divide parts or units within buildings into compartments. A typical example is division of a building into individual living units, e.g. flats. The dividing compartment walls and floors have fire resistances specified in accordance with the building size and function.

Where pipes penetrate a compartment interface, they must have a means of preventing the spread of fire, smoke and hot gases through the void they occupy. Non-combustible pipe materials may be acceptably sealed with cement and sand mortar, but the most vulnerable are plastic pipes of low heat resistance. The void through which they pass can be sleeved in a non-combustible material for at least 1 m each side. One of the most successful methods for plastic pipes is to fit an intumescent collar at the abutment with, or within, the compartment wall or floor. Under heat, these become a carbonaceous char, expand and compress the warm plastic to close the void for up to four hours.

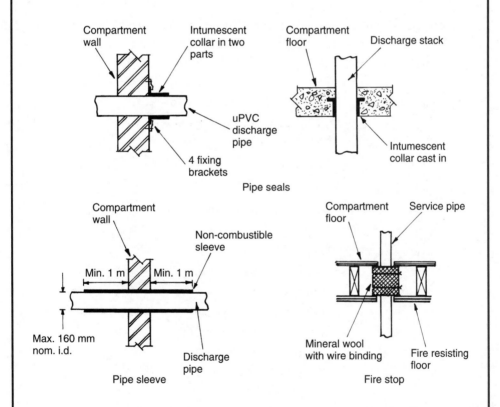

Ref: Building Regulations, Approved Document B3: Fire spread.

Note: See also page 468.

Sanitation Flow Rate – Formula

Simultaneous demand process – considers the number of appliances likely to be used at any one time, relative to the total number installed on a discharge stack.

Formula:

$$m = np + 1\cdot8 \sqrt{2np(1-p)}$$

where: m = no of appliances discharging simultaneously
   n = no. of appliances installed on the stack
   p = appliance discharge time (t) ÷ intervals between use (T).

Average time for an appliance to discharge = 10 seconds (t)
Intervals between use (commercial premises) = 600 seconds (T)
              (public premises)   = 300 seconds (T)

Commercial premises, e.g. offices, factories, etc., p = 10 ÷ 600 = 0·017.
Public premises, e.g. cinemas, stadium, etc., p = 10 ÷ 300 = 0·033.

E.g. an office building of ten floors with four WCs, four urinals, four basins and one sink on each floor.

Total number of appliances (n) = 13 × 10 floors = 130

Substituting factors for p and n in the formula:

$$m = (130 \times 0\cdot017) + 1\cdot8\sqrt{2 \times 130 \times 0\cdot017 \, (1-0\cdot017)}$$
$$m = 2\cdot21 + (1\cdot8 \times 2\cdot08) = 5\cdot96$$

Simultaneous demand factor = m ÷ n
              = 5·96 ÷ 130 = 0·045 or 4·5%

Flow rates (see page 279):

Four WCs at 2·3 l/s	= 9·2
Four urinals at 0·15 l/s	= 0·6
Four basins at 0·6 l/s	= 2·4
One sink at 0·9 l/s	= 0.9
Total per floor	= 13·1 l/s
Total for ten floors	= 131 l/s

Allowing 4·5% simultaneous demand = 131 × 4·5% = 5·9 l/s.

The use of discharge units for drain and sewer design is shown on pages 233 and 234. The same data can be used to ascertain the size of discharge stacks and pipes.

Using the example from the previous page:

Four WCs at 14 DUs	=	56
Four urinals at 0·3 DUs	=	1·2
Four basins at 3 DUs	=	12
One sink at 14 DUs	=	14
Total per floor	=	83·2
Total for ten floors	=	832 discharge units

Discharge units can be converted to flow in litres per second from the chart:

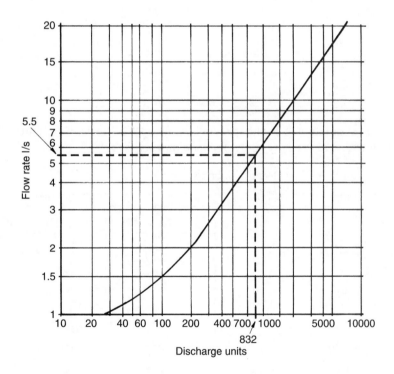

From the chart, a total loading of 832 discharge units can be seen to approximate to 5·5 l/s. A fair comparison with the 5·9 l/s calculated by formula on the preceding page.

Formula:

$$q = K \sqrt[3]{d^8}$$

where: q = discharge or flow rate in l/s
K = constant of 32×10^{-6}
d = diameter of stack in mm.

Transposing the formula to make d the subject:

$$d = \sqrt[8]{(q \div K)^3} \qquad q = 5 \cdot 5 \text{ l/s (see previous page)}$$

$$d = \sqrt[8]{(5 \cdot 5 \div 32 \times 10^{-6})^3}$$

= 91·9 mm, i.e. a 100 mm nom. i.d. stack.

Discharge units on stacks:

Discharge stack nom. i.d. (mm)	Max. No. of DUs
50	20
65	80
75	200
90	400
100	850
150	6400

Using the example from the preceding page, 832 discharge units can be adequately served by a 100 mm diameter stack.

Discharge units on discharge branch pipes:

Discharge pipe, nom. i.d. (mm)	Branch gradient		
	1 in 100	1 in 50	1 in 25
32		1	1
40		2	8
50		10	26
65		35	95
75	40	100	230
90	120	230	460
100	230	430	1050
150	2000	3500	7500

Ref: BS EN 12056-2: Gravity drainage systems inside buildings.

9 GAS INSTALLATION, COMPONENTS AND CONTROLS

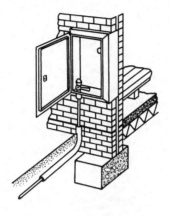

NATURAL GAS – COMBUSTION

MAINS GAS SUPPLY AND INSTALLATION

GAS SERVICE PIPE INTAKE

METERS

GAS CONTROLS AND SAFETY FEATURES

GAS IGNITION DEVICES AND BURNERS

PURGING AND TESTING

GAS APPLIANCES

BALANCED FLUE APPLIANCES

OPEN FLUE APPLIANCES

FLUE BLOCKS

FLUE TERMINALS

FLUE LINING

SHARED FLUES

FAN ASSISTED GAS FLUES

VENTILATION REQUIREMENTS

FLUE GAS ANALYSIS

GAS CONSUMPTION

GAS PIPE SIZING

Properties of natural gas are considered on page 126. Some further features include:

- Ignition temperature, 700°C.

- Stoichiometric mixture – the quantity of air required to achieve complete combustion of gas. For combustion, the ratio of air volume to natural gas volume is about 10·6:1. Therefore, about 10% gas to air mixture is required to achieve complete combustion. As air contains about 20% oxygen, the ratio of oxygen to gas is approximately 2:1. Developing this a little further – natural gas is about 90% methane, therefore:

$$CH_4 + 2O_2 = CO_2 + 2H_2O$$

1 part methane + 2 parts oxygen = 1 part carbon dioxide + 2 parts water

If there is insufficient air supply to a gas burner, incomplete combustion will result. This produces an excess of carbon monoxide in the flue; a toxic and potentially deadly gas.

- Products of complete combustion – water vapour, carbon dioxide and the nitrogen already contained in the air. Correct combustion can be measured by simple tests to determine the percentage of carbon dioxide in flue gases. The Draeger and Fyrite analysers shown on page 328 are suitable means for this assessment.

- Flues – these are necessary to discharge the products of combustion safely and to enhance the combustion process. The application of flues is considered in more detail later in this chapter. Some gas appliances such as small water heaters and cookers are flueless. Provided they are correctly installed, they will produce no ill-effects to users. The room in which they are installed must be adequately ventilated, otherwise the room air could become vitiated (oxygen depleted). For a gas cooker, this means an openable window or ventilator. A room of less than 10 m^3 requires a permanent vent of 5000 mm^2.

BG Group Plc (formerly British Gas Plc) supply gas to communities through a network of mains, installed and maintained by Lattice Group plc (Transco). Gas marketing and after-sales services are provided by a number of commercial franchisees for the consumer's choice.

Some of the underground service pipes have been in place for a considerable time. These are manufactured from steel and although protected with bitumen, PVC or grease tape (Denso), they are being progressively replaced with non-corrosive yellow uPVC for mains and polyethylene for the branch supplies to buildings. The colour coding provides for recognition and to avoid confusion with other utilities in future excavation work.

Mains gas pressure is low compared with mains water. It is unlikely to exceed 75 mbar (750 mm water gauge or 7·5 kPa) and this is reduced by a preset pressure governor at the consumer's meter to about 20 mbar.

A service pipe of 25 mm nominal bore is sufficient for normal domestic installations. For multi-installations such as a block of flats, the following can be used as a guide:

Nominal bore (mm)	No. of flats
32	2–3
38	4–6
50*	>6

* Note: Supplies of 50 mm nom. bore may be provided with a service valve after the junction with the main. Where commercial premises are supplied and the risk of fire is greater than normal, e.g. a garage, a service pipe valve will be provided regardless of the pipe size and its location will be clearly indicated. Pipes in excess of 50 mm nom. bore have a valve fitted as standard.

Gas mains should be protected by at least 375 mm ground cover (450 mm in public areas).

Refs: The Gas Act.
      The Gas Safety (Installation and Use) Regulations.

The details shown below represent two different established installations. Some of these may still be found, but unless there are exceptional circumstances, the meter is no longer located within a building. An exception may be a communal lobby to offices or a block of flats. The preferred meter location for the convenience of meter readers and security of building occupants is on the outside of a building. This can be in a plastic cupboard housing on the external wall or in a plastic box with hinged lid sunken into the ground at the building periphery.

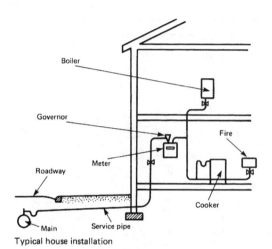

Typical house installation

Prior to conversion to natural gas in the 1960s, a condensate receiver was used to trap moisture from town or coal gas where it was impractical to incline the service pipe back to the main.

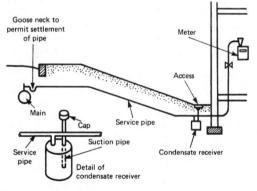

Use of condensate receiver

A service pipe is the term given to the pipe between the gas main and the primary meter control. A polyethylene pipe is used underground and steel or copper pipe where it is exposed. Wherever possible, the service pipe should enter the building on the side facing the gas main. This is to simplify excavations and to avoid the pipe having to pass through parts of the substructure which could be subject to settlement. The service pipe must not:

- pass under the base of a wall or foundations to a building
- be installed within a wall cavity or pass through it except by the shortest possible route
- be installed in an unventilated void space – suspended and raised floors with cross-ventilation may be an exception
- have electrical cables taped to it
- be near any heat source.

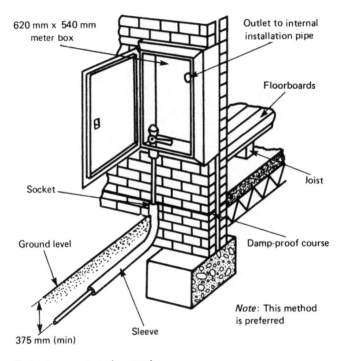

620 mm x 540 mm meter box

Outlet to internal installation pipe

Floorboards

Joist

Socket

Ground level

Damp-proof course

375 mm (min)

Sleeve

Note: This method is preferred

Entry to an external meter box

Where there is insufficient space or construction difficulties preclude the use of an external meter box or external riser, with certain provisions, the service pipe may be installed under a solid concrete floor or through a suspended floor.

For a solid floor, a sleeve or duct should be provided and built into the wall to extend to a pit of approximately 300 × 300 mm plan dimensions. The service pipe is passed through the duct, into the pit and terminated at the meter position with a control valve. The duct should be as short as possible, preferably not more than 2 m. The space between the duct and the service pipe is sealed at both ends with mastic and the pit filled with sand. The floor surface is made good to match the floor finish. If the floor is exposed concrete, e.g. a garage, then the duct will have to bend with the service pipe to terminate at floor level and be mastic sealed at this point.

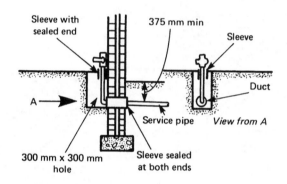

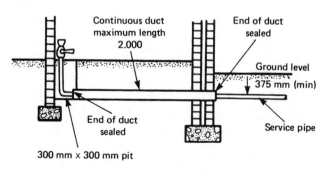

Service pipe entry into solid floor

Where a service pipe passes through a wall or a solid concrete floor, it must be enclosed by a sleeve of slightly larger diameter pipe to provide space to accommodate any building settlement or differential movement. The outside of the sleeve should be sealed with cement mortar and the space between the sleeve and service pipe provided with an intumescent (fire resistant) mastic sealant.

If an internal meter is used, the space or compartment allocated for its installation must be well ventilated. A purpose-made void or air brick to the outside air is adequate. The surrounding construction should be of at least 30 minutes' fire resistance. In commercial and public buildings the period of fire resistance will depend on the building size and purpose grouping.

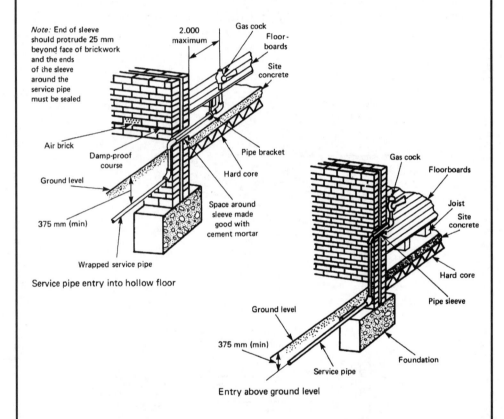

Note: End of sleeve should protrude 25 mm beyond face of brickwork and the ends of the sleeve around the service pipe must be sealed

2.000 maximum

Gas cock
Floor-boards
Site concrete

Air brick
Damp-proof course
Ground level
375 mm (min)
Wrapped service pipe
Pipe bracket
Hard core
Space around sleeve made good with cement mortar

Service pipe entry into hollow floor

Gas cock
Floorboards
Joist
Site concrete
Hard core
Pipe sleeve
Ground level
375 mm (min)
Service pipe
Foundation

Entry above ground level

Ref: Building Regulations, Approved Document B: Fire safety.

Gas service pipe risers must be installed in fire protected shafts constructed in accordance with the Building Regulations, Approved Document B: Fire safety. Possible methods for constructing a shaft include:

- A continuous shaft ventilated to the outside at top and bottom. In this situation a fire protected sleeve is required where a horizontal pipe passes through the shaft wall.
- A shaft which is fire stopped at each floor level. Ventilation to the outside air is required at both high and low levels in each isolated section.

Shafts are required to have a minimum fire resistance of 60 minutes and the access door or panel a minimum fire resistance of 30 minutes. The gas riser pipe must be of screwed or welded steel and be well supported throughout with a purpose-made plate at its base. Movement joints or flexible pipes and a service valve are provided at each branch.

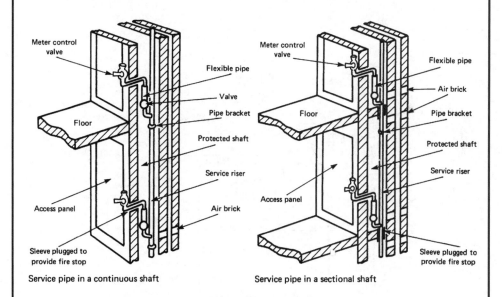

Service pipe in a continuous shaft

Service pipe in a sectional shaft

Refs: Building Regulations, Approved Document B3: Section 9, Compartmentation.
BS 8313: Code of practice for accommodation of building services in ducts.

Installation of Gas Meters

The gas meter and its associated controls are the property of the gas authority. It should be sited as close as possible to the service pipe entry to the building, ideally in a purpose-made meter cupboard on the external wall. The cupboard should be positioned to provide easy access for meter maintenance, reading and inspection. The immediate area around the meter must be well ventilated and the meter must be protected from damage, corrosion and heat. A constant pressure governor is fitted to the inlet pipework to regulate the pressure at about 20 mbar (2 kPa or 200 mm w.g.).

Electricity and gas meters should not share the same compartment. If this is unavoidable, a fire resistant partition must separate them and no electrical conduit or cable should be closer than 50 mm to the gas meter and its installation pipework. One exception is the earth equi-potential bond cable. This must be located on the secondary pipework and within 600 mm of the gas meter.

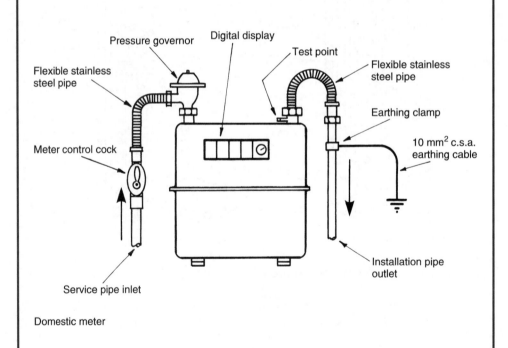

Gas meters measure the volume of gas in cubic feet or cubic metres consumed within a building. The charge is converted to kilowatt-hours (kWh). 100 cubic feet or 2·83 cubic metres is approximately 31 kWh, (see page 329). Some older meters have dials but these have been largely superseded by digital displays which are easier to read.

There are basically three categories of meter:

1. Domestic credit.
2. Domestic pre-payment.
3. Industrial credit.

Credit meters measure the fuel consumed and it is paid for after use at 3-monthly billing intervals. Monthly payments can be made based on an estimate, with an annual adjustment made to balance the account.

Pre-payment meters require payment for the fuel in advance by means of coins, cards, key or tokens. Tokens are the preferred method and these are purchased at energy showrooms, post offices and some newsagents. A variation known as the Quantum meter uses a card to record payment. These cards are purchased at designated outlets and can be recharged with various purchase values.

Industrial meters have flanged connections for steel pipework. Flexible connections are unnecessary due to the pipe strength and a firm support base for the meter. A by-pass pipe is installed with a sealed valve. With the supply authority's approval this may be used during repair or maintenance of the meter.

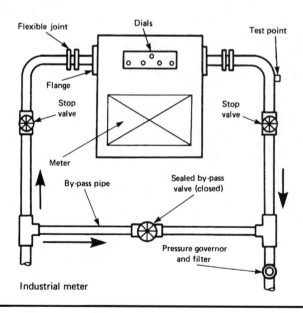

Industrial meter

A constant pressure governor is fitted at the meter to regulate pressure into the system. It is secured with a lead seal to prevent unqualified tampering. Individual appliances may also have factory fitted pressure governors, located just before the burners. Gas passes through the valve and also through the by-pass to the space between the two diaphragms. The main diaphragm is loaded by a spring and the upward and downward forces acting upon this diaphragm are balanced. The compensating diaphragm stabilises the valve. Any fluctuation of inlet pressure inflates or deflates the main diaphragm, raising or lowering the valve to maintain a constant outlet pressure.

A meter control cock has a tapered plug which fits into a tapered body. As gas pressures are very low, the valve can operate by a simple 90° turn to align a hole in the plug with the bore of the valve body, and vice versa. The drop-fan safety cock prevents the valve being turned accidently.

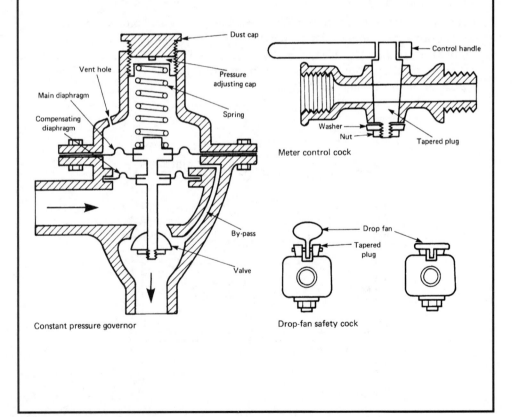

Constant pressure governor

Meter control cock

Drop-fan safety cock

For correct combustion of natural gas, burner design must allow for the velocity of the gas-air mixture to be about the same as the flame velocity. Natural gas has a very slow burning velocity, therefore there is a tendency for a flame to lift off the burner. This must be prevented as it will allow gas to escape, possibly exploding elsewhere! Correct combustion will occur when the gas pressure and injector bore are correct and sufficient air is drawn in, provided the gas-air velocity is not too high to encourage lift off. Some control over lift-off can be achieved by a retention flame fitted to the burner. Flame lift-off may also be prevented by increasing the number of burner ports to effect a decrease in the velocity of the gas-air mixture. A box-type of burner tray is used for this purpose.

If the gas pressure is too low, or the injector bore too large, insufficient air is drawn into the burner. This can be recognised by a smoky and unstable flame, indicating incomplete combustion and an excess of carbon monoxide. At the extreme, light-back can occur. This is where the flame passes back through the burner to ignite on the injector.

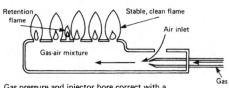

Gas pressure too low or injector bore too large

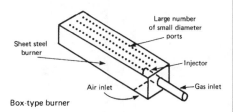

Gas pressure and injector bore correct but with no retention flame

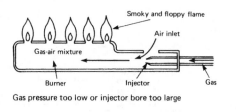

Gas pressure and injector bore correct with a retention flame

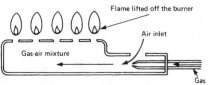

Box-type burner

Gas Thermostats

A thermostat is a temperature sensitive device which operates a gas valve in response to a predetermined setting. Hot water heaters and boilers may be fitted with two thermostats:

1. Working thermostat – controls the water flow temperature from the boiler. It has a regulated scale and is set manually to the user's convenience. It engages or disengages the gas valve at a water temperature of about 80°C.

2. High limit thermostat – normally preset by the boiler manufacturer to function at a water temperature of about 90°C. It is a thermal cut-out safety device which will isolate the gas supply if the working thermostat fails.

The rod-type thermostat operates by a difference in thermal response between brass and invar steel. When water surrounding a brass tube becomes hot, the tube expands. This draws the steel rod with it until a valve attached to the rod closes off the fuel supply. The reverse process occurs as the water cools.

The vapour expansion thermostat has a bellows, capillary tube and probe filled with ether. When water surrounding the probe becomes hot, the vapour expands causing the bellows to respond by closing the fuel valve. Cooling water reverses the process.

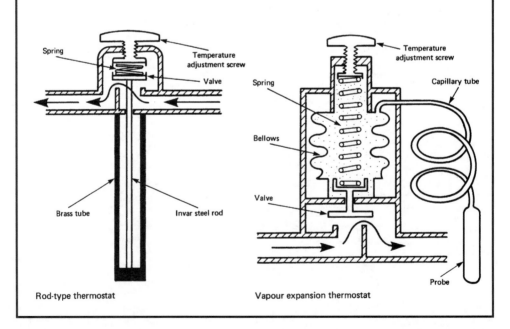

Rod-type thermostat

Vapour expansion thermostat

A rod-type thermostat is often connected to a relay valve to
control gas supply to the burner. When the boiler is operational, gas
flows to the burner because valves A and B are open. Gas pressure
above and below the diaphragm are equal. When the water reaches
the required temperature, the brass casing of the rod thermostat
expands and draws the invar steel rod with it to close valve A. This
prevents gas from flowing to the underside of the diaphragm. Gas
pressure above the diaphragm increases, allowing valve B to fall
under its own weight to close the gas supply to the burner. As the
boiler water temperature falls, the brass casing of the thermostat
contracts to release valve A which reopens the gas supply.

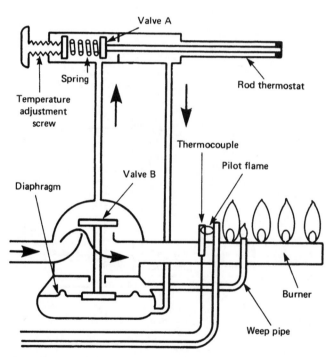

Operating principles of rod thermostat
and gas relay valve

Gas water heaters/boilers and other heat producing appliances such as air heaters must be fitted with a safety device to prevent gas flowing in event of the pilot light extinguishing. Whilst functional, the pilot light plays on a thermo-couple suspended in the gas flame. The hot thermo-couple energises an electromagnetic or solenoid valve to open and allow gas to flow. This is otherwise known as a thermo-electric pilot flame failure safety device. The drawing below shows the interrelationship of controls and the next page illustrates and explains the safety device in greater detail.

To commission the boiler from cold, the thermo-electric valve is operated manually by depressing a push button to allow gas flow to the pilot flame. A spark igniter (see page 304) illuminates the flame whilst the button is kept depressed for a few seconds, until the thermo-couple is sufficiently warm to automatically activate the valve.

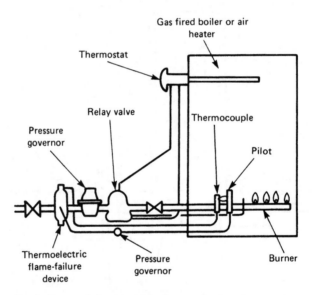

Gas boiler or air heater controls

Ref: BS 5258: Safety of domestic gas appliances.

Thermo-electric – has an ancillary thermo-couple sensing element consisting of two dissimilar metals joined together at each end to form an electrical circuit. When the thermo-couple is heated by the gas pilot flame, a small electric current is generated. This energises an electromagnet in the gas valve which is retained permanently in the open position allowing gas to pass to the relay valve. If the pilot flame is extinguished, the thermo-couple cools and the electric current is no longer produced to energise the solenoid. In the absence of a magnetic force, a spring closes the gas valve.

Bi-metallic strip – has a bonded element of brass and invar steel, each metal having a different rate of expansion and contraction. The strip is bent into a U shape with the brass on the outside. One end is anchored and the other attached to a valve. The valve responds to thermal reaction on the strip. If the pilot flame is extinguished, the bent bi-metallic strip contracts opening to its original position and closing the gas supply and vice versa.

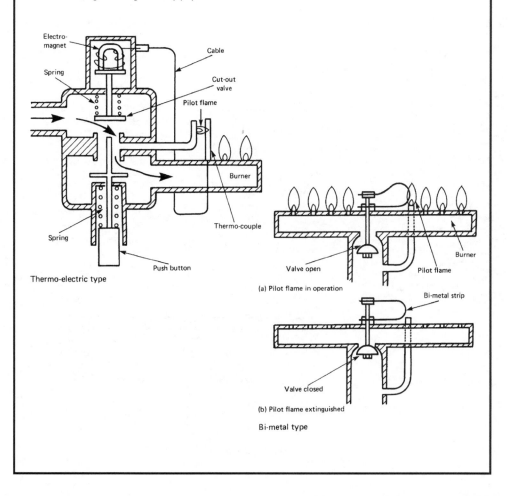

Thermo-electric type

(a) Pilot flame in operation

(b) Pilot flame extinguished

Bi-metal type

Gas Ignition Devices

Lighting the pilot flame with matches or tapers is unsatisfactory. It is also difficult to effect whilst operating the push button control on the gas valve. An integral spark igniter is far more efficient. These are usually operated by mains electricity. An electric charge is compounded in a capacitor, until a trigger mechanism effects its rapid discharge. This electrical energy passes through a step-up transformer to create a voltage of 10 or 15 kV to produce a spark. The spark is sufficient to ignite the pilot flame. Spark generation of this type is used in appliances with a non-permanent pilot flame. This is more fuel economic than a permanent flame. The spark operation is effected when the system thermostat engages an automatic switch in place of the manual push switch shown below and a gas supply to the pilot.

A piezoelectric spark igniter contains two crystals. By pressurising them through a cam and lever mechanism from a push button, a large electric voltage potential releases a spark to ignite the gas.

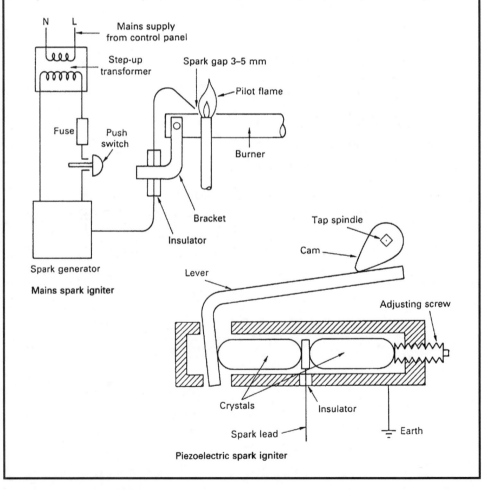

Mains spark igniter

Piezoelectric spark igniter

It is very important that new gas installations are thoroughly purged of air and debris that may remain in the completed pipework. This also applies to existing installations that have been the subject of significant changes. If air is not removed, it is possible that when attempting to ignite the gas, a gas-air mixture will cause a blow back and an explosion. Before purging, the system should be pressure tested for leakages - see next page.

Procedure:

- Ensure ample ventilation where gas and air will escape from the system.

- Prohibit smoking, use of electrical switches, power tools, etc. in the vicinity of the process.

- Close the main gas control valve at the meter.

- Disconnect the secondary pipework at the furthest fitting.
  Note: if the last appliance has a flame failure safety device, no gas will pass beyond it, therefore remove its test nipple screw.

- Turn on the main gas control valve until the meter is completely purged.

- Purging the meter is achieved by passing through it a volume of gas at least equal to five times its capacity per revolution of the meter mechanism. Most domestic meters show 0.071 cu. ft. (0.002 m^3) per dial revolution, so: $5 \times 0.071 = 0.355$ cu. ft. (0.010 m^3) of gas is required.

- Turn off the main gas control valve and reconnect the open end or replace the last appliance test nipple.

- Turn on the main gas control valve and purge any remaining air to branch appliances until gas is smelt.

- Test any previous disconnections by applying soap solution to the joint. Leakage will be apparent by foaming of the solution.

- When all the air in the system has been removed, appliances may be commissioned.

Ref: BS 6891: Specification for installation of low pressure gas pipework of up to 28 mm in domestic premises.

Testing Gas Installations for Soundness

Testing a new installation:

- Cap all open pipe ends and turn appliances off.

- Close the main control valve at the meter. If the meter is not fitted, blank off the connecting pipe with a specially prepared cap and test nipple.

- Remove the test nipple screw from the meter or blanking cap and attach the test apparatus by the rubber tubing.

- Level the water in the manometer at zero.

- Pump or blow air through the test cock to displace 300 mm water gauge (30 mbar) in the manometer. This is approximately one and a half times normal domestic system pressure.

- Wait 1 minute for air stabilisation, then if there is no further pressure drop at the manometer for a further 2 minutes the system is considered sound.

- If leakage is apparent, insecure joints are the most likely source. These are painted with soap solution which foams up in the presence of air seepage.

Testing an existing system:

- Close all appliance valves and the main control valve at the meter.

- Remove the test nipple screw on the meter and attach the test apparatus.

- Open the main control valve at the meter to record a few millimetres water gauge.

- Close the valve immediately and observe the manometer. If the pressure rises this indicates a faulty valve.

- If the valve is serviceable, continue the test by opening the valve fully to record a normal pressure of 200 to 250 mm w.g. Anything else suggests that the pressure governor is faulty.

- With the correct pressure recorded, turn off the main valve, allow 1 minute for air stabilisation and for a further 2 minutes there should be no pressure fluctuation.

- Check for any leakages as previously described.

When used with a flexible tube, hand bellows and control cock, this equipment is suitable for measuring gas installation pressure and testing for leakage. It is also suitable for air testing drains and discharge stacks.

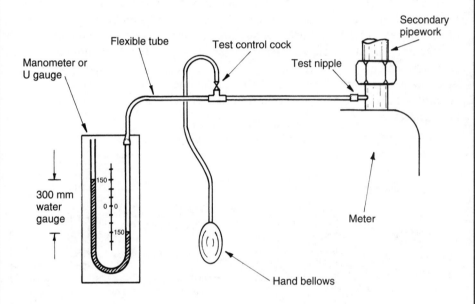

The glass tube is contained in a protective metal or wooden box. It is mounted against a scale graduated in millibars or millimetres. 1 mbar is the pressure exerted by a 9·81 mm (10 mm is close enough) head of water. Water is levelled in the tube to zero on the scale. Care must be taken to note the scale calibration. Some manometers are half scale, which means the measures are in mbar or mm but they are double this to give a direct reading. Others are indirect, as shown. With these, the water displacements either side of the zero must be added.

Fires – these have a relatively low energy rating, usually no more than 3 kW. They are set in a fire recess and use the lined flue for extraction of burnt gases. Air from the room is sufficient for gas combustion, as appliances up to 7 kW net input do not normally require special provision for ventilation. Heat is emitted by convection and radiation.

Decorative fuel effect fires – these are a popular alternative to the traditional gas fire. They burn gas freely and rely on displacement of heat by the colder air for combustion to encourage burnt gas extraction indirectly into the flue. Sufficient air must be available from a purpose-made air inlet to ensure correct combustion of the gas and extraction of burnt gases. An air brick with permanent ventilation of at least 10,000 mm^2 is sufficient for fires up to 12·7 kW net input rating. Log and coal effect fires are designed as a visual enhancement to a grate by resembling a real fire, but as a radiant heater they compare unfavourably with other forms of gas heat emitters.

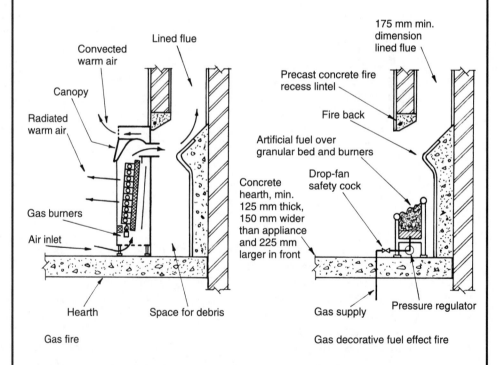

Ref: BS 5871: Specification for installation of gas fires, convector heaters, fire/back boilers and decorative fuel effect fires.

Radiant heaters – in tube format these are simple and effective heat emitters, most suited to high ceiling situations such as industrial units, warehouses and factories. They suspend above the work area and provide a very efficient downward radiation of up to 40 kW. Gas is fired into one end of a tube and the combustion gases extracted by fan assisted flue at the other. The tube may be straight or return in a U shape to increase heat output. A polished stainless steel back plate functions as a heat shield and reflector.

The control box houses an air intake, electronic controls, gas regulators and safety cut-out mechanisms. This includes a gas isolator in event of fan failure. To moderate burning, the end of the tube has a spiral steel baffle to maintain even temperature along the tube.

Advantages over other forms of heating include a rapid heat response, low capital cost, easy maintenance and high efficiency.

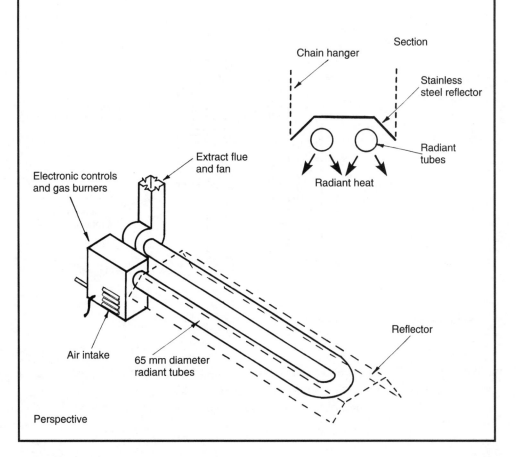

309

Gas Appliances – Convector Heater

Convector – a wall mounted, balanced flue appliance rated up to about 7 kW. They are compact units, room sealed and therefore independent of natural draught from the room in which they are installed. The flue is integral with the appliance and must be installed on an external wall. An exception is when the flue is fan assisted, as this will permit a short length of horizontal flue to the outside wall.

Air for combustion of gas is drawn from outside, through a different pathway in the same terminal as the discharging combusted gases. Correct installation will ensure that the balance of air movement through the terminal is not contaminated by exhaust gases.

About 90% of the heat emitted is by convection, the remainder radiated. Some convectors incorporate a fan, so that virtually all the heat is convected.

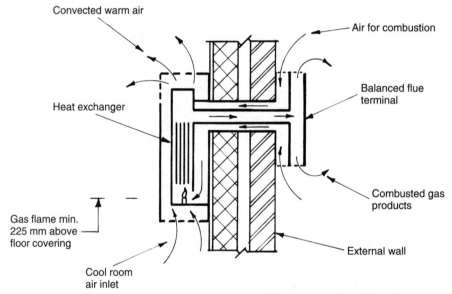

Room sealed gas convector heater

Refs: Building Regulations, Approved Document J: Combustion appliances and fuel storage systems.
BS 5440-1: Specifiation for installation and maintenance of flues.

The balanced flue appliance has the air inlet and flue outlet sealed from the room in which it is installed. It is more efficient than a conventional open flue pipe as there are less heat losses in and from the flue. As it is independent of room ventilation there are no draughts associated with combustion and there is less risk of combustion products entering the room. It is also less disruptive to the structure and relatively inexpensive to install.

A balanced flue is designed to draw in the air required for gas combustion from an area adjacent to where it discharges its combusted gases. These inlets and outlets must be inside a windproof terminal sited outside the room in which the appliance is installed. Gas appliances in a bath or shower room, or in a garage must have balanced flues.*

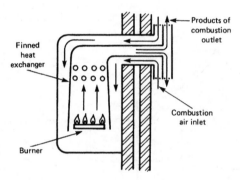

Balanced flue water heater

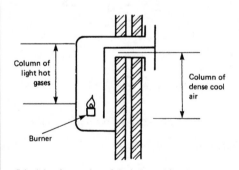

Principle of operation of the balanced flue heater

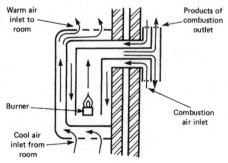

Balanced flue convector heater

* Ref: Gas Safety (Installation and Use) Regulations.

Balanced flue terminals must be positioned to ensure a free intake of air and safe dispersal of combustion products. Generally, they should be located on a clear expanse of wall, not less than 600 mm from internal or external corners and not less than 300 mm below openable windows, air vents, grilles, gutters or eaves.

A terminal less than 2 m from ground level should be fitted with a wire mesh guard to prevent people contacting with the hot surface. Where a terminal is within 600 mm below a plastic gutter, an aluminium shield 1·5 m long should be fitted to the underside of the gutter immediately above the terminal.

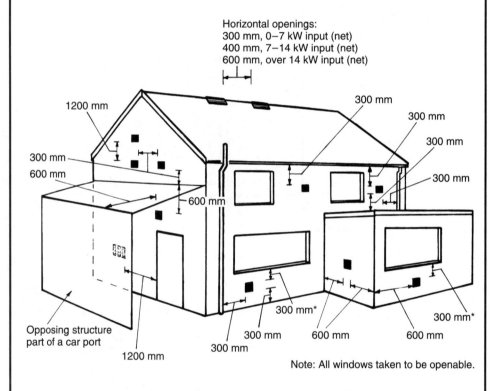

Horizontal openings:
300 mm, 0–7 kW input (net)
400 mm, 7–14 kW input (net)
600 mm, over 14 kW input (net)

1200 mm

300 mm

300 mm

300 mm

300 mm

300 mm

600 mm

300 mm

600 mm

Opposing structure part of a car port

300 mm*

300 mm*

300 mm

600 mm

600 mm

1200 mm

300 mm

Note: All windows taken to be openable.

*0–7 kW input (net) – 300 mm, 7–14 kW input (net) – 600 mm, 14–32 kW input (net) – 1500 mm, over 32 kW input (net) – 2000 mm
Balanced flue and ridge natural draught terminal positions (min. dimensions)

Refs: Building Regulations, Approved Document J: Section J3, Protection of Building.
BS 5440 Installation and maintenance of flues and ventilation for gas appliances of rated input not exceeding 70 kW net.

Natural draught flues – appliances discharging flue gases by natural convection are located on an external wall. There must be some regard for the adjacent construction as unsatisfactory location may result in:

- inefficient combustion of fuel
- risk of fire
- combustion products staining the wall
- combustion gases entering the building.

Fan assisted flues – appliances fitted with these can be located a short distance from an external wall. Smaller terminals are possible due to the more positive extraction of the flue gases. Terminal location is not as critical as for natural draught flues, but due regard must still be given to adjacent construction.

Location of balanced flue terminals (min. distance in mm):

Location of terminal	Natural draught	Fan assisted
Directly under an openable window or a ventilator	300*	300
Under guttering or sanitation pipework	300	75
Under eaves	300	200
Under a balcony or a car port roof	600	200
Horizontally to an opening window	As ridge openings shown previous page	300
Opening in a car port	1200	1200
Horizontally from vertical drain and discharge pipes	300	150 75 < 5 kW input (net)
Horizontally from internal or external corners	600	300
Above ground, balcony or flat roof	300	300
From an opposing wall, other surface or boundary	600	600
Opposite another terminal	600	1200
Vertically from a terminal on the same wall	1200	1500
Horizontally from a terminal on the same wall	300	300

* See note on previous page.

A gas appliance may be situated in a fire recess and the chimney structure used for the flue. The chimney should have clay flue linings to BS EN 1457: Chimneys — Clay/ceramic flue liners. A stainless steel flexible flue lining may be installed where the chimney was built before 1 February 1966, provided the lining complies with BS 715: Specification for metal flue pipes, fittings, terminals and accessories for gas fired appliances with a rated input not exceeding 60 kW.

Other suitable flue materials include:

- precast hollow concrete flue blocks to BS 1289
- pipes made from stainless steel (BS 1449), enamelled steel (BS 6999), cast iron (BS 41) or fibre cement (BS 567 or 835)
- products that satisfy an acceptable quality standard, such as that awarded by the British Board of Agrément.

Flues must be correctly sized from appliance manufacturer's data. If a flue is too large or too long, overcooling of the flue gases will produce condensation. This occurs at about 60°C when the gases cool to the dew point of water. The following factors will determine the flue size:

- heat input to the appliance
- resistance to the flow of combustion gases caused by bends and the terminal
- length of the flue.

Spigot and socket flue pipes are installed socket uppermost and joints made with fire cement. For the efficient conveyance of combusted gases, flue pipes should be vertical wherever possible. Where they pass through a floor or other combustible parts of the structure they should be fitted with a non-combustible sleeve.

A ventilation opening (air brick) for combustion air is required in the external wall of the room containing the appliance. As a guide, for large boilers in their own plant room a ventilation-free area of at least twice the flue area is required. For domestic appliances, 500 mm^2 for each kilowatt of input rating over 7 kW net is adequate.

Ref: Building Regulations, Approved Document J: Heat producing appliances.

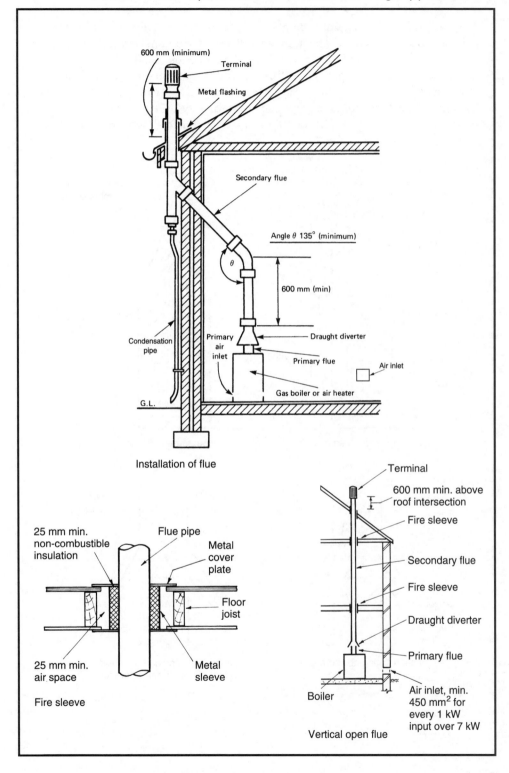

600 mm (minimum)

Terminal

Metal flashing

Secondary flue

Angle θ 135° (minimum)

θ

600 mm (min)

Draught diverter

Condensation pipe

Primary air inlet

Primary flue

Air inlet

Gas boiler or air heater

G.L.

Installation of flue

25 mm min. non-combustible insulation

Flue pipe

Metal cover plate

Floor joist

Metal sleeve

25 mm min. air space

Fire sleeve

Terminal

600 mm min. above roof intersection

Fire sleeve

Secondary flue

Fire sleeve

Draught diverter

Primary flue

Boiler

Air inlet, min. 450 mm² for every 1 kW input over 7 kW

Vertical open flue

Draught Diverter

The purpose of a draught diverter is to admit diluting air into the primary flue to reduce the concentration of combustion gases and to reduce their temperature in the flue. The draught diverter, as the name suggests, also prevents flue downdraughts from extinguishing the gas pilot flame by diverting the draughts outside of the burners. Draught diverters can be provided in two ways. Either as an open lower end to the flue (integral) or an attachment (separate) to the primary flue.

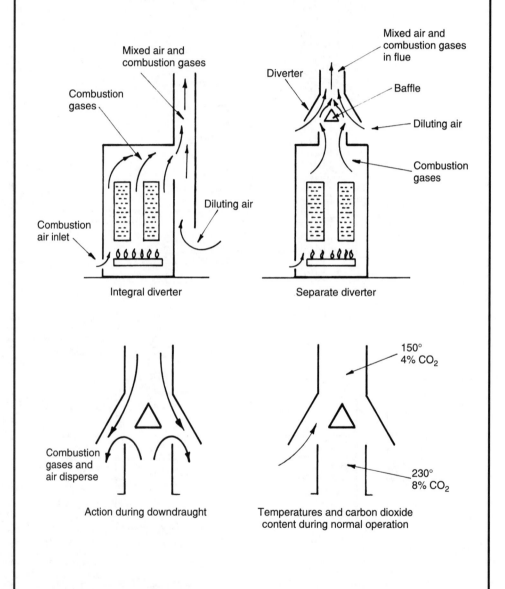

Precast concrete flue blocks are manufactured from high alumina cement and dense aggregagtes, to resist the effects of toxic flue gases and condensation. They are jointed with high alumina cement mortar and laid alternately and integrally with the inner leaf of concrete blockwork in a cavity wall. This optimises space and appearance, as there is no chimney structure projecting into the room or unsightly flue pipe. The void in the blocks is continuous until it joins a twin wall insulated flue pipe in the roof space to terminate at ridge level.

These flue blocks are specifically for gas fires and convectors of relatively low rating. Whilst a conventional circular flue to a gas fire must be at least 12 000 mm^2 cross-sectional area, these rectangular flue blocks must have a minimum flue dimension of 90 mm and cross-sectional area of 16 500 mm^2.

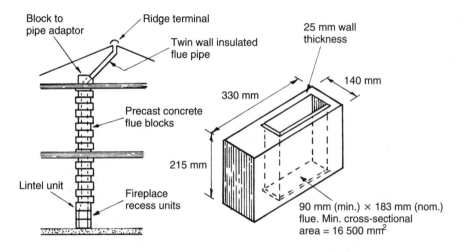

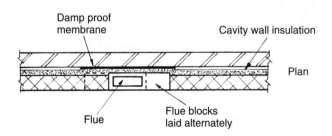

Refs: BS 1289: Flue blocks and masonry terminals for gas appliances.
BS 4543: Factory made (insulated) chimneys.

A flue terminal has several functions:

- to prevent entry of birds, squirrels, etc.
- to prevent entry of rain and snow
- to resist the effects of down draughts
- to promote flue pull and extraction of combusted gases.

Location – should be with regard to positive and negative wind pressures acting on a roof to permit free wind flow across the terminal and not be too close to windows and other ventilation voids. The preferred location is at or above the ridge of a pitched roof. Elsewhere, the following can be used as guidance:

Location	Min. height (mm) to lowest part of outlet
Within 1·5 m horizontally of a vertical surface, e.g. dormer	600 above top of structure
Pitched roof <45°	600 from roof intersection
Pitched roof >45°	1000 " " "
Flat roof	250 " " "
Flat roof with parapet*	600 " " "

* Note: if horizontal distance of flue from parapet is greater than 10 × parapet height, min. flue height = 250 mm.

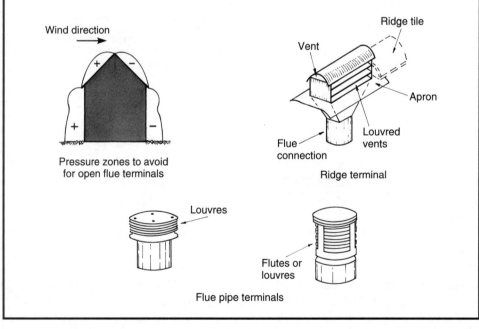

Wind direction

Pressure zones to avoid for open flue terminals

Ridge tile

Vent

Apron

Flue connection

Louvred vents

Ridge terminal

Louvres

Flue pipe terminals

Flutes or louvres

Pitched roof:

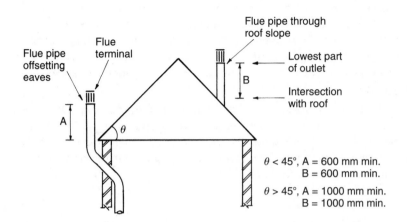

Flue pipe through roof slope

Flue terminal

Flue pipe offsetting eaves

Lowest part of outlet

B

Intersection with roof

θ < 45°, A = 600 mm min.
B = 600 mm min.

θ > 45°, A = 1000 mm min.
B = 1000 mm min.

Flat roof:

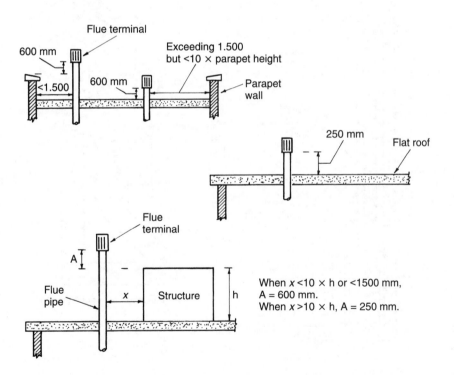

Flue terminal

600 mm

<1.500

600 mm

Exceeding 1.500 but <10 × parapet height

Parapet wall

250 mm

Flat roof

Flue terminal

A

Flue pipe

x

Structure

h

When x <10 × h or <1500 mm, A = 600 mm.
When x >10 × h, A = 250 mm.

Ref: BS 5440: Installation and maintenance of flues and ventilation for gas appliances of rated input not exceeding 70 kW net.

Stainless Steel Flue Lining

Traditional brick chimneys have unnecessarily large flues when used with gas burning appliances. If an existing unlined chimney is to be used, a flexible stainless steel lining should be installed to prevent the combustion products and condensation from breaking down the old mortar joints. By reducing the flue area with a lining, this will accelerate the discharge of gases (efflux velocity), preventing them from lowering sufficiently in temperature to generate excessive condensation.

Coils of stainless steel lining material are available in 100, 125 and 150 mm diameters to suit various boiler connections. The existing chimney pot and flaunching are removed to permit the lining to be lowered and then made good with a clamping plate, new flaunching and purpose-made terminal.

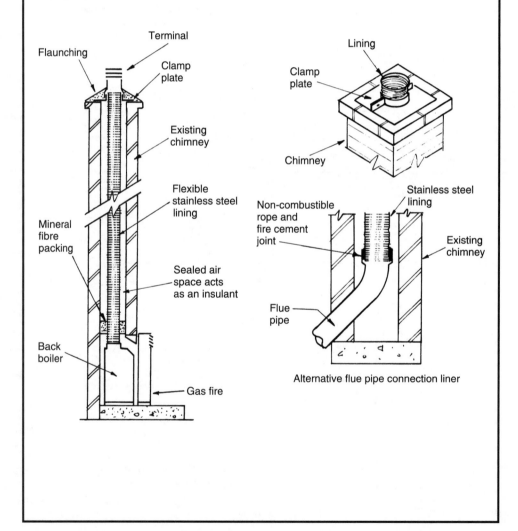

Alternative flue pipe connection liner

This is a cost-effective alternative to providing a separate flue for each gas appliance installed in a multi-storey/multi-unit building. It was originally developed by the South-east Gas Board to utilise balanced flues attached to a central ventilated void. Appliances use a central duct for air supply to the gas burners and to discharge their products of combustion. The dilution of burnt gases must be sufficient to prevent the carbon dioxide content exceeding 1·5% at the uppermost appliance. The size of central void depends on the number of appliances connected. Tables for guidance are provided in BS 5440: Installation and maintenance of flues and ventilation for gas appliances of rated input not exceeding 70 kW net.

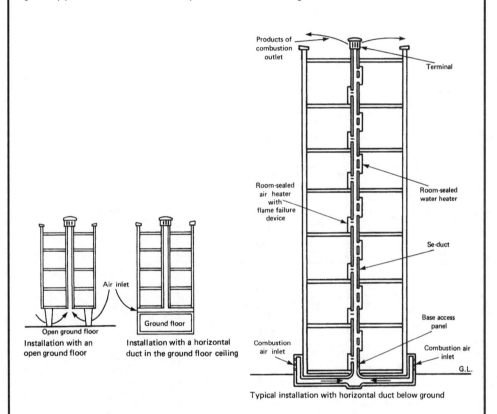

Installation with an open ground floor

Installation with a horizontal duct in the ground floor ceiling

Typical installation with horizontal duct below ground

Note: A flame failure device is otherwise known as a flame supervision device.

The U duct system is similar in concept to the Se-duct, but used where it is impractical to supply air for combustion at low level. The U duct has the benefits of the Se-duct, but it will require two vertical voids which occupy a greater area. The downflow duct provides combustion air from the roof level to appliances. Appliances of the room sealed-type are fitted with a flame failure/supervision device to prevent the build-up of unburnt gases in the duct. They can only connect to the upflow side of the duct. Stable air flow under all wind conditions is achieved by using a balanced flue terminal, designed to provide identical inlet and outlet exposure. As with the Se-duct, the maximum amount of carbon dioxide at the uppermost appliance inlet must be limited to 1·5%.

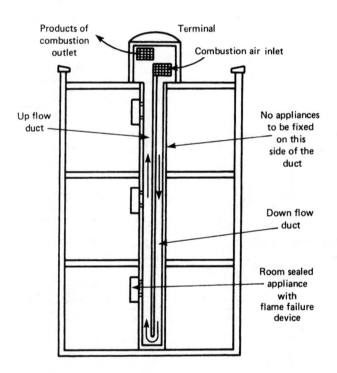

Products of combustion outlet

Terminal

Combustion air inlet

Up flow duct

No appliances to be fixed on this side of the duct

Down flow duct

Room sealed appliance with flame failure device

Typical installation of U duct

The shunt duct system is applicable to installation of several conventional appliances with open flues in the same building. It economises in space and installation costs when compared with providing each appliance with an individual flue. It is limited to ten consecutive storeys due to the effects of varying wind pressures and each appliance must be fitted with a draught diverter and a flame failure/supervision device. Gas fires and water heaters may be connected to this system, provided the subsidiary flue from each is at least 1·2 m and 3 m long respectively, to ensure sufficient draught.

Other shared flue situations may be acceptable where conventional open flued appliances occupy the same room. Consultation with the local gas authority is essential, as there are limitations. An exception is connection of several gas fires to a common flue. Also, a subsidiary branch flue connection to the main flue must be at least 600 mm long measured vertically from its draught diverter.

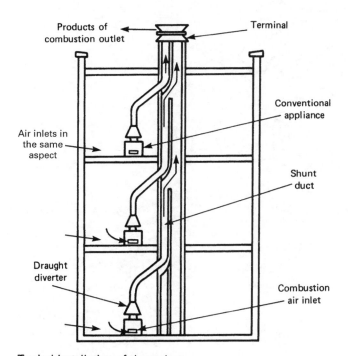

Typical installation of shunt duct

Note: Shared flues sized in accordance with BS 5440.

Fan Assisted Gas Flues

With high rise shops, office buildings and flats sharing the same boiler, problems can arise in providing a flue from ground floor plant rooms. Instead of extending a vertical flue from ground level to the top of a building, it is possible to air dilute the flue gases and discharge them at relatively low level by installing an extract fan in the flue. As the boiler fires, the fan draws fresh air into the flue to mix with the products of gas combustion and to discharge them to the external air. The mixed combustion gases and diluting air outlet terminal must be at least 3 m above ground level and the carbon dioxide content of the gases must not exceed 1%. A draught sensor in the flue functions to detect fan operation. In the event of fan failure, the sensor shuts off the gas supply to the boilers.

The plant room is permanently ventilated with air bricks or louvred vents to ensure adequate air for combustion. Ventilation voids should be at least equivalent to twice the primary flue area.

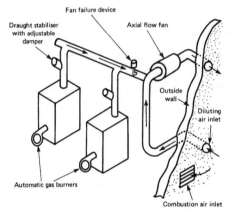

Installation using one outside wall and boilers with automatic burners

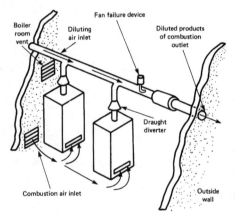

Installation, using two outside walls and boilers with draught diverters

Fan assistance with the dilution and removal of combustion products has progressed from commercial and industrial applications in open flues, to domestic appliance balanced flues. In addition to diluting the CO_2 content at the flue gases point of discharge, fanned draught balanced flue systems have the following advantages over standard balanced flues:

- Positive control of flue gas removal without regard for wind conditions.
- Location of flue terminal is less critical – see page 313.
- Flue size (inlet and outlet) may be smaller.
- Flue length may be longer, therefore the boiler need not be mounted on an external wall.
- Heat exchanger may be smaller due to more efficient burning of gas. Overall size of boiler is reduced.

The disadvantages are, noise from the fan and the additional features could make the appliance more expensive to purchase and maintain.

If the fan fails, the air becomes vitiated due to lack of oxygen and the flames smother. The flame failure/protection device then closes the gas valve.

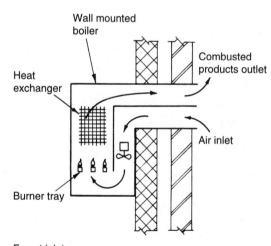

Fan at inlet

Note: Fan may deliver air only to combustion chamber, or an air/gas mixture

Fan at outlet

Note: Fan will be specifically designed to withstand high flue gas temperatures

Location of fan

Room sealed balanced flue appliances do not require a purpose-made air vent for combustion as the air supply is integral with the terminal. Where installed in a compartment or in an enclosure such as a cupboard an air vent is necessary to remove excess heat. With open or conventional flue appliances, access must be made for combustion air if the appliance input rating is in excess of 7 kW (net). This equates to at least 500 mm^2 of free area per kW over 7 kW (net), e.g. the ventilation area required for an open flued boiler of 20 kW (net) input rating will be at least $20 - 7 = 13 \times 500 = 6500$ mm^2.

Conventionally flued appliances will also require air for cooling if they are installed in a compartment. This may be by natural air circulation through an air brick or with fan enhancement.

Flueless appliances such as a cooker or instantaneous water heater require an openable window direct to outside air, plus the following ventilation grille requirements:

Oven, hotplate or grill:

Room volume (m^3)	Ventilation area (mm^2)
<5	10 000
5–10	5000 (non-required if a door opens directly to outside air)
>10	non-required

Instantaneous water heater (max. input 11 kW (net)):

Room volume (m^3)	Ventilation area (mm^2)
<5	not permitted
5–10	10 000
10–20	5000
>20	non-required

Vents should be sited where they cannot be obstructed. At high level they should be as close as possible to the ceiling and at low level, not more than 450 mm above floor level. When installed between internal walls, vents should be as low as possible to reduce the spread of smoke in the event of a fire.

Open flued gas fires rated below 7 kW (net) require no permanent ventilation, but decorative fuel effect fires will require a vent of at least 10 000 mm^2 free area.

The next page illustrates requirements for room sealed and open flued appliances.

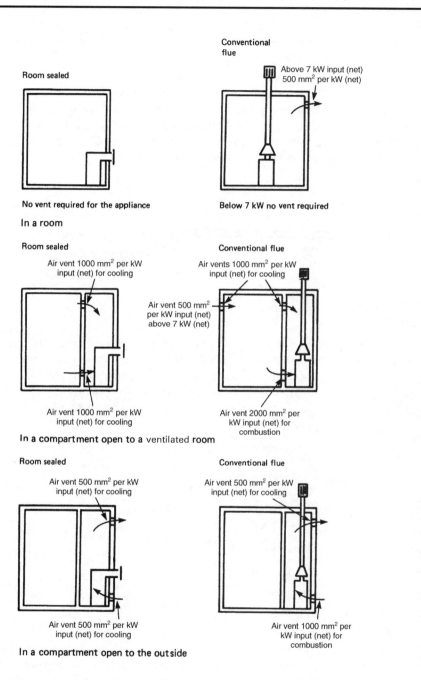

Room sealed

No vent required for the appliance

In a room

Conventional
flue

Above 7 kW input (net)
500 mm² per kW (net)

Below 7 kW no vent required

Room sealed

Air vent 1000 mm² per kW
input (net) for cooling

Air vent 1000 mm² per kW
input (net) for cooling

Conventional flue

Air vents 1000 mm² per kW
input (net) for cooling

Air vent 500 mm²
per kW input (net)
above 7 kW (net)

Air vent 2000 mm² per
kW input (net) for
combustion

In a compartment open to a ventilated room

Room sealed

Air vent 500 mm² per kW
input (net) for cooling

Air vent 500 mm² per kW
input (net) for cooling

Conventional flue

Air vent 500 mm² per kW
input (net) for cooling

Air vent 1000 mm² per
kW input (net) for
combustion

In a compartment open to the outside

Refs: Building Regulations, Approved Document J: Combustion
appliances and fuel storage systems.
BS 5440: Installation and maintenance of flues and ventilation
for gas appliances of rated input not exceeding 70 kW net.

Combusted Gas Analysis

Simple field tests are available to assess the efficiency of gas combustion with regard to the percentage of carbon monoxide and carbon dioxide in the flue gases.

Draeger analyser – hand bellows, gas sampler tube and a probe. The tube is filled with crystals corresponding to whether carbon monoxide or carbon dioxide is to be measured. The probe is inserted into the flue gases and the bellows pumped to create a vacuum. The crystals absorb different gases and change colour accordingly. Colours correspond with a percentage volume.

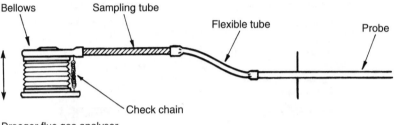

Draeger flue gas analyser

Fyrite analyser – hand bellows, container of liquid reactant and a probe. Flue gases are pumped into the container which is inverted so that the liquid reactant absorbs the gas in solution. The liquid rises to show the percentage carbon dioxide corresponding to a scale on the container. Oxygen content can also be measured using an alternative solution.

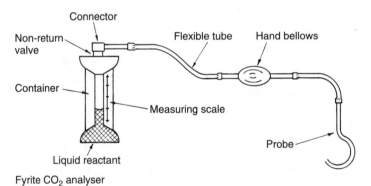

Fyrite CO_2 analyser

Note: Flue gas samples can be taken by inserting the probe below the draught diverter or through the access plate on top of the appliance combustion chamber. Samples can also be taken at the terminal.

Ref: BS 1756: Methods for sampling and analysis of flue gases.

Typical natural gas consumption figures for domestic appliances:

Boiler $1 \cdot 6 \, m^3/hour$
Cooker $1 \cdot 0$ "   "
Fire $0 \cdot 5$ "   "

Exact gas consumption rate (Q) can be calculated from the following formula:

$$Q = \frac{\text{Appliance rating} \times 3600}{\text{Calorific value of gas}}$$

Given that the calorific values for natural gas and propane (LPG) are $38\,500 \, kJ/m^3$ and $96\,000 \, kJ/m^3$ respectively, the value of Q for a 20 kW input boiler is:

$$\text{Nat. gas}: Q = \frac{20 \times 3600}{38\,500} = 1 \cdot 87 \, m^3/h$$

$$\text{Propane}: Q = \frac{20 \times 3600}{96\,000} = 0 \cdot 75 \, m^3/h$$

Operating costs – fuel tariffs can be obtained from the various gas suppliers. A typical charge for natural gas is $1 \cdot 3$ pence per kWh. If the 20 kW input boiler consumes gas for 5 hours per day, the operating cost will be:

$$1 \cdot 3 \times 20 \times 5 = £1 \cdot 30 \text{ per day or } £9 \cdot 10 \text{ per week}$$

To convert gas metered in units of cubic feet, multiply by 0·0283, i.e. 1 cu. ft. = 0·0283 m^3.

Gas consumed in kWh:

$$\frac{m^3 \times \text{volume conversion factor } (1 \cdot 02264) \times \text{calorific value } (MJ/m^3)}{3 \cdot 6}$$

where: 1 kWh = 3·6 MJ.

e.g. 100 cu. ft at 2·83 m^3

$$\frac{2 \cdot 83 \times 1 \cdot 02264 \times 38 \cdot 5}{3 \cdot 6} = 31 \, kWh$$

Gas Pipe Sizing

To determine the size of pipework, two factors must be established:

1. The gas consumption (Q).
2. The effective length of pipework.

Effective length of pipework is taken as the actual length plus the following allowances for fittings in installations up to 28 mm outside diameter copper tube:

Fitting	Equivalent length (m)
elbow	0·5
tee	0·5
bend (90°)	0·3

The gas discharge in m^3/hour for copper tube for varying effective lengths is as follows:

Tube diam. (mm o.d)	Effective pipe length (m)							
	3	6	9	12	15	20	25	30
8	0·52	0·26	0·17	0·13	0·10	0·07		
10	0·86	0·57	0·50	0·37	0·30	0·22	0·18	0·15
12	1·50	1·00	0·85	0·82	0·69	0·52	0·41	0·34
15	2·90	1·90	1·50	1·30	1·10	0·95	0·92	0·88
22	8·70	5·80	4·60	3·90	3·40	2·90	2·50	2·30
28	18·00	12·00	9·40	8·00	7·00	5·90	5·20	4·70

This table is appropriate for 1 mb (10 mm w.g.) pressure drop for gas of relative density 0.6.

Example:

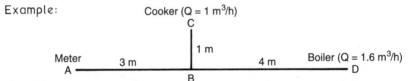

Note: A to B contains 3 elbows and 1 tee
  B to C contains 3 elbows
  B to D contains 4 elbows

Pipe A to B, gas flow = 1 m^3/h + 1.6 m^3/h = 2.6 m^3/h

Actual pipe length = 3 m

Effective pipe length = $3 + (3 \times 0·5) + (1 \times 0·5) = 5$ m

From the table, a 22 mm o.d. copper tube can supply 2.6 m^3/h for up to 23·75 metres (by interpolating between 20 and 25 m).

Pressure drop over only 5 m will be: $5 \div 23·75 = 0·21$ mb (2·1 mm w.g.).

Pipes B to C and B to D can be calculated similarly.

Ref: BS 6891: Specification for installation of low pressure gas pipework of up to 28 mm in domestic premises.

10 ELECTRICAL SUPPLY AND INSTALLATIONS

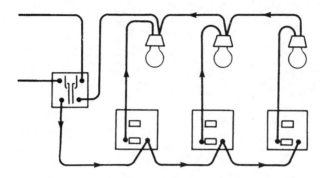

THREE-PHASE GENERATION AND SUPPLY

ELECTRICITY DISTRIBUTION

INTAKE TO A BUILDING

EARTHING SYSTEMS AND BONDING

CONSUMER UNIT

POWER AND LIGHTING CIRCUITS

OVERLOAD PROTECTION

ELECTRIC WIRING

TESTING COMPLETED INSTALLATION

CABLE RATING

DIVERSITY

DOMESTIC AND INDUSTRIAL INSTALLATIONS

ELECTRIC SPACE HEATING

SPACE HEATING CONTROLS

CONSTRUCTION SITE ELECTRICITY

LIGHT SOURCES, LAMPS AND LUMINAIRES

LIGHTING CONTROLS

EXTRA-LOW-VOLTAGE LIGHTING

LIGHTING DESIGN

DAYLIGHTING

TELECOMMUNICATIONS INSTALLATION

In 1831 Michael Faraday succeeded in producing electricity by plunging a bar magnet into a coil of wire. This is credited as being the elementary process by which we produce electricity today, but the coils of wire are cut by a magnetic field as the magnet rotates. These coils of wire (or stator windings) have an angular spacing of 120° and the voltages produced are out of phase by this angle for every revolution of the magnets. Thus generating a three-phase supply.

A three-phase supply provides 73% more power than a single-phase supply for the addition of a wire. With a three-phase supply, the voltage between two line or phase cables is 1·73 times that between the neutral and any one of the line cables, i.e. 230 volts × 173 = 400 volts, where 1·73 is derived from the square root of the three phases.

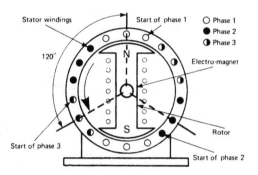

Simplified detail of three-phase generator or alternator

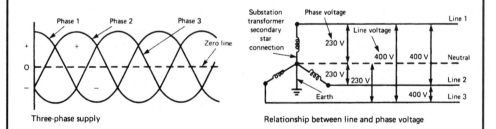

Three-phase supply

Relationship between line and phase voltage

Note: The following section on electrical systems should be read with regard to:

Building Regulations, Approved Document P: Electrical safety, and BS 7671: Requirements for Electrical Installations, the IEE Wiring Regulations 16th edition.

In the UK electricity is produced at power generating stations at 25 kilovolt (kV) potential, in three-phase supply at 50 cycles per second or hertz (Hz). Thereafter it is processed by step-up transformers to 132, 275 or 400 kV before connecting to the national grid. Power to large towns and cities is by overhead lines at 132 kV or 33 kV where it is transformed to an 11 kV underground supply to sub-stations. From these sub-stations the supply is again transformed to the lower potential of 400 volts, three-phase supply and 230 volts, single-phase supply for general distribution.

The supply to houses and other small buildings is by an underground ring circuit from local sub-stations. Supplies to factories and other large buildings or complexes are taken from the 132 or 33 kV main supply. Larger buildings and developments will require their own transformer, which normally features a delta-star connection to provide a four-wire, three-phase supply to the building.

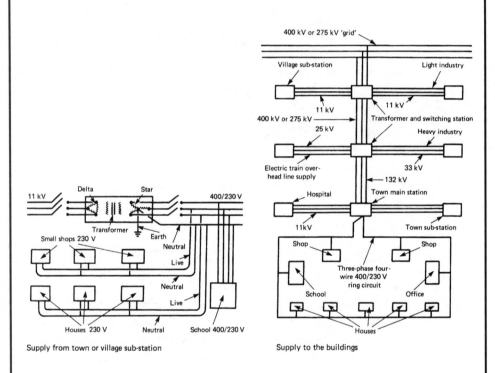

Supply from town or village sub-station

Supply to the buildings

Note: For easy identification, each phase cable has colour coded plastic insulation of red, yellow or blue. The neutral is colour coded black. An outer sheathing of red or black provides for future identification.

A sub-station is required for the conversion, transformation and control of electrical power. It is used where large buildings or complexes of buildings require greater power than the standard low or medium potential of 230 and 400 volts. A sub-station must be constructed on the customer's premises. It is supplied by high voltage cables from the electricity authority's nearest switching station. The requirements for a sub-station depend upon the number and size of transformers and switchgear.

A transformer is basically two electric windings, magnetically inter-linked by an iron core. An alternating electromotive force applied to one of the windings produces an electromagnetic induction corresponding to an electromotive force in the other winding.

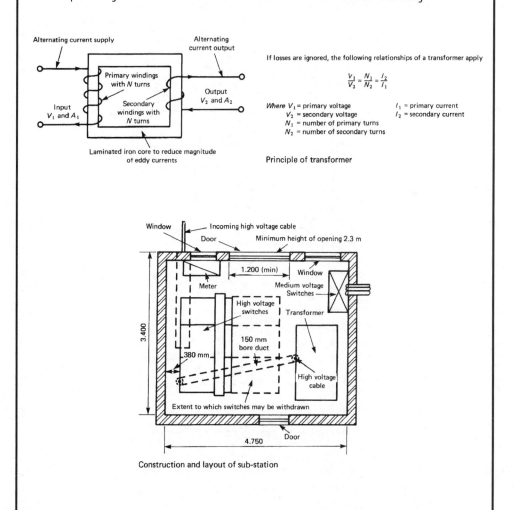

If losses are ignored, the following relationships of a transformer apply

$$\frac{V_1}{V_2} = \frac{N_1}{N_2} = \frac{I_2}{I_1}$$

Where V_1 = primary voltage $\qquad I_1$ = primary current
V_2 = secondary voltage $\qquad I_2$ = secondary current
N_1 = number of primary turns
N_2 = number of secondary turns

Principle of transformer

Construction and layout of sub-station

The termination and metering of services cables to buildings is determined by the electricity authority's supply arrangements. Most domestic supplies are underground with the service cable terminating at the meter cupboard, as shown. Depth of cover to underground cables should be at least 750 mm below roads and 450 mm below open ground. In remote areas the supply may be overhead. Whatever method is used, it is essential that a safety electrical earthing facility is provided and these are considered on the next page. All equipment up to and including the meter is the property and responsibility of the supplier. This also includes a fusible cut-out, neutral link and in some situations a transformer. Meters are preferably sited in a purpose-made reinforced plastic compartment set in or on the external wall of a building.

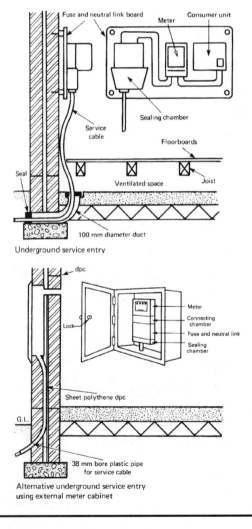

Underground service entry

Alternative underground service entry using external meter cabinet

Supply systems require a safety electrical earthing facility. The manner in which this is effected will depend on whether the supply is overhead or underground and the conductive property of the ground surrounding the installation. Systems are classified in accordance with a letter coding:

First letter – type of earthing:

T – at least one point of the supply is directly earthed.

I – the supply is not directly earthed, but connected to earth through a current limiting impedance. Not acceptable for public supplies in the UK.

Second letter – installation earthing arrangement:

T – all exposed conductive metalwork is directly earthed.

N – all exposed conductive metalwork is connected to an earth provided by the supply company.

Third and fourth letters – earth conductor arrangement:

S – earth and neutral conductors separate.

C – earth and neutral conductors combined.

Common supply and earthing arrangements are:

TT (shown below).

TN-S and TN-C-S (shown next page).

TT system:

Most used in rural areas where the supply is overhead. An earth terminal and electrode is provided on site by the consumer. As an extra safety feature, a residual current device (RCD), generally known as a trip switch, is located between the meter and consumer unit. The RCD in this situation should be of the time delayed type – see page 348.

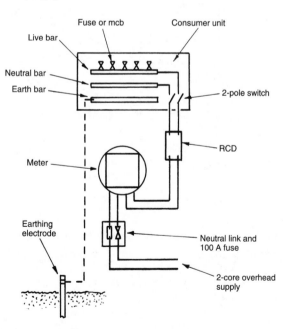

Fuse or mcb   Consumer unit
Live bar
Neutral bar
Earth bar
2-pole switch
RCD
Meter
Earthing electrode
Neutral link and 100 A fuse
2-core overhead supply

TN-S system – this is widely used in the UK, with the electricity supply company providing an earth terminal with the intake cable. This is usually the metal sheathing around the cable, otherwise known as the supply protective conductor. It connects back to the star point at the area transformer, where it is effectively earthed.

TN-C-S system – this is as the TN-S system, but a common conductor is used for neutral and earth supply. The supply is therefore TN-C, but with a separated neutral and earth in the consumer's installation it becomes TN-C-S. This system is also known as protective multiple earth (PME). The advantage is that a fault to earth is also a fault to neutral, which creates a high fault current. This will operate the overload protection (fuse or circuit breaker) rapidly.

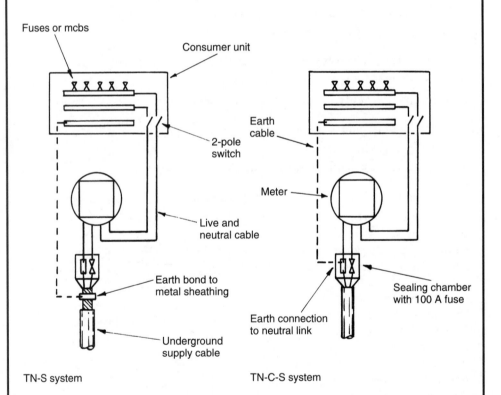

Fuses or mcbs

Consumer unit

2-pole switch

Earth cable

Meter

Live and neutral cable

Earth bond to metal sheathing

Sealing chamber with 100 A fuse

Earth connection to neutral link

Underground supply cable

TN-S system

TN-C-S system

Note: Specification of installation cable between supply company's sealing chamber and consumer's unit – phase/live and neutral 25 mm^2, earth 10 mm^2 cross-sectional area.

The Institution of Electrical Engineers (IEE) Wiring Regulations require the metal sheaths and armour of all cables operating at low and medium voltage to be cross-bonded to ensure the same potential as the electrical installation. This includes all metal trunking and ducts for the conveyance and support of electrical services and any other bare earth continuity conductors and metalwork used in conjunction with electrical appliances. The bonding of the services shall be as close as possible to the point of entry of the services into a building. Other fixed metalwork shall be supplementary earth bonded.

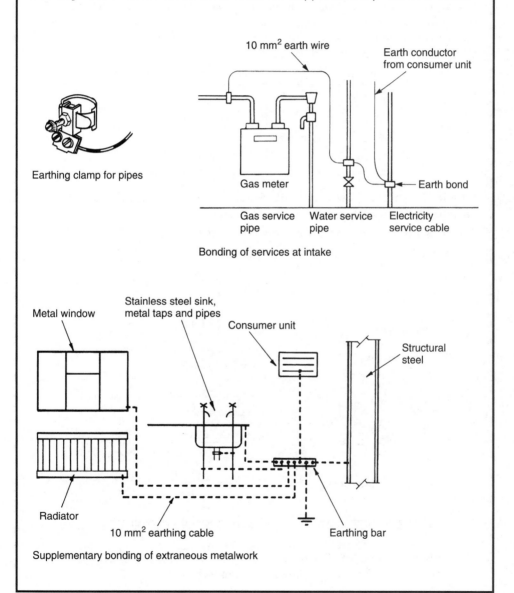

Earthing clamp for pipes

10 mm^2 earth wire

Earth conductor from consumer unit

Earth bond

Gas meter

Gas service pipe   Water service pipe   Electricity service cable

Bonding of services at intake

Metal window

Stainless steel sink, metal taps and pipes

Consumer unit

Structural steel

Radiator

10 mm^2 earthing cable   Earthing bar

Supplementary bonding of extraneous metalwork

Historically, electrical installations required a separate fuse and isolator for each circuit. Modern practice is to rationalise this into one 'fuse box', known as a consumer's power supply control unit or consumer unit for short. This unit contains a two-pole switch isolator for the phase/live and neutral supply cables and three bars for the live, neutral and earth terminals. The live bar is provided with several fuse ways or miniature circuit breakers (up to 16 in number for domestic use) to protect individual circuits from overload. Each fuse or mcb is selected with a rating in accordance with its circuit function. Traditional fuses are rated at 5, 15, 20, 30 and 45 amps whilst the more modern mcbs are rated in accordance with BS EN 60898: Circuit breakers for over current protection ... at 6, 10, 16, 20, 32, 40, 45 and 50 amps.

Circuit	Mcb rating (amps)
Lighting	6
Immersion heater	16 or 20*
Socket ring main	32
Cooker	40 or 45*
Shower	40 or 45*

* Depends on the power rating of appliance. A suitable mcb can be calculated from: Amps = Watts ÷ Voltage.

E.g. A 3 kW immersion heater: Amps = 3000 ÷ 230 = 13.

Therefore a 16 amp rated mcb is adequate.

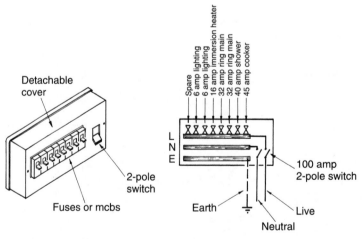

Eight-way consumer unit   Typical contents of consumer unit

Refs: BS 5486 and BS EN 60439: Specification for low-voltage switchgear and controlgear assemblies.

A ring circuit is used for single-phase power supply to three-pin sockets. It consists of PVC sheathed cable containing live and neutral conductors in PVC insulation and an exposed earth looped into each socket outlet. In a domestic building a ring circuit may serve an unlimited number of sockets up to a maximum floor area of 100 m². A separate circuit is also provided solely for the kitchen, as this contains relatively high rated appliances. Plug connections to the ring have small cartridge fuses up to 13 amp rating to suit the appliance wired to the plug. The number of socket outlets from a spur should not exceed the number of socket outlets and fixed appliances on the ring.

Cable rating:
2·5 mm² c.s.a.

Consumer unit:
BS 5486 and
BS EN 60439.

3-pin plugs and
sockets: BS 1363.

Plug cartridge
fuses: BS 1362.

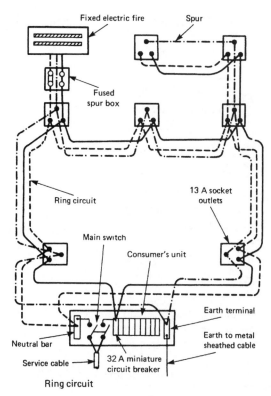

Note: Fixed appliances such as fires, heating controls and low powered water heaters can be connected to a fused spur from a ring socket. Appliances and installations with a load factor above 3 kW, e.g. immersion heater, cooker, extension to an outbuilding, etc. must not be connected to any part of a ring circuit. These are supplied from a separate radial circuit from the consumer unit.

Power Sockets

Power sockets should be positioned between 150 mm and 250 mm above floor levels and work surfaces. An exception is in buildings designed for the elderly or infirm, where socket heights should be between 750 and 900 mm above the floor. Every socket terminal should be fitted with a double outlet to reduce the need for adaptors. Disposition of sockets would limit the need for lead lengths to no more than 2 m.

The following provides guidance on the minimum provision for power sockets in domestic accommodation:

Location	Minimum quantity of sockets
Living rooms	8
Kitchen	6
Master bedroom	6
Dining room	4
Study bedroom	4
Utility room	4
Single bedrooms	4
Hall and landing	2
Garage/workshop	2
Bathroom	1 – double insulated shaver socket

Maximum appliance load (watts) and plug cartridge fuse (BS 1362) selection for 230 volt supply:

Maximum load (W)	Plug fuse rating (amp)
230	1
460	2
690	3
1150	5
1610	7
2300	10
2900	13

Calculated from: Watts = Amps × Voltage.

A radial circuit may be used as an alternative to a ring circuit to supply any number of power sockets, provided the following limitations are effected:

Cable c.s.a. (mm²)	Minimum overload protection (amps)	Remarks
2.5	20	Max. 20 m² floor area, 17 m cable
4.0	30	Max. 50 m² floor area, 21 m cable

With 2·5 mm² cable length limitation of 17 m over 20 m² floor area for a radial supply to sockets, a ring main with a maximum cable length of 54 m over 100 m² will usually prove to be more effective. Therefore radial circuits are more suited to the following:

Application	Cable c.s.a. (mm²)	Minimum overload protection (amps)	Remarks
Lighting	1·5	5	Max. 10 light fittings
Immersion heater	2·5	15	Butyl rubber flex from 2-pole control switch
Cooker	6	30	Cable and fuse
	10	45	ratings to suit cooker rating
Shower	4, 6 or 10	30 to 45	See page 253
Storage radiator	2·5	20	See page 360
Outside extension	2·5	20	Nominal light and power
	4	30	Max. five sockets and 3 amp light circuit (next page)

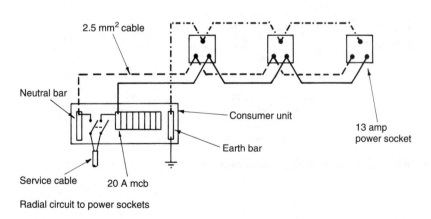

Radial circuit to power sockets

Radial Extension to an Outbuilding

An electricity supply to an outside building may be overhead at a height not less than 3·5 m. It may be supported in a conduit or from a catenary suspension wire. An underground supply is less obtrusive and should be at least 500 mm below the surface. The cable should be armoured PVC sheathed or copper sheathed mineral insulated (MICC). Standard PVC insulated cable may be used, provided it is enclosed in a protective conduit. Fused isolators are required in the supply building and the outside building, and a residual current device (RCD) 'trip switch' should also be installed after the fused switch control from the consumer unit. Two-point-five mm^2 c.s.a. cable is adequate for limited installations containing no more than a power socket and lighting. In excess of this, a 4 mm^2 c.s.a. cable is preferred particularly if the outbuilding is some distance to overcome the voltage drop.

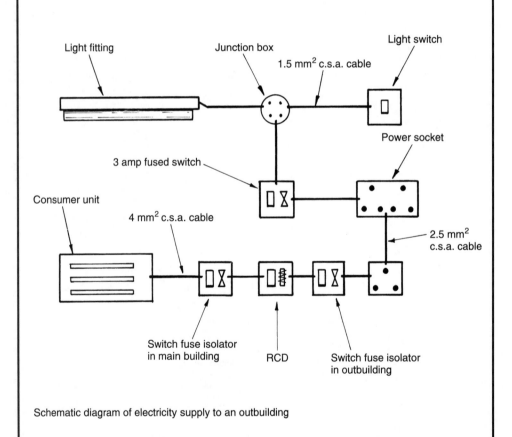

Schematic diagram of electricity supply to an outbuilding

Lighting circuits can incorporate various switching arrangements. In a one-way switch circuit the single-pole switch must be connected to the live conductor. To ensure that both live and neutral conductors are isolated from the supply a double-pole switch may be used, although these are generally limited to installations in larger buildings where the number and type of light fittings demand a relatively high current flow. Provided the voltage drop (4% max., see page 355) is not exceeded, two or more lamps may be controlled by a one-way single-pole switch.

In principle, the two-way switch is a single-pole changeover switch interconnected in pairs. Two switches provide control of one or more lamps from two positions, such as that found in stair/landing, bedroom and corridor situations. In large buildings, every access point should have its own lighting control switch. Any number of these may be incorporated into a two-way switch circuit. These additional controls are known as intermediate switches.

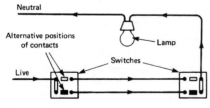

One-way single pole switch circuit controlling one lamp.

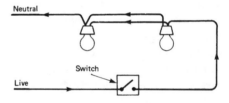

One-way single pole switch circuit controlling two or more lamps

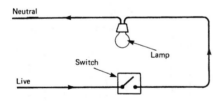

Two-way switching

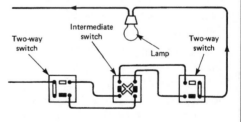

Two-way switching with one intermediate switch

The purpose of a 'master' switch is to limit or vary the scope of control afforded by other switches in the same circuit. If a 'master' switch (possibly one with a detachable key option) is fixed near the main door of a house or flat, the householder is provided with a means of controlling all the lights from one position.

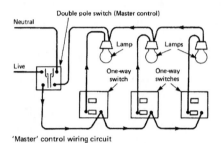

'Master' control wiring circuit

A sub-circuit for lighting is generally limited to a total load of 10, 100 watt light fittings. It requires a 5 amp fuse or 6 amp mcb overload protection at the consumer unit. The importance of not exceeding these ratings can be seen from the simple relationship between current (amps), power (watts) and potential (voltage), i.e. Amps = Watts ÷ Volts. To avoid overloading the fuse or mcb, the limit of 10 lamps @ 100 watts becomes:

$$Amps = (10 \times 100) \div 230 = 4.3$$

i.e. <5 amps fuse protection.

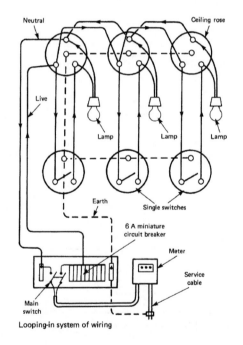

Looping-in system of wiring

In large buildings higher rated overload protection is often used due to the greater load.

Wiring for lighting is usually undertaken using the 'looping-in' system, although it is possible to use junction boxes instead of ceiling roses for connections to switches and light fittings.

Electrical installations must be protected from current overload, otherwise appliances, cables and people using the equipment could be damaged. Protection devices can be considered in three categories:

1. Semi-enclosed (rewirable) fuses.
2. High breaking or rupturing capacity (HBC or HRC) cartridge fuses.
3. Miniature circuit breakers (mcb).

None of these devices necessarily operate instantly. Their efficiency depends on the degree of overload. Rewirable fuses can have a fusing factor of up to twice their current rating and cartridge fuses up to about 1·6. Mcbs can carry some overload, but will be instantaneous (0·01 seconds) at very high currents.

Characteristics:
Semi-enclosed rewirable fuse:
Inexpensive.
Simple, i.e. no moving parts.
Prone to abuse (wrong wire could be used).
Age deterioration.
Unreliable with temperature variations.
Cannot be tested.

Cartridge fuse:
Compact.
Fairly inexpensive, but cost more than rewirable.
No moving parts.
Not repairable.
Could be abused.

Miniature circuit breaker:
Relatively expensive.
Factory tested.
Instantaneous in high current flow.
Unlikely to be misused.

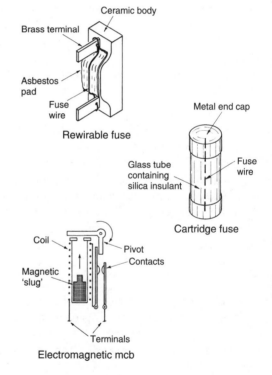

Rewirable fuse

Cartridge fuse

Electromagnetic mcb

Refs: BS 88: Cartridge fuses for voltages up to and including 1000 V a.c. and 1500 V d.c.
BS 1361: Specification for cartridge fuses for a.c. circuits in domestic and similar premises.
BS EN 60269: Low voltage fuses.
BS EN 60898: Specification for circuit breakers for overcurrent protection for household and similar installations.

Residual Current Devices (RCD) are required where a fault to earth may not produce sufficient current to operate an overload protection device (fuse or mcb), e.g. an overhead supply. If the impedance of the earth fault is too high to enable enough current to effect the overload protection, it is possible that current flowing to earth may generate enough heat to start a fire. Also, the metalwork affected may have a high potential relative to earth and if touched could produce a severe shock.

An RCD has the load current supplied through two equal and opposing coils, wound on a common transformer core. When the live and neutral currents are balanced (as they should be in a normal circuit), they produce equal and opposing fluxes in the transformer or magnetic coil. This means that no electromotive force is generated in the fault detector coil. If an earth fault occurs, more current flows in the live coil than the neutral and an alternating magnetic flux is produced to induce an electromotive force in the fault detector coil. The current generated in this coil activates a circuit breaker.

Whilst a complete system can be protected by a 100 mA (milliamp) RCD, it is possible to fit specially equipped sockets with a 30 mA RCD where these are intended for use with outside equipment. Plug-in RCDs are also available for this purpose. Where both are installed it is important that discrimination comes into effect. Lack of discrimination could effect both circuit breakers simultaneously, isolating the whole system unnecessarily. Therefore the device with the larger operating current should be specified with a time delay mechanism.

The test resistor provides extra current to effect the circuit breaker. This should be operated periodically to ensure that the mechanics of the circuit breaker have not become ineffective due to dirt or age deterioration. A notice to this effect is attached to the RCD.

Ref: BS ENs 61008 and 61009: Electrical accessories – residual current operated circuit-breakers for household and similar uses.

An RCD is not appropriate for use with a TN-C system, i.e. combined neutral and earth used for the supply, as there will be no residual current when an earth fault occurs as there is no separate earth pathway.

They are used primarily in the following situations:

- Where the electricity supply company do not provide on earth terminal, e.g. a TT overhead supply system.
- In bedrooms containing a shower cubicle.
- For socket outlets supplying outdoor portable equipment.

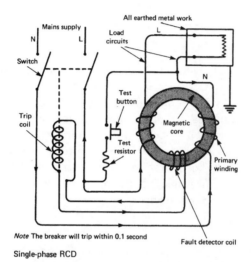

Note The breaker will trip within 0.1 second

Single-phase RCD

Note The breaker will trip within 0.1 second

Three-phase RCD

A three-phase device operates on the same principle as a single-phase RCD, but with three equal and opposing coils.

Armoured cable is used for mains and sub-mains. The cable is laid below ground level, breaking the surface where it enters sub-stations or transformers and other buildings. High voltage cable is protected below ground by precast concrete 'tiles'.

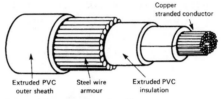

Armoured three phase four wire cable
for laying below ground level

Conduit for electrical services is produced in steel (galvanised or painted black) or plastic tube into which insulated cables are drawn. The conduit protects the cable from physical damage and heat. It also provides continuous support and if it is metal, it may be used as an earth conductor. Standard outside diameters are 20, 25, 32 and 40 mm. Steel is produced in either light or heavy gauge. Light gauge is connected by grip fittings, whilst the thicker walled heavy gauge can be screw threaded to fittings and couplings. Plastic conduit has push-fit connections.

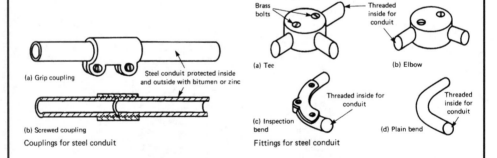

(a) Grip coupling

(b) Screwed coupling

Couplings for steel conduit

(a) Tee

(b) Elbow

(c) Inspection bend

(d) Plain bend

Fittings for steel conduit

Refs: BS 6346: Specification for 600/1000 V and 1900/3300 V armoured electric cables having PVC insulation.
BS EN 50086: Specification for conduit systems for electrical installations.
BS 7846: Specification for 600/1000 V armoured fire-resistant electric cables having low emission of smoke and corrosive gases when affected by fire.

Mineral insulated copper covered cable (MICC) has copper conductors insulated with highly compressed magnesium oxide powder inside a copper tube. When installing the cable, it is essential that the hygroscopic insulant does not come into contact with a damp atmosphere. Cutting the cable involves special procedures which are used to seal the insulant from penetration of atmospheric dampness. The cable provides an excellent earth conductor; it is also resistant to most corrosive atmospheres and is unaffected by extremes of heat.

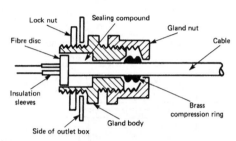

Section of termination joint for mineral insulated copper covered cable (MICC)

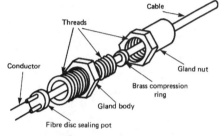

Exploded view of termination joint for mineral insulated copper covered cable

PVC and rubber insulated cables are relatively inexpensive and simple to install, requiring clipped support at regular intervals. PVC cables are in general use, but they have a temperature limitation between 0°C and 70°C. Below zero they become brittle and are easily damaged and at the higher temperature they become soft, which could encourage the conductor to migrate through the PVC. Outside of these temperatures, the cable must be protected or an appropriate rubber insulant specified. Cables usually contain one, two or three conductors. In three-core cable the live and neutral are insulated with red and black colour coding respectively. The earth is bare and must be protected with green and yellow sleeving where exposed at junction boxes, sockets, etc. Blue and yellow insulated conductors are occasionally used where an additional facility is required, e.g. two-way lighting.

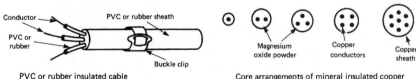

PVC or rubber insulated cable

Core arrangements of mineral insulated copper covered cables

Refs: BS 6004 and 6007: Specifications for PVC and rubber insulated cables for electric power and lighting, respectively.

There are a variety of wiring schemes depending on the degree of sophistication required and the extent of controls, i.e. thermostats, motorised valves, etc. Boiler and control equipment manufacturers provide installation manuals to complement their products. From these the installer can select a control system and wiring diagram to suit their client's requirements.

The schematic diagrams shown relate to a gravity or convected primary flow and return and pumped heating system (see page 69) and a fully pumped hot water and heating system using a three-way motorised valve (see page 91).

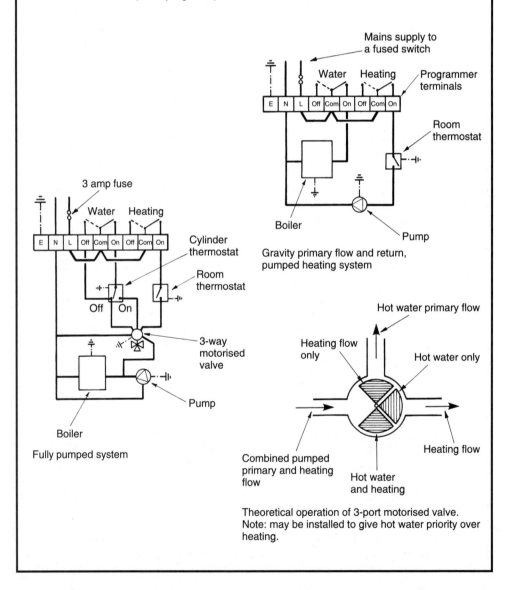

Gravity primary flow and return, pumped heating system

Fully pumped system

Theoretical operation of 3-port motorised valve. Note: may be installed to give hot water priority over heating.

Electrical installations must be tested on completion to verify that the system will operate efficiently and safely. The tests are extensive, as defined in the Institution of Electrical Engineers Regulations. They can only be carried out by a competent person, i.e. a qualified electrician or electrical engineer. The following tests are an essential part of the proceedings:

- Continuity.
- Insulation.
- Polarity.

Testing is undertaken by visual inspection and the use of a multipurpose meter (multimeter) or an instrument specifically for recording resistance, i.e. an ohmmeter.

Continuity – there are several types of continuity test for ring mains. Each is to ensure integrity of the live, neutral and earth conductors without bridging (shorting out) of connections. The following is one established test to be applied to each conductor:

- Record the resistance between the ends of the ring circuit (A).
- Record the resistance between closed ends of the circuit and a point mid-way in the circuit (B).
- Check the resistance of the test lead (C).
- Circuit integrity is indicated by: A ÷ 4 approx. = B – C.

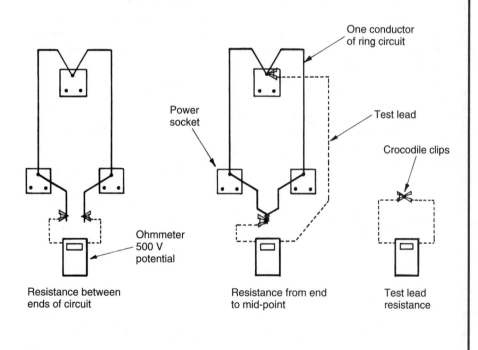

One conductor of ring circuit

Power socket

Test lead

Crocodile clips

Ohmmeter 500 V potential

Resistance between ends of circuit

Resistance from end to mid-point

Test lead resistance

Insulation – this test is to ensure that there is a high resistance between live and neutral conductors and these conductors and earth. A low resistance will result in current leakage and energy waste which could deteriorate the insulation and be a potential fire hazard. The test to earth requires all lamps and other equipment to be disconnected, all switches and circuit breakers closed and fuses left in. Ohmmeter readings should be at least 1 MΩ.

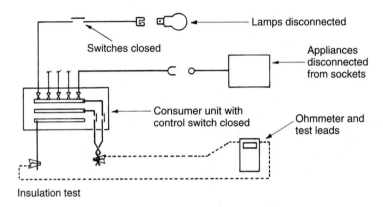

Insulation test

Polarity – this is to ensure that all switches and circuit breakers are connected in the phase or live conductor. An inadvertant connection of switchgear to a neutral conductor would lead to a very dangerous situation where apparent isolation of equipment would still leave it live! The test leads connect the live bar in the disconnected consumer unit to live terminals at switches. A very low resistance reading indicates the polarity is correct and operation of the switches will give a fluctuation on the ohmmeter.

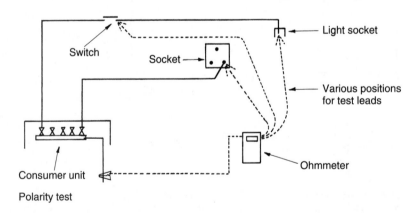

Polarity test

Ref: BS EN 61010-1: Safety requirements for electrical equipment for measurement, control and laboratory use.

Standard applications	Cable specification (mm² c.s.a.)
Lighting	1 or 1·5
Immersion heater	1·5 or 2·5
Sockets (ring)	2·5
Sockets (radial)	2·5 or 4 (see page 343)
Cooker	6 or 10
Shower	4, 6 or 10 (see page 253)

Some variations occur as the specification will depend on the appliance or circuit loading – see calculation below. Where non-standard circuits or special installations are necessary, the cable specification must be calculated in the following stages:

- Determine the current flowing.
- Select an appropriate cable (see table below).
- Check that the voltage drop is not greater than 4%.

Current ratings and voltage reduction for PVC insulated cables:

c.s.a. (mm²)	Current carrying capacity (amps)		Voltage drop (mV/amp/m)
	In conduit	Clipped	
1	13·5	15·5	44
1.5	17·5	20	29
2.5	24	27	18
4	32	37	11
6	41	47	7·3
10	57	65	4·4

E.g. a 7.2 kW shower with a cable length of 10 m in conduit:

Amps = Watts ÷ Volts = 7200 ÷ 230 = 31·3
From table, select 4 mm² c.s.a. (32 amps)

Voltage drop = (mV × Current flowing × Cable length) ÷ 1000
= (11 × 31·3 × 10) ÷ 1000 = 3·44 volts

Maximum voltage drop = 230 × 4% = 9·2 volts.
Therefore, 4 mm² c.s.a. cable is satisfactory.

Note: Correction factors may need to be applied, e.g., when cables are grouped, insulated or in an unusual temperature. The IEE regulations should be consulted to determine where corrections are necessary.

Diversity

Diversity in electrical installations permits specification of cables and overload protection devices with regard to a sensible assessment of the maximum likely demand on a circuit. For instance, a ring circuit is protected by a 30 amp fuse or 32 amp mcb, although every socket is rated at 13 amps. Therefore if only three sockets were used at full rating, the fuse/mcb would be overloaded. In practice this does not occur, so some diversity can be incorporated into calculations.

Guidance for diversity in domestic installations:

Circuit	Diversity factor
Lighting	66% of the total current demand.
Power sockets	100% of the largest circuit full load current + 40% of the remainder.
Cooker	10 amps + 30% full load + 5 amps if a socket outlet is provided.
Immersion heater	100%.
Shower	100% of highest rated + 100% of second highest + 25% of any remaining.
Storage radiators	100%.

E.g. a house with 7·2 kW shower, 3 kW immersion heater, three ring circuits and three lighting circuits of 800 W each:

Appliance/circuit	Current demand (amps)	Diversity allowance (amps)
Shower	$\dfrac{7200}{230} = 31\cdot3$	$31\cdot3 \times 100\% = 31\cdot3$
Ring circuit-1	30	$30 \times 100\% = 30$
Ring circuit-2	30	$30 \times 40\% = 12$
Ring circuit-3	30	$30 \times 40\% = 12$
Lighting	$3 \times 800 = \dfrac{2400}{230} = 10\cdot4$	$10\cdot4 \times 66\% = 6\cdot9$
		Total = 92·2 amps

For a factory of modest size where the electrical load is not too high, a three-phase, four-wire, 400 volts supply will be sufficient. The distribution to three-phase motors is through exposed copper busbars in steel trunking running around the periphery of the building. Supply to individual motors is through steel conduit via push button switchgear. In addition to providing protection and support, the trunking and conduit can be used as earth continuity.

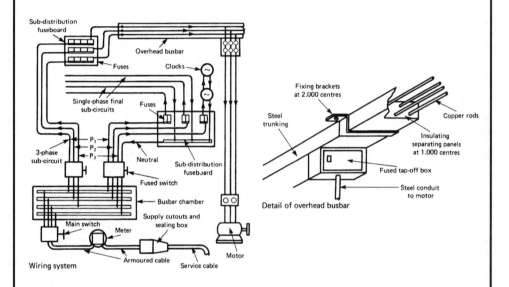

Switches must be within easy reach of machinery operators and contain a device to prevent restarting of the motor after a power failure stoppage.

Overhead busbars provide an easily accessible means of connecting supplies to machinery by bolting the cable to the busbars.

Lighting and other single-phase circuits are supplied through separate distribution fuse boards.

Refs: BS EN 60439: Specification for low-voltage switchgear and control assemblies.
BS EN 60439-2: Particular requirements for busbar trunking systems (busways).

For large developments containing several buildings, either radial or ring distribution systems may be used.

Radial system – separate underground cables are laid from the sub-station to each building. The system uses more cable than the ring system, but only one fused switch is required below the distribution boards in each building.

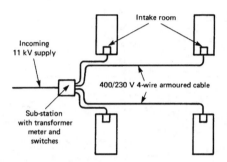

Radial distribution (block plan)

Ring circuit system – an underground cable is laid from the sub-station to loop in to each building. To isolate the supply, two fused switches are required below the distribution boards in each building. Current flows in both directions from the intake, to provide a better balance than the radial system. If the cable on the ring is damaged at any point, it can be isolated for repair without loss of supply to any of the buildings.

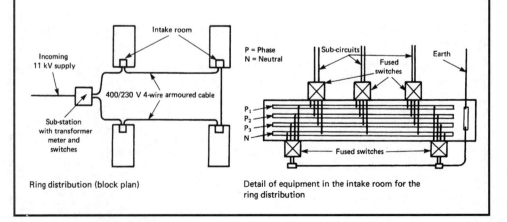

Ring distribution (block plan)

Detail of equipment in the intake room for the ring distribution

The rising main supply system is used in high rise offices and flats. Copper busbars run vertically inside trunking and are given support by insulated bars across the trunking chamber. The supply to each floor is connected to the rising main by means of tap-off units. To balance electrical distribution across the phases, connections at each floor should be spread between the phase bars. If a six-storey building has the same loading on each floor, two floors would be supplied from separate phases. Flats and apartments will require a meter at each tap-off unit.

To prevent the spread of fire and smoke, fire barriers are incorporated with the busbar chamber at each compartment floor level. The chamber must also be fire stopped to the full depth of the floor.

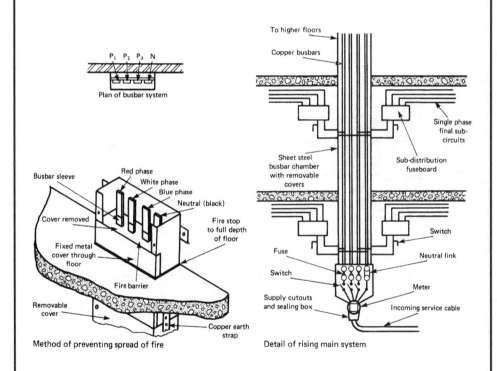

Plan of busbar system

Method of preventing spread of fire

Detail of rising main system

Ref: Building Regulations, Approved Document B3: Internal fire spread (structure).

It is uneconomic to shut down electricity generating plant over night, even though there is considerably less demand. To encourage the use of off-peak energy, the electricity supply companies offer it at an inexpensive tariff. A timer and white meter or economy 7 (midnight to 0700) meter controls the supply to an energy storage facility.

Underfloor – makes use of the thermal storage properties of a concrete floor. High resisting insulated conductors are embedded in the floor screed at 100 to 200 mm spacing, depending on the desired output. This is about 10 to 20 W/m of cable. To be fully effective the underside of the screed should be completely insulated and thermostatic regulators set in the floor and the room.

Block heaters – these are rated between 1 kW and 6 kW and incorporate concrete blocks to absorb the off-peak energy (see next page).

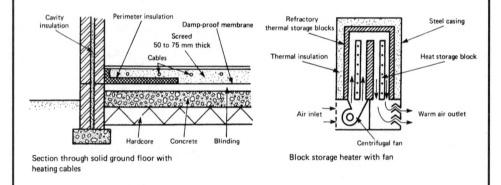

Section through solid ground floor with heating cables

Block storage heater with fan

Electrically heated ceilings use standard tariff electricity supply. The heating element is flexible glasscloth with a conducting silicone elastomer.

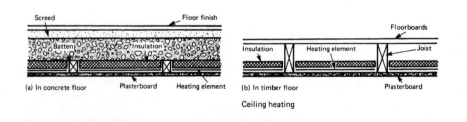

(a) In concrete floor

(b) In timber floor

Ceiling heating

Night storage heaters – these have developed from very bulky cabinets containing concrete blocks which effectively absorb the overnight electrical energy and dissipate it gradually during the next day. Improvements in storage block material have considerably reduced the size of these units to compare favourably with conventional hot water radiators. They contain a number of controls, including a manually set input thermostat on each heater, an internal thermostat to prevent overheating and a time programmed fan. Manufacturers provide design tables to establish unit size. As a rough guide, a modern house will require about 200 W output per square metre of floor area. Storage heaters are individually wired on radial circuits from the off-peak time controlled consumer unit.

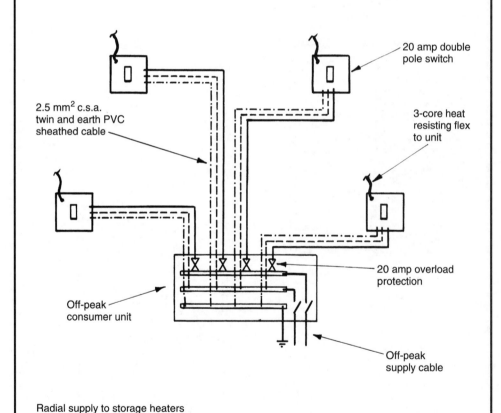

2.5 mm² c.s.a. twin and earth PVC sheathed cable

20 amp double pole switch

3-core heat resisting flex to unit

20 amp overload protection

Off-peak consumer unit

Off-peak supply cable

Radial supply to storage heaters

Electrically heated warm air systems are a development of the storage heater concept – see previous two pages. A central unit rated from 6 kW to 12 kW absorbs electrical energy off-peak and during the day delivers this by fan to various rooms through a system of insulated ducting. A room thermostat controls the fan to maintain the air temperature at the desired level. Air volume to individual rooms is controlled through an outlet register or diffuser.

Stub duct system – the unit is located centrally and warm air conveyed to rooms by short ducts with attached outlets.

Radial duct system – warm air from the unit is supplied through several radial ducts designated to specific rooms. Outlet registers are located at the periphery of rooms to create a balanced heat distribution.

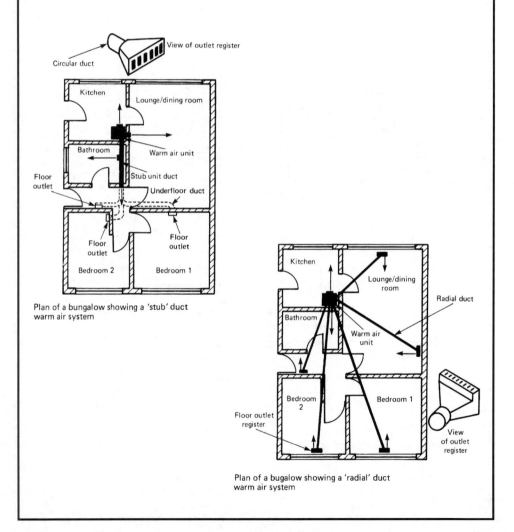

Plan of a bungalow showing a 'stub' duct warm air system

Plan of a bugalow showing a 'radial' duct warm air system

There are numerous types of independent heat emitters for use with 13 amp power sockets or fused spur sockets.

Panel heater – the heat output is mainly radiant from a surface operating temperature of between 204°C and 240°C. For safety reasons it is mounted at high level and may be guarded with a mesh screen.

Infra-red heater – contains an iconel-sheathed element or nickel chrome spiral element in a glass tube, backed by a curved reflector. May be used at high level in a bathroom and controlled with a string pull.

Oil-filled heater – similar in appearance to steel hot water radiators, they use oil as a heat absorbing medium from one or two electrical elements. Heat is emitted by radiant and convected energy. An integral thermostat allows for manual adjustment of output.

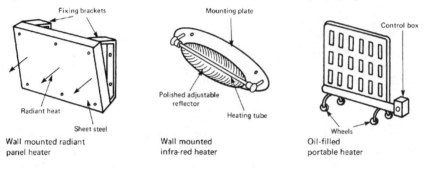

Wall mounted radiant panel heater

Wall mounted infra-red heater

Oil-filled portable heater

Convector heater – usually has two electrical elements with independent control to vary the output. May be used where a constant level of background warmth is required.

Parabolic reflector fire – has the heating element in the focal point to create efficient radiant heat output.

Wall mounted fan heaters – usually provided with a two-speed fan to deliver air through a bank of electrical elements at varying velocities. Direction is determined by adjustable louvres.

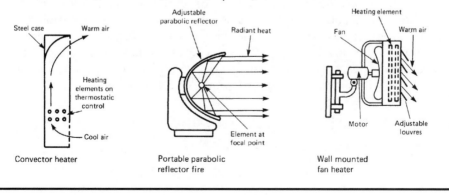

Convector heater

Portable parabolic reflector fire

Wall mounted fan heater

Controls For Electric Night Storage Space Heaters

Controls vary from simple switches and sensors integrated with appliances, to overall system management programmed through time switches and optimisers:

- Manual charge control – set by the user to regulate energy input and output. The effect can be variable and unreliable as it does not take into account inconsistencies such as daily variations in temperature.

- Automatic charge control – sensors within the heater and room are pre-set to regulate the electrical input charge. When room temperature is high, the sensor in the heater reduces the energy input. Conversely, the energy input is increased when the room temperature is low.

- Heat output control – this is a damper within the heater casing. It can be adjusted manually to regulate heat emission and prevent a room overheating. A variable speed fan can be used to similar effect or to vary the amount of heat emission and its distribution.

- Time switch/programmer and room thermostat – the simplest type of programmed automatic control applied individually to each heater or as a means of system or group control. Where applied to a system of several emitters, individual heaters should still have some means of manual or preferably automatic regulation. This type of programmed timing is also appropriate for use with direct acting thermostatically switched panel-type heaters.

- 'CELECT-type' controls – this is a type of optimiser control which responds to pre-programmed times and settings, in addition to unknown external influences such as variations in the weather. Zones or rooms have sensors which relate room information to the controller or system manager, which in turn automatically adjusts individual storage heater charge periods and amount of energy input to suit the room criteria. This type of control can also be used for switching of panel heaters.

A temporary supply of electricity for construction work may be obtained from portable generators. This may be adequate for small sites but most developments will require a mains supply, possibly up to 400 volts in three phases for operating hoists and cranes. Application must be made in good time to the local electricity authority to ascertain the type of supply and the total load. The incoming metered supply provided by the electricity company will be housed in a temporary structure constructed to the authority's approval. Thereafter, site distribution and installation of reduced voltage transformers is undertaken by the developer's electrical contractor subject to the supply company's inspection and testing.

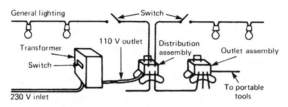

Reduced voltage distribution

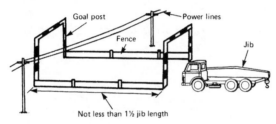

Goal posts (or barrier fences) give protection against contact with overhead power lines

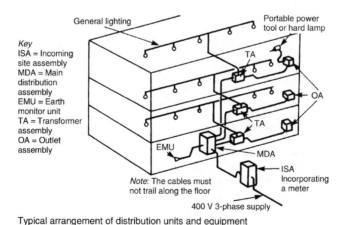

Key
ISA = Incoming site assembly
MDA = Main distribution assembly
EMU = Earth monitor unit
TA = Transformer assembly
OA = Outlet assembly

Note: The cables must not trail along the floor

Typical arrangement of distribution units and equipment

Equipment:

Incoming site assembly (ISA) – provided by the local electricity supply company. It contains their switchgear, overload protection, transformers and meters for a 400 volt, three-phase supply at 300, 200 and 100 amps.

Main distribution assembly (MDA) – contains three-phase and single-phase distribution boards, overload protection and lockable switchgear. May be combined with the ISA to become an ISDA.

Transformer assembly (TA) – supplied from the MDA to transform voltage down to 110 V, 50 V and possibly 25 V for use in very damp situations.

Earth monitor unit (EMU) – used where mobile plant requires flexible cables at mains voltage. A very low-voltage current is conducted between plant and EMU and earth conductor, so that if this is interrupted by a fault a monitoring unit disconnects the supply.

Socket outlet assembly (SOA) – a 110 volt supply source at 32 amps with switchgear and miniature circuit breakers for up to eight 16 amp double pole sockets to portable tools.

Cable colour codes and corresponding operating voltage:

Colour	Voltage
Violet	25
White	50
Yellow	110
Blue	230
Red	400
Black	500/650

Refs: BS 4363: Specification for distribution assemblies for electricity supplies for construction and building sites.
BS 7375: Code of practice for distribution of electricity on construction and building sites.
BS EN 60439-4: Specification for low-voltage switchgear and control assemblies. Particular requirements for assemblies for construction sites.

Light is a form of electromagnetic radiation. It is similar in nature and behaviour to radio waves at one end of the frequency spectrum and X-rays at the other. Light is reflected from a polished (specular) surface at the same angle that strikes it. A matt surface reflects in a number of directions and a semi-matt surface responds somewhere between a polished and a matt surface.

Angle of incidence θ_1 =
Angle of reflection θ_2

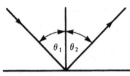

Light reflected
from a polished surface

Light is reflected in all
directions

Light reflected
from a matt surface

Some light is scattered and some
light is reflected directionally

Light scattered and
reflected from a semi-matt
surface

Light is scattered in all directions
(diffusion)

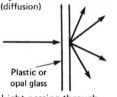

Plastic or
opal glass

Light passing through
a diffusing screen

Light is bent or refracted when
passing through a surface between
two media

Intensity of light
and lux

Illumination produced from a light source perpendicular to the surface:

$$E = I \div d^2$$

E = illumination on surface (lux)

I = Illumination intensity from source (cd)

d = distance from light source to surface (m).

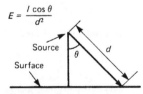

Illumination produced
from a light source not
perpendicular to the surface

Definitions and units of measurement:

- Luminous intensity – candela (cd), a measurement of the magnitude of luminance or light reflected from a surface, i.e. cd/m^2.
- Luminous flux – lumen (lm), a measurement of the visible light energy emitted.
- Illuminance – Lumens per square metre (lm/m^2) or lux (lx), a measure of the light falling on a surface.
- Efficacy – efficiency of lamps in lumens per watt (lm/W). Luminous efficacy = Luminous flux output ÷ Electrical power input.
- Glare index – a numerical comparison ranging from about 10 for shaded light to about 30 for an exposed lamp. Calculated by considering the light source size, location, luminances and effect of its surroundings.

Examples of illumination levels and limiting glare indices for different activities:

Activity/location	Illuminance (lux)	Limiting glare index
Assembly work: (general)	250	25
(fine)	1000	22
Computer room	300	16
House	50 to 300*	n/a
Laboratory	500	16
Lecture/classroom	300	16
Offices: (general)	500	19
(drawing)	750	16
Public house bar	150	22
Shops/supermarkets	500	22
Restaurant	100	22

* Varies from 50 in bedrooms to 300 in kitchen and study.

The Building Regulations, Approved Document L2 requires that non-domestic buildings have reasonably efficient lighting systems and make use of daylight where appropriate.

Filament lamps – the tungsten iodine lamp is used for floodlighting. Evaporation from the filament is controlled by the presence of iodine vapour. The gas-filled, general-purpose filament lamp has a fine tungsten wire sealed within a glass bulb. The wire is heated to incandescence (white heat) by the passage of an electric current.

Discharge lamps – these do not have a filament, but produce light by excitation of a gas. When voltage is applied to the two electrodes, ionisation occurs until a critical value is reached when current flows between them. As the temperature rises, the mercury vaporises and electrical discharge between the main electrodes causes light to be emitted.

Fluorescent tube – this is a low pressure variation of the mercury discharge lamp. Energised mercury atoms emit ultra-violet radiation and a blue/green light. The tube is coated internally with a fluorescent powder which absorbs the ultra-violet light and re-radiates it as visible light.

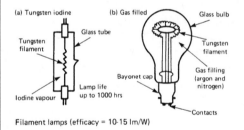

Filament lamps (efficacy = 10-15 lm/W)

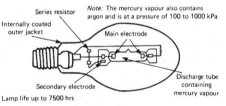

Mercury-vapour discharge lamp (efficacy = 50 lm/W)

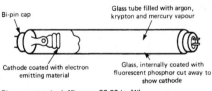

Fluorescent tube (efficacy = 20-60 lm/W)

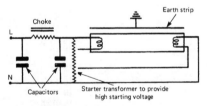

Control gear is needed to start the discharge and to keep the light steady during operation. A transformer provides a quick start.

Fluorescent strip lamps have many applications. The fittings and reflectors shown are appropriate for use in industrial locations, with a variation which creates an illuminated ceiling more suited to shops and offices. A false ceiling of thermaluscent panels provides well-diffused illumination without glare and contributes to the insulation of the ceiling. Other services should not be installed in the void as they will cast shadows on to the ceiling. Tubes are mounted on batten fittings and the inside of the void should be painted white to maximise effect.

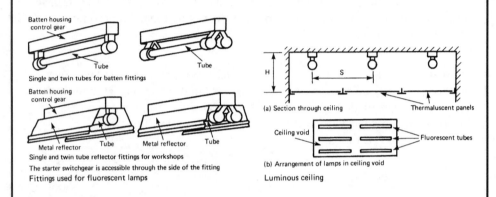

Single and twin tubes for batten fittings

Single and twin tube reflector fittings for workshops
The starter switchgear is accessible through the side of the fitting
Fittings used for fluorescent lamps

(a) Section through ceiling

(b) Arrangement of lamps in ceiling void

Luminous ceiling

High pressure sodium discharge lamps produce a consistent golden white light in which it is possible to distinguish colours. They are suitable for floodlighting, commercial and industrial lighting and illumination of highways. The low pressure variant produces light that is virtually monochromatic. The colour rendering is poor when compared to the high pressure lamp. Sodium vapour pressure for high and low pressure lamps is 0·5 Pa and 33 kPa, and typical efficacy is 125 and 180 lm/W respectively.

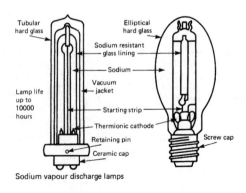

Sodium vapour discharge lamps

Fittings for lighting may be considered in three categories:

1. General utility – designed to be effective, functional and economic.
2. Special – usually provided with optical arrangements such as lenses or reflectors to give directional lighting.
3. Decorative – designed to be aesthetically pleasing or to provide a feature, rather than to be functional.

From an optical perspective, the fitting should obscure the lamp from the discomfort of direct vision to reduce the impact of glare.

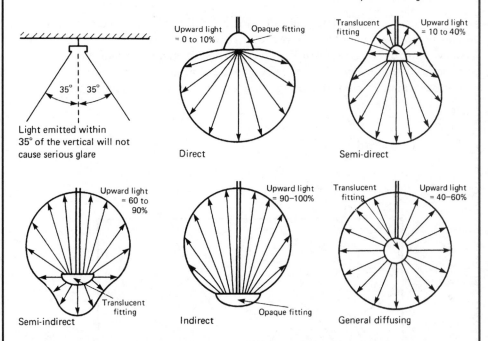

Light emitted within 35° of the vertical will not cause serious glare

Direct

Semi-direct

Semi-indirect

Indirect

General diffusing

Ventilated fittings allow the heat produced by the lamps to be recirculated through a ceiling void to supplement a warm air ventilation system. The cooling effect on the lamp will also improve its efficiency.

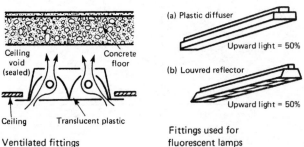

Ventilated fittings

Fittings used for fluorescent lamps

Luminaires and Polar Curves

Luminaire – a word to describe the complete lighting unit including the lamp. When selecting a lamp type, it is important to select a luminaire to complement the lamp both functionally and aesthetically. A luminaire has several functions: it defines the lamp position, protects the lamp and may contain the lamp control mechanism. In the interests of safety it must be well insulated, in some circumstances resistant to moisture, have adequate appearance for purpose and be durable.

Polar curve – shows the directional qualities of light from a lamp and luminaire by graphical representation, as shown in outline on the previous page. A detailed plot can be produced on polar co-ordinated paper from data obtained by photometer readings at various angles from the lamp. The co-ordinates are joined to produce a curve.

Typical representation:

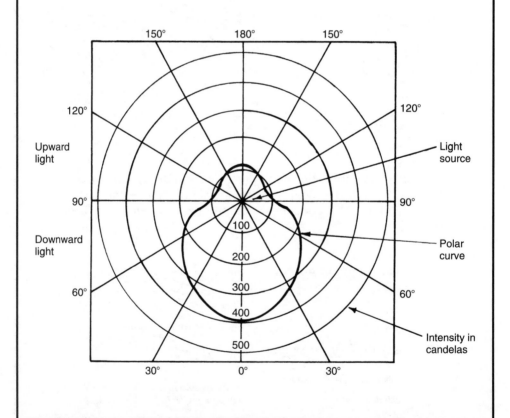

Compact fluorescent lamps are a smaller variation and development of the standard fluorescent tube fitting. They are manufactured with conventional bayonet or screw fittings. Unit cost is higher than tungsten filament bulbs but will last over 8000 hours, consuming only about 25% of the energy of a conventional bulb. Tungsten filament bulbs have a life expectancy of about 1000 hours.

The comfort type produces gentle diffused light and is suitable where continuous illumination is required. The prismatic types are more robust and are suitable for application to workshops and commercial premises. Electronic types are the most efficient, consuming only 20% of the energy that would be used in a tungsten filament bulb. Compact fluorescent lamps are not appropriate for use with dimmer switches.

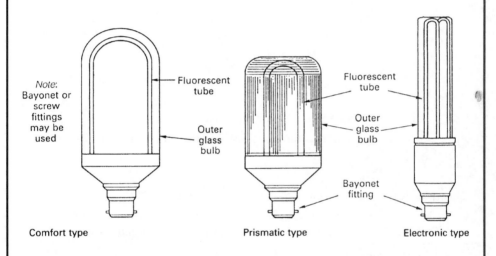

Note: Bayonet or screw fittings may be used

Fluorescent tube

Outer glass bulb

Comfort type

Fluorescent tube

Outer glass bulb

Bayonet fitting

Prismatic type

Fluorescent tube

Outer glass bulb

Electronic type

The Buildings Regulations, Approved Document L, lists compact fluorescent lamps as an acceptable means for lighting non-domestic buildings.

Energy Saving Chart

Energy saver	Ordinary light bulb	Energy saving	Over 8000 hours save up to
25 W	100 W	80%	47·70
18 W	75 W	73%	36·25
11 W	60 W	80%	31·16
9 W	40 W	72%	19·72

Domestic energy costed at 7·95p/kWh

Interior lighting – the energy consumed by lighting in dwellings depends on the overall performance and efficiency of luminaires, lamps and control gear. The Building Regulations require that fixed lighting in a reasonable number of locations where lighting has most use (see table), be fitted with lamps having a luminous efficacy in excess of 40 lumens per circuit-watt. The term circuit-watt is used instead of watt, as this includes the power used by the lamp plus the installation and control gear.

Guidance on number of locations where efficient lighting should be provided:

Rooms created in a dwelling	Minimum number of locations
1–3	1
4–6	2
7–9	3
10–12	4

Hall, stairs and landing are regarded as one room.

An integral (attached to the building) conservatory is considered a room.

Garages, loft and outbuildings are not included.

Exterior lighting – reasonable provisions are required for economic use. This could include any of the following or a combination of:

- efficient lamps
- automatic timed switching control
- photo-electric switching control

Note: Lamps that satisfy the criteria of efficiency include fluorescent tubes and compact fluorescent lamps. Special socket fittings can be made to prevent interchange with unsuitable standard tungsten lamps.

Refs. Building Regulations, Approved Document L1: Conservation of fuel and power in dwellings.
Low energy domestic lighting – ref. GIL 20, BRESCU publications.

Lighting efficiency is expressed as the initial (100 hour) efficacy averaged over the whole building –

Offices, industrial and storage buildings,
not less than 40 luminaire-lumens per circuit-watt.
Other buildings,
not less than 50 lamp-lumens per circuit-watt.
Display lighting,
not less than 15 lamp-lumens per circuit-watt.

A formula and tables for establishing conformity with these criteria are provided in the Building Regulations, Approved Document.

Lighting control objectives:
• to maximise daylight.
• to avoid unnecessary use of artificial lighting when spaces are unoccupied.

Control facilities:
• Local easily accessible manual switches or remote devices including infra-red transmitters, sonic, ultra-sonic and telecommunication controls.
• Plan distance from switch to luminaire, maximum 8 metres or 3 times fitting height above floor (take greater).
• Time switches as appropriate to occupancy.
• Photo-electric light metering switches.
• Automatic infra-red sensor switches which detect the absence or presence of occupants.

Controls specific to display lighting include dedicated circuits that can be manually switched off when exhibits or merchandise presentations are not required. Timed switching that automatically switches off when premises are closed.

Refs. Building Regulations, Approved Document L2: Conservation of fuel and power in buildings other than dwellings.
BRE Information Paper 2/99, Photo-electric controls of lighting: design, set-up and installation issues.

Extra-low-voltage lighting has application to display lighting for shops and exhibitions. It is also used as feature lighting in domestic premises where set in the ceiling in kitchens and bathrooms. These situations benefit from the low heat emission, good colour rendering and very low running costs of this form of lighting. System potential is only 12 volts a.c., through a transformed 230 volt mains supply. High performance 50 watt tungsten halogen dichroic lamps are compact and fit flush with the mounting surface.

Electricity is supplied from the transformer through a fused splitter to provide a fairly uniform short length of cable to each lamp. Similarity in cable lengths is important to maintain equivalent voltage drop and a short length of cable will minimise voltage drop. Lamps are very sensitive to change in voltage, therefore correct selection of transformer is essential. A voltage drop of 6% (approx. 0·7 volts) will reduce the illuminating effect by about 30%. Cable sizing is also critical with regard to voltage drop. The low voltage creates a high current, i.e. just one 50 watt bulb at 12 volts = 4·17 amps (see page 355 for cable sizing).

Schematic ELV lighting:

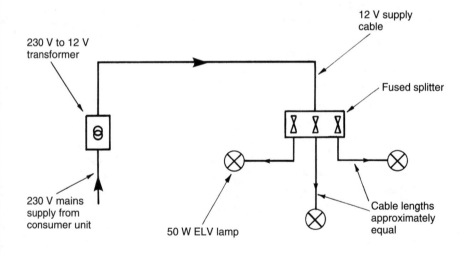

Note: A variation is the use of individual low-voltage lamps which contain their own transformer. However, these are relatively expensive items and are attached to special fittings.

The lumen method of lighting design is used to determine a lighting layout that will provide a design maintained illuminance. It is valid if the luminaires are mounted above the working plane in a regular pattern. The method uses the formula: $N = (E \times A) \div (F \times U \times M)$.

N = number of lamps

E = average illuminance on the working plane (lux)

A = area of the working plane (m²)

F = flux from one lamp (lumens)

U = utilisation factor

M = maintenance factor.

The utilisation factor (U) is the ratio of the lumens received on the working plane to the total flux output of lamps in the scheme. The maintenance factor (M) is a ratio which takes into account the light lost due to an average expectation of dirtiness of light fittings and surfaces.

Spacing-to-height ratio (SHR) is the centre-to-centre (S) distance between adjacent luminaires to their mounting height (H) above the working plane. Manufacturers' catalogues can be consulted to determine maximum SHRs, e.g. a luminaire with trough reflector is about 1·65 and an enclosed diffuser about 1·4.

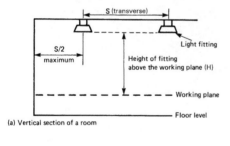

(a) Vertical section of a room

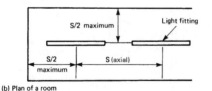

(b) Plan of a room

Method of spacing fluorescent tubes

Example. An office 8 m long by 7 m wide requires an illumination level of 400 lux on the working plane. It is proposed to use 80 W fluorescent fittings having a rated output of 7375 lumens each. Assuming a utilisation factor of 0.5 and a maintenance factor of 0.8 design the lighting scheme.

$$N = \frac{E \times A}{F \times U \times M} \quad \therefore N = \frac{400 \times 8 \times 7}{7375 \times 0.5 \times 0.8} \quad N = 7.59, \text{ use 8 fittings}$$

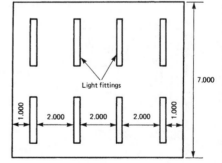

Layout of fluorescent tubes for the office

Permanent Supplementary Lighting of Interiors

Illumination of building interiors is a very important factor for designers. This will relate to user convenience and visual impact of the building. Overall considerations fall into three categories:

A – daylighting alone, in which the window area occupies about 80% of the façades

B – permanent supplementary artificial lighting of interiors, in which the window area is about 20% of the façades

C – permanent artificial lighting of interiors in which there are no windows.

Occupants of buildings usually prefer a view to the outside. Therefore the choice of lighting for most buildings is from type A or B. With type B the building may be wider, because artificial lighting is used to supplement daylighting. Although the volume is the same as type A the building perimeter is less, thus saving in wall construction. Type B building also has lower heat gains and energy losses through the glazing, less noise from outside and less maintenance of windows.

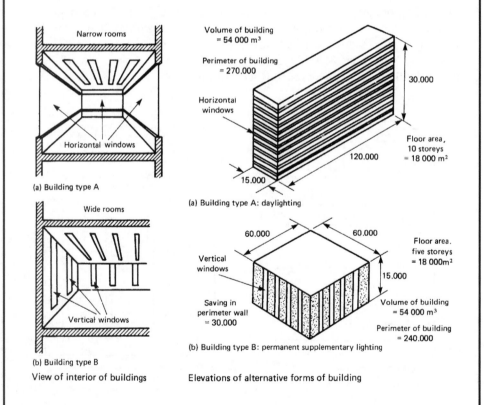

(a) Building type A

(b) Building type B

View of interior of buildings

(a) Building type A: daylighting

(b) Building type B: permanent supplementary lighting

Elevations of alternative forms of building

Ref: BS 8206-1: Code of practice for artificial lighting.

378

The daylight received inside a building can be expressed as 'the ratio of the illumination at the working point indoors, to the total light available simultaneously outdoors'. This can also be expressed as a percentage and it is known as the 'daylight factor'.

The daylight factor includes light from:

- Sky component – light received directly from the sky; excluding direct sunlight.
- External reflected component – light received from exterior reflecting surfaces.
- Internal reflected component – light received from internal reflecting surfaces.

If equal daylight factor contours are drawn for a room, they will indicate how daylighting falls as distance increases from a window.

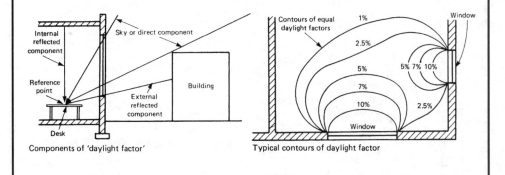

Components of 'daylight factor'

Typical contours of daylight factor

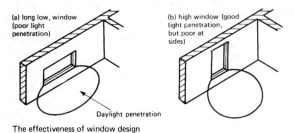

The effectiveness of window design

Refs: BRE Digests 309 and 310: Estimating daylight in buildings.
BS 8206-2: Code of practice for daylighting.

The effect of daylight in a room can be studied by using scaled models. Providing that textures and colours of a room surface are the same, an approximate result may be obtained.

An estimate of the effect of daylight in a room may also be made from daylight factor protractors and associated tables of data. These were developed by the Building Research Establishment for use with scaled drawings to determine the sky component from a sky of uniform luminance.

There are pairs of protractors to suit different window types. Protractor No. 1 is placed on the cross-section as shown. Readings are taken where the sight lines intersect the protractor scale.

In the diagram, the sky component $= 8 \cdot 5 - 4 = 4 \cdot 5\%$ and an altitude angle of 30°. The sky component of $4 \cdot 5\%$ must be corrected by using protractor No. 2. This is placed on the plan as shown. Readings from protractor No. 2 are $0 \cdot 25$ and $0 \cdot 1$, giving a total correction factor of $0 \cdot 35$. Therefore $4 \cdot 5 \times 0 \cdot 35 = 1 \cdot 6\%$.

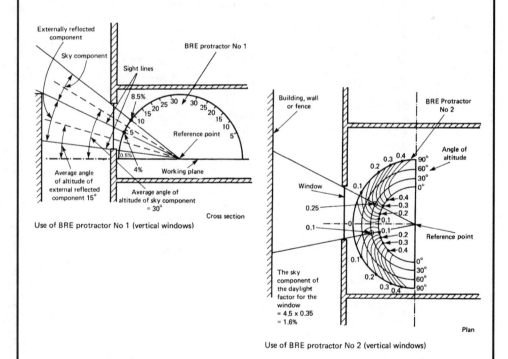

Use of BRE protractor No 1 (vertical windows)

Use of BRE protractor No 2 (vertical windows)

Note: Daylight protractors number 1 to 10. They are available with a guide from the Building Research Establishment, ref. Publication code AP80

The external reflected component of the daylight factor for a uniform sky may be taken as approximately 0·1 x the equivalent sky component. Using the diagrams shown in Daylighting – 2, the value may be found as follows:

- Readings from protractor No. 1 are 4% and 0·5%.

- Equivalent sky component = 4% – 0·5% = 3·5%.

- Average angle of altitude = 15°.

- Readings on protractor No. are 0·27 and 0·09 (for 15°).

- Correction factor = 0·27 + 0·09 = 0·36.

- Equivalent uniform sky component = 3·5% x 0·36 = 1·26%.

- Externally reflected component = 0·1 x 1·26% = 0·126%.

To establish the daylight factor, the internal reflected component is calculated and added to both the sky and externally reflected components – see example.

Example: Find the minimum internally reflected component of the daylight factor for a room measuring 10 m x 8 m x 2·5 m high, having a window in one wall with an area of 20 m². The floor has an average reflection factor of 20% and the walls and ceiling average reflection factors of 60% and 70% respectively.

$$\text{Window area as a percentage of floor area} = \frac{20}{80} \times \frac{100}{1} = 25\%$$

Referring to Table 2 (p. 382) the minimum internally reflected component = 1·3%.

Allowing a maintenance factor of 0·9 for dirt on the windows the value will be modified to 1·3 x 0·9 = 1·17%.

For the example given in daylighting 2 and 3 the daylight factor will be the addition of the three components = 1·6 + 0·126 + 1·17 = 2·9%

Table 1 Reflection factors

Reflection factors (%)		Reflection factors (%)	
White	75–88	Golden yellow	62
Light stone	53	Orange	36
Middle stone	37	Eau-de-nil	48
Light buff	60	Sky blue	47
Middle buff	43	Turquoise	27
Light grey	44	Light brown	30
Dark grey	26	Middle brown	20
Pale cream	73	Salmon pink	42

Table 2 Minimum internally reflected component of the daylight factor (%)

Ratio of window area to floor area	Window area as a percentage of floor area	Floor reflection factor (%)											
		10				20				40			
		Wall reflection factor (per cent)											
		20	40	60	80	20	40	60	80	20	40	60	80
		%	%	%	%	%	%	%	%	%	%	%	%
1:50	2			0·1	0·2		0·1	0·1	0·2		0·1	0·2	0·2
1:20	5	0·1	0·1	0·2	0·4	0·1	0·2	0·3	0·5	0·1	0·2	0·4	0·6
1:14	7	0·1	0·2	0·3	0·5	0·1	0·2	0·4	0·6	0·2	0·3	0·6	0·8
1:10	10	0·1	0·2	0·4	0·7	0·2	0·3	0·6	0·9	0·3	0·5	0·8	1·2
1:6·7	15	0·2	0·4	0·6	1·0	0·2	0·5	0·8	1·3	0·4	0·7	1·1	1·7
1:5	20	0·2	0·5	0·8	1·4	0·3	0·6	1·1	1·7	0·5	0·9	1·5	2·3
1:4	25	0·3	0·6	1·0	1·7	0·4	0·8	1·3	2·0	0·6	1·1	1·8	2·8
1:3·3	30	0·3	0·7	1·2	2·0	0·5	0·9	1·5	2·4	0·8	1·3	2·1	3·3
1:2·9	35	0·4	0·8	1·4	2·3	0·5	1·0	1·8	2·8	0·9	1·5	2·4	3·8
1:2·5	40	0·5	0·9	1·6	2·6	0·6	1·2	2·0	3·1	1·0	1·7	2·7	4·2
1:2·2	45	0·5	1·0	1·8	2·9	0·7	1·3	2·2	3·4	1·2	1·9	3·0	4·6
1:2	50	0·6	1·1	1·9	3·1	0·8	1·4	2·3	3·7	1·3	2·1	3·2	4·9

Note: The ceiling reflection factor is assumed to be 70%

There are other methods for determining daylight factor. Some are simple rules of thumb and others more detailed formulae. An example of each are shown below.

- Rule of thumb – D = 0·1 × P
  where: D = daylight factor

  P = percentage of glazing relative to floor area.

E.g. a room 80 m² floor area with 15 m² of glazing.

$$D = 0·1 × 15/80 × 100/1 = 1·875\%$$

- Formula –

$$D = \frac{T × G × \theta × M}{A(1-R^2)}$$

where: D = average daylight factor

T = transmittance of light through glass
(clear single glazing = 0·85, clear double glazing = 0·75)

G = glazed area (m²)

θ = angle of sky component

M = maintenance factor (see page 377)

A = total area of interior surfaces, inc. windows (m²)

R = reflection factors (see page 382).

E.g. using the data from the example on page 381 and assuming a 50% reflection factor, double glazing and a sky component angle of 35°.

$$D = \frac{0·75 × 20 × 35 × 0·9}{250 \ (1-[50/100]^2)} = 2·52\%$$

All calculations and estimates of daylight factor and glazing area must conform with the basic allowances defined in the Building Regulations, Approved Document L – Conservation of Fuel and Power:

Dwellings: Windows, doors and rooflights – maximum 25% of the total floor area.

Non-domestic buildings:

Residential – windows and personnel doors, maximum 30% of exposed wall area.

Industrial and storage buildings – windows and personnel doors, maximum 15% of exposed wall area.

Places of assembly, shops and offices – windows and personnel doors, maximum 40% of exposed wall area (excludes display windows).

Note: Rooflights in non-domestic, max. 20% of exposed roof area.

Telecommunications Installation

Cabling systems that were originally used solely for telephone communications now have many other applications. These include fire alarms, security/intruder alarms, computer networking, teleprinters, facsimile machines, etc. The voltage and current are very low and have no direct connection to the mains electricity in a building. Therefore, telecommunications and mains cabling should be distinctly separated in independent conduits and trunking for reasons of safety and to prevent interference.

External telecommunications cables may supply a building from overhead or underground, the latter being standard for new building work. The intake is below surface level at a point agreed with the cable supplier. In large buildings the incoming cable supplies a main distribution unit which has connections for the various parts of the building. Cables supply both switchboards and individual telephones from vertical risers. There may be limitations on the number of cables supplied from risers and early consultation with the cable supplier is essential to determine this and any other restrictions.

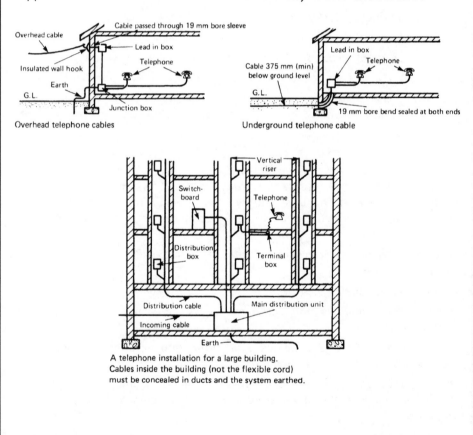

Overhead telephone cables

Underground telephone cable

A telephone installation for a large building.
Cables inside the building (not the flexible cord)
must be concealed in ducts and the system earthed.

11 MECHANICAL CONVEYORS – LIFTS, ESCALATORS AND TRAVELATORS

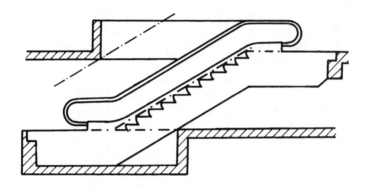

To function efficiently and to provide access for the elderly and disabled, modern offices and public buildings are provided with suitably designed lift installations. Planning (as with all services) should commence early in the design programme. Priority must be given to locating lifts centrally within a building to minimise horizontal travel distance. Consideration must also be given to position, relative to entrances and stairs. Where the building size justifies several passenger lifts, they should be grouped together. In large buildings it is usual to provide a group of lifts near the main entrance and single lifts at the ends of the building. The lift lobby must be wide enough to allow pedestrian traffic to circulate and pass through the lift area without causing congestion. For tall buildings in excess of 15 storeys, high speed express lifts may be used which by-pass the lower floors.

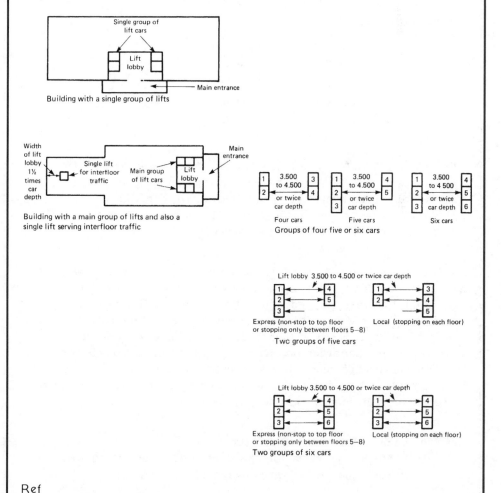

Building with a single group of lifts

Building with a main group of lifts and also a single lift serving interfloor traffic

Groups of four five or six cars

Two groups of five cars

Express (non-stop to top floor or stopping only between floors 5–8)

Local (stopping on each floor)

Two groups of six cars

Express (non-stop to top floor or stopping only between floors 5–8)

Local (stopping on each floor)

Ref

Requirements:

- Necessary in all buildings over three storeys high.
- Essential in all buildings over a single storey if they are accessed by the elderly or disabled.
- Minimum standard – one lift per four storeys.
- Minimum walking distance to access a lift – 45 m.
- Floor space and lift car capacity can be estimated at $0 \cdot 2m^2$ per person.

Lift speed:

Type	Car speed (m/s)
Goods (electric or hydraulic)	$0 \cdot 2$–1
Electric passenger <4 floors	$0 \cdot 3$–$0 \cdot 8$
4–6 floors	$0 \cdot 8$–$1 \cdot 2$
6–9 floors	$1 \cdot 2$–1.5
9–15 floors*	5–7
paternoster	$<0 \cdot 4$
Hydraulic passenger[†]	$0 \cdot 1$–$1 \cdot 0$

* Express lift that does not stop at the lower floor levels. The upper speed limit is 7 m/s because of the inability of the human ear to adapt to rapid changing atmospheric conditions.
[†] Overall theoretical maximum travel distance is 21 m vertically, therefore limited to four or five storeys.

Electric motor – low speed lifts operate quite comfortably with an a.c. motor to drive the traction sheave through a worm gear (see page 395). For faster speed applications a d.c. motor is preferable. This is supplied via a mains generator for each lift motor. D.c. motors have historically provided better variable voltage controls, more rapid and smoother acceleration, quieter operation, better floor levelling and greater durability in resisting variable demands. Recent developments with a.c. motors have made them more acceptable and these are now becoming more widely used.

Refs: BS 5655: Lifts and service lifts.
BS 5656: Safety rules for the construction and installation of escalators and passenger conveyors.

High tensile steel ropes are used to suspend lift cars. They have a design factor of safety of 10 and are usually at least four in number. Ropes travel over grooved driving or traction sheaves and pulleys. A counterweight balances the load on the electric motor and traction gear.

Methods for roping vary:

Single wrap 1:1 – the most economical and efficient of roping systems but is limited in use to small capacity cars.

Single wrap 1:1 with diverter pulley – required for larger capacity cars. It diverts the counterweight away from the car. To prevent rope slip, the sheave and pulley may be double wrapped.

Single wrap 2:1 – an alternative for use with larger cars. This system doubles the load carrying capacity of the machinery but requires more rope and also reduces the car speed by 50%.

Double wrap – used to improve traction between the counterweight, driving sheave and steel ropes.

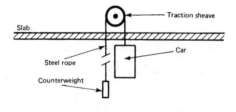

Single wrap 1:1 roped

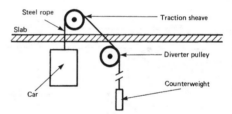

Single wrap 1:1 roped with diverter pulley

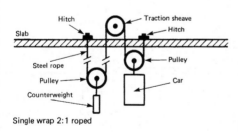

Single wrap 2:1 roped

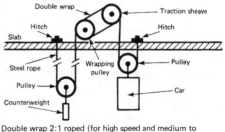

Double wrap 2:1 roped (for high speed and medium to heavy duty loads)

Single wrap 3:1 – used for heavy goods lifts where it is necessary to reduce the force acting upon the machinery bearings and counterweight. The load carrying capacity is increased by up to three times that of uniform ratio, but the capital costs are higher with increased pulleys and greater length of rope. By comparison, the car speed is also reduced to one-third.

Drum drive – a system with one set of ropes wound clockwise around the drum and another set anti-clockwise. It is equally balanced, as one set unwinds the other winds. The disadvantage of the drum drive is that as height increases, the drum becomes less controllable, limiting its application to rises of about 30 m.

Compensating rope and pulley – used in tall buildings where the weight of the ropes in suspension will cause an imbalance on the driving gear and also a possible bouncing effect on the car. The compensating ropes attach to the underside of car and counterweight to pass around a large compensating pulley at low level.

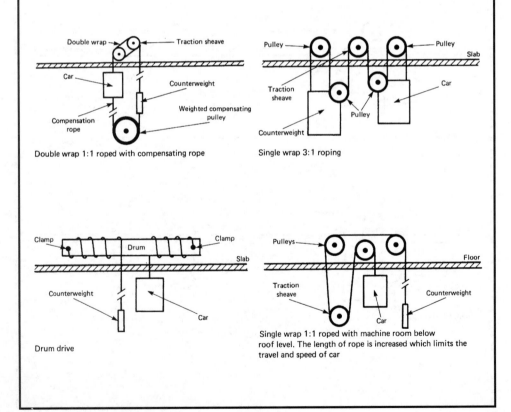

Double wrap 1:1 roped with compensating rope

Single wrap 3:1 roping

Drum drive

Single wrap 1:1 roped with machine room below roof level. The length of rope is increased which limits the travel and speed of car

The single automatic push button system is the simplest and least sophisticated of controls. The lift car can be called and used by only one person or group of people at a time. When the lift car is called to a floor, the signal lights engraved 'in use' are illuminated on every floor. The car will not respond to any subsequent landing calls, nor will these calls be recorded and stored. The car is under complete control of the occupants until they reach the required floor and have departed the lift. The 'in use' indicator is now switched off and the car is available to respond to the next landing call. Although the control system is simple and inexpensive by comparison with other systems, it has its limitations for user convenience. It is most suited to light traffic conditions in low rise buildings such as nursing homes, small hospitals and flats.

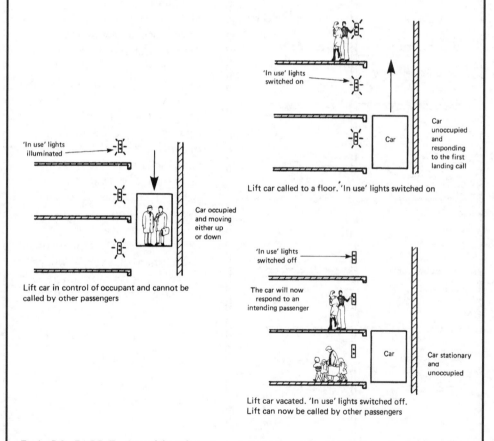

'In use' lights switched on

Car unoccupied and responding to the first landing call

Lift car called to a floor. 'In use' lights switched on

'In use' lights illuminated

Car occupied and moving either up or down

Lift car in control of occupant and cannot be called by other passengers

'In use' lights switched off

The car will now respond to an intending passenger

Car stationary and unoccupied

Lift car vacated. 'In use' lights switched off. Lift can now be called by other passengers

Ref. BS 5655-7: Specification for manual control devices, indicators and additional fittings.

Down collective – stores calls made by passengers in the car and those made from the landings. As the car descends, landing calls are answered in floor sequence to optimise car movement. If the car is moving upwards, the lift responds to calls made inside the car in floor sequence. After satisfying the highest registered call, the car automatically descends to answer all the landing calls in floor sequence. Ony one call button is provided at landings. This system is most suited to flats and small hotels, where the traffic is mainly between the entrance lobby and specific floors.

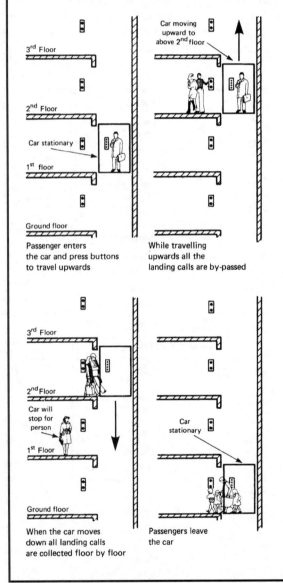

3rd Floor

2nd Floor

Car stationary

1st floor

Ground floor

Passenger enters
the car and press buttons
to travel upwards

Car moving
upward to
above 2nd floor

While travelling
upwards all the
landing calls are by-passed

3rd Floor

2nd Floor

Car will
stop for
person

1st Floor

Ground floor

When the car moves
down all landing calls
are collected floor by floor

Car
stationary

Passengers leave
the car

Full or directional collective – a variation in which car and landing calls are immediately stored in any number. Upward and downward intermediate landing calls are registered from one of two directional buttons. The uppermost and lowest floors only require one button. The lift responds to calls in floor order independent of call sequence, first in one direction and then the other. It has greater flexibility than the down collective system and is appropriate for offices and departmental stores where there is more movement between intermediate floors.

Two cars may be co-ordinated by a central processor to optimise efficiency of the lifts. Each car operates individually on a full or down collective control system. When the cars are at rest, one is stationed at the main entrance lobby and the other, which has call priority, at a mid point within the building or at another convenient floor level. The priority car will answer landing calls from any floor except the entrance lobby. If the priority car is unable to answer all call demands within a specific time, the other car if available will respond. A similar system may also apply to three cars, with two stationary at the entrance lobby and one available at mid point or the top floor.

With the supervisory control system, each car operates on full collective control and will respond to calls within a dedicated zone. A micro-processor determines traffic demand and locates cars accordingly to each operating zone.

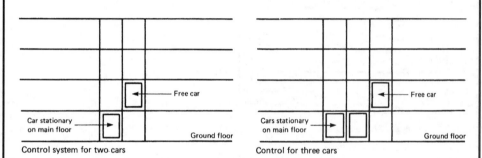

Control system for two cars   Control for three cars

Supervisory control for three or more cars

Lift Doors

Door operation is by an electric motor through a speed reduction unit, clutch drive and connecting mechanism. The type of entrance and doors form a vital part of the lift installation. The average lift car will spend more time at a floor during passenger transfer time than it will during travel. For general passenger service, either side opening, two-speed or even triple-speed side opening doors are preferred. The most efficient in terms of passenger handling is the two-speed centre opening. The clear opening may be greater and usable clear space becomes more rapidly available to the passengers. Vertical centre-bi-parting doors are suitable for very wide openings, typical of industrial applications.

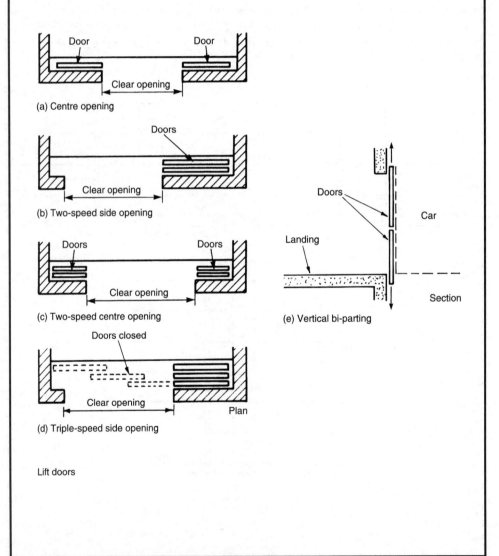

(a) Centre opening

(b) Two-speed side opening

(c) Two-speed centre opening

(d) Triple-speed side opening

(e) Vertical bi-parting

Lift doors

Wherever possible the machine room should be sited above the lift shaft. This location minimises the length of ropes and optimises efficiency. The room should be ventilated, but the vent opening must not be over the equipment. Machinery must be well secured to a concrete base. To reduce sound transmission and vibration, compressed cork, dense rubber or a composite layer is used as an intermediate mounting.

A steel lifting beam is built into the structure above the machinery for positioning or removing equipment for maintenance and repair. Sufficient floor space is necessary for the inspection and repair of equipment – see BS 5655: Lifts and service lifts, for guidance on machine room dimensions relative to capacity.

To prevent condensation the room must be well insulated and heated to provide a design air temperature between 10°C and 40°C. Walls, ceiling and floor should be smooth finished and painted to reduce dust formation. A regular pattern of room cleaning and machinery maintenance should be scheduled.

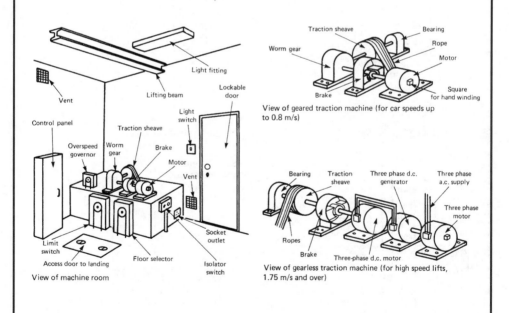

View of geared traction machine (for car speeds up to 0.8 m/s)

View of machine room

View of gearless traction machine (for high speed lifts, 1.75 m/s and over)

Buffers – located at the base of the shaft. They are usually oil loaded for lift speeds >1·5 m/s and otherwise spring loaded. Some variations use compressible plastics.

Overspeed governor – a steel rope passes round a tension pulley in the pit and a governor pulley in the machine room. It also attaches to the lift car's emergency braking system. Overspeeding locks the governor as it responds to spring loaded fly-weight inertia from the centrifugal force in its accelerating pulley. This also switches off power to the lift. The tightening governor rope actuates the safety braking gear.

Safety gear – hardened steel wedges are arranged in pairs each side of the lift car to slow down and stop the car by frictional contact with the car guide rail. Slow- and medium-speed lifts have pairs of hardened steel cams which instantaneously contact a steel channel secured to the lift wall.

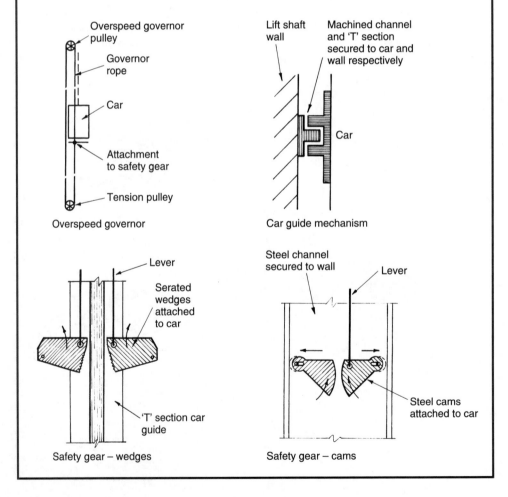

Overspeed governor pulley
Governor rope
Car
Attachment to safety gear
Tension pulley

Overspeed governor

Lift shaft wall
Machined channel and 'T' section secured to car and wall respectively
Car

Car guide mechanism

Lever
Serated wedges attached to car
'T' section car guide

Safety gear – wedges

Steel channel secured to wall
Lever
Steel cams attached to car

Safety gear – cams

To satisfy the need for economies in lift manufacturing processes, BS 5655 provides a limited range of dimensions. Therefore, architects will have to establish passenger transport requirements as a preliminary design priority. The size of lift shaft will depend upon the car capacity and the space required for the counterweight, guides and landing door. The shaft extends below the lowest level served to provide a pit. This permits a margin for car overtravel and a location for car and counterweight buffers. The pit must be watertight and have drainage facilities. Shaft and pit must be plumb and the internal surfaces finished smooth and painted to minimise dust collection. A smoke vent with an unobstructed area of $0 \cdot 1$ m^2 is located at the top of the shaft. The shaft is of fire resistant construction as defined for 'protected shafts' in the Building Regulations. This will be at least 30 minutes and is determined by building function and size. No pipes, ventilating ducts or cables (other than those specifically for the lift) must be fitted within the shaft. A clearance is required at the top of the lift for car overtravel. Counterweight location is at the back or side of the car.

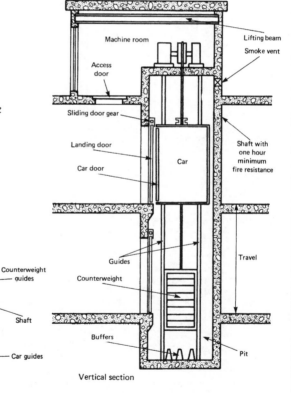

Refs: BS 5655: Lifts and service lifts.
Building Regulations, Approved Document B3: Internal fire spread.

Typical Single Lift Dimensions

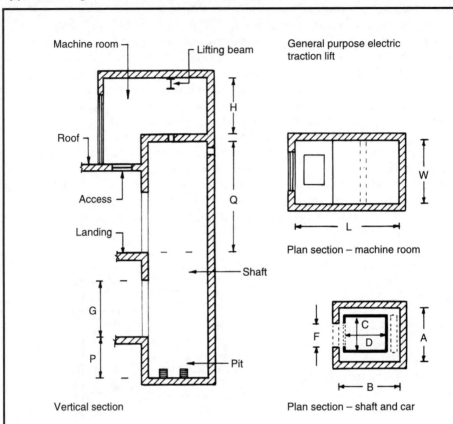

Machine room — Lifting beam

General purpose electric traction lift

Roof —

Access —

Landing —

H

Q

G

P

Shaft

Pit

Vertical section

Plan section – machine room

W

L

Plan section – shaft and car

C
D
F
A
B

All dimensions in metres:

Shaft size		Car size			Door size		Pit	Machine room			
A	B	C	D	E	F	G	P	Q	H	L	W
1·8	2·1	1·1	1·4	2·2	0·8	2·0	1·7	4/4·2	2·6	3·7	2·5
1·9	2·3	1·35	1·4	2·2	0·8	2·0	1·7	4/4·2	2·6	3·7	2·5
2·4	2·3	1·6	1·4	2·3	1·1	2·1	1·8	4·2	2·7	4·9	3·2
2·6	2·3	1·95	1·4	2·3	1·1	2·1	1·9	4·4	2·7	4·9	3·2
2·6	2·6	1·95	1·75	2·3	1·1	2·1	1·9	4·4	2·8	5·5	3·2

Note: Dimension E refers to the car door height.

A paternoster consists of a series of open fronted two-person cars suspended from hoisting chains. Chains run over sprocket wheels at the top and bottom of the lift shaft. The lift is continuously moving and provides for both upward and downward transportation of people in one shaft. Passengers enter or leave the car while it is moving, therefore waiting time is minimal. Passengers will have to be fairly agile, which limits this type of installation to factories, offices, universities, etc. It is not suitable in buildings that accommodate the infirm or elderly! When a car reaches its limit of travel in one direction, it moves across to the adjacent set of hoisting chains to engage with car guides and travel in the other direction. In the interests of safety, car speed must not exceed 0·4 m/s.

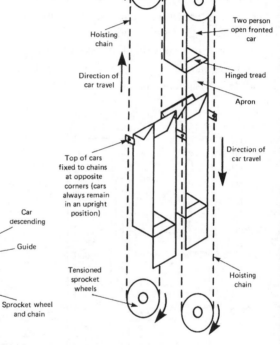

Sprocket wheels driven by an electric motor

Hinged hood

Hoisting chain

Two person open fronted car

Direction of car travel

Hinged tread

Apron

Direction of car travel

Top of cars fixed to chains at opposite corners (cars always remain in an upright position)

Tensioned sprocket wheels

Hoisting chain

Bearing

Car rising

Car descending

Guide

Car moving across

Sprocket wheel and chain

Bearing

Plan of lift at top changeover

View of installation

Paternosters convey about 600 persons per hour. This type of lift has the advantage of allowing passengers to begin their journeys undelayed, regardless of travel direction. Simplicity of control gear adds to the advantages, resulting in fewer breakdowns by eliminating normal processes of stopping, starting, accelerating and decelerating. They are most suited to medium-rise buildings.

Direct acting – the simplest and most effective method, but it requires a borehole below the pit to accommodate the hydraulic ram. The ram may be one piece or telescopic. In the absence of a counterweight, the shaft width is minimised. This will save considerably on construction costs and leave more space for general use.

Side acting – the ram is connected to the side of the car. For large capacity cars and heavy goods lifts, two rams may be required, one each side of the car. A borehole is not necessary, but due to the cantilever design and eccentric loading of a single ram arrangement, there are limitations on car size and load capacity.

Direct side acting – the car is cantilevered and suspended by a steel rope. As with side acting, limitations of cantilever designs restrict car size and payload. Car speed may be increased.

Indirect side acting – the car is centrally suspended by a steep rope and the hydraulic system is inverted.

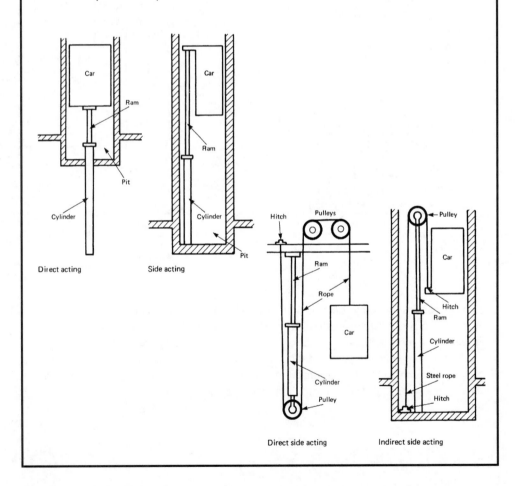

Direct acting   Side acting   Direct side acting   Indirect side acting

Originally, hydraulic lifts used mains water supply as the operating medium. The main was pressurised from a central pumping station to service lift installations in several buildings. The oil-hydraulic system has oil pressure fed by a pump into a cylinder to raise the ram and lift car. Each lift has its own pumping unit and controller. These units are usually sited at or near to the lowest level served, no more than 10 m from the shaft. The lift is ideal in lower rise buildings where moderate speed and smooth acceleration is preferred. Car speed ranges from 0·1 to 1 m/s and the maximum travel is limited to about 21 m. The lift is particularly suitable for goods lifts and for hospitals and old people's homes. Most hydraulic lifts carry the load directly to the ground, therefore as the shaft does not bear the loads, construction is less expensive than for a comparable electric lift installation.

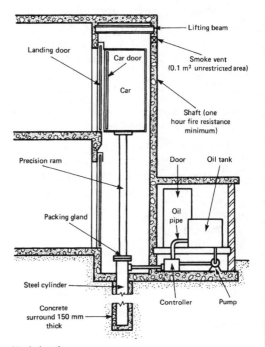

Vertical section

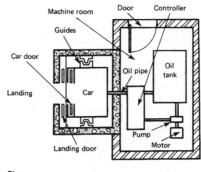

Plan

BS 5655-10·2 provides specific guidance for the testing and examination of hydraulic lifts. See also BS 5655-2: for safety rules applied to constructing and installing hydraulic lifts.

Oil-hydraulic Lift Pumping Unit and Packing Gland

Upward movement – the oil pressure must be gradually increased. The up solenoid valve is energised by an electric current and opens to allow oil to enter above piston D. as the area of piston D is greater than valve C, the oil pressure closes the valve and allows high pressure oil to flow to the cylinder and lift the ram and the car.

Downward movement – the oil pressure must be gradually decreased. The lowering solenoid valve is energised by an electric current and opens allowing oil to flow back to the tank through the by-pass. As the area of piston A is greater than valve B, the reduced oil pressure behind the piston allows valve B to open. Oil flows into the tank and the car moves downwards.

A special packing gland with several seals is required between the cylinder and ram.

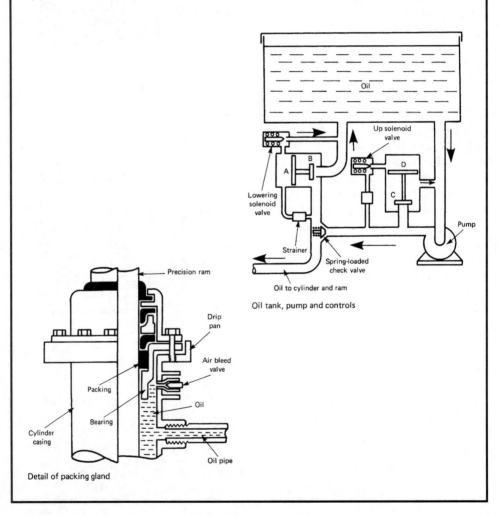

Oil tank, pump and controls

Detail of packing gland

Lift performance depends on:

- acceleration;

- retardation,

- car speed;

- speed of door operation; and

- stability of speed and performance with variations of car load.

The assessment of population density may be found by allowing between one person per 9·5 m^2 of floor area to 11·25 m^2 of floor area. For unified starting and finishing times 17% of the population per five minutes may be used. For staggered starting and finishing times 12% of the population may be used.

The number of lifts will have an effect on the quality of service. Four 18-person lifts provide the same capacity as three 24-person lifts but the waiting time will be about twice as long with the three-car group.

The quality of service may be found from the interval of the group. 25–35 seconds interval is excellent, 35–45 seconds is acceptable for offices, 60 seconds for hotels and 90 seconds for flats.

Further criteria for the comfort and convenience of lift users:

- Directional indication of location of the lift lobby for people unfamiliar with the building.

- Call buttons at landings and in the car positioned for ease of use with unambiguous definition for up and down directions.

- Call buttons to be at a level appropriate for use by people with disabilities and small children.

- Call display/car location display at landings to be favourably positioned for a group of people to watch the position of all cars and for them to move efficiently to the first car arriving.

- Call lights and indicators with an audible facility to show which car is first available and in which direction it is travelling.

- Lobby space of sufficient area to avoid congestion by lift users and general pedestrian traffic in the vicinity.

Estimating the Number of Lifts Required

Example: An office block with 20 storeys above ground floor having unified starting and stopping times is to have a floor area above the ground floor of 8000 m² and floor pitch of 3 m. A group of four lifts, each car having a capacity of 20 persons and a car speed of 2·5 m/s are specified. The clear door width is to be 1·1 m and the doors are to open at a speed of 0·4 m/s. Estimate the interval and quality of service that is to be provided.

1 Peak demand for a 5-minute period $= \dfrac{8000\text{m}^2 \times 17\%}{11\,\text{m}^2/\text{person} \times 100}$

$= 124$ persons

2 Car travel $= 20 \times 3\text{ m} = 60\text{ m}$

3 Probable number of stops $= S - S\left(\dfrac{S-1}{S}\right)^n$

(where S = maximum number of stops)

$\therefore$ Probable number of stops $= 20 - 20\left(\dfrac{20-1}{20}\right)^{16}$

$= 11$

(where n = number of passengers usually approximately 80% of capacity)

4 Upward journey time $= S_1\left(\dfrac{L}{S_1 V} + 2V\right)$

where S_1 = probable number of stops L = travel V = speed

$\therefore$ Upward journey time $= 11\left(\dfrac{60}{11 \times 2\cdot5} + 2 \times 2\cdot5\right)$

$= 79$ seconds

5 Downward journey time $= \left(\dfrac{L}{V} + 2V\right)$

$= \dfrac{60}{2\cdot5} + 2 \times 2\cdot5$

$= 29$ seconds

6 Door operating time $= 2\,(S_1 + 1)\,\dfrac{W}{Vd}$

where W = width of door opening; Vd = opening speed

$\therefore$ Door operating time $= 2\,(11+1)\,\dfrac{1\cdot1}{0\cdot4} = 66$ seconds

7 The average time taken for each person to get into and out of a lift car may be taken as 2 seconds

$\therefore$ Transfer time $= 2n = 2 \times 16 = 32$ seconds

8 Round trip time $= 79 + 29 + 66 + 32 = 206$ seconds

9 Capacity of group $= \dfrac{5\text{ mins} \times 60 \times 4 \times 20 \times 0\cdot8}{206}$

$= 93$ persons per 5 minutes

10 Interval for the group $= \dfrac{206}{4} = 51\cdot5$ seconds

The capacity of the group of lifts and the interval for the group are satisfactory. (Note: Cars less than 12 capacity are not satisfactory)

During the early part of the 20th century, it became apparent that the growing number of high rise buildings would require special provisions for fire control. The firefighting lift was conceived as a means of rapidly accessing the upper floors. Early innovations prioritised the passenger lift by means of a 'break-glass' key switch which brought the lift to the ground floor immediately. This is now unlikely to be acceptable to building insurers and the fire authorities. It is also contrary to current building standards which specify a separate lift installation specifically for firefighting purposes.

Special provisions for firefighting lifts:

- Minimum duty load of 630 kg.
- Minimum internal dimensions of 1·1 m wide × 1·4 m deep × 2·0 m high.
- Provision of an emergency escape hatch in the car roof.
- Top floor access time − maximum 60 seconds.
- Manufactured from non-combustible material.
- A two-way intercommunications system installed.
- Door dimensions at least 0·8 m wide × 2·0 m high of fire resisting construction.
- Two power supplies − mains and emergency generator.

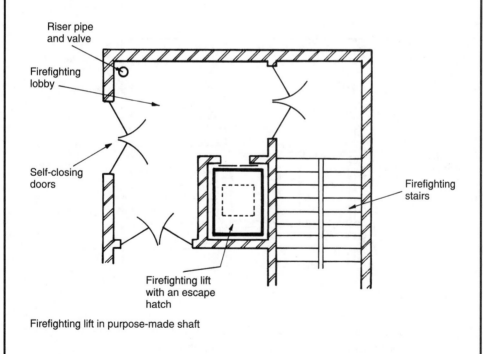

Riser pipe and valve

Firefighting lobby

Self-closing doors

Firefighting stairs

Firefighting lift with an escape hatch

Firefighting lift in purpose-made shaft

Building Regulations – structures with floors at a height greater than 18 m above fire service vehicle access (usually ground level), or with a basement greater than 10 m below fire service vehicle access, should have accessibility from a purpose-made firefighting lift. All intermediate floors should be served by the lift. Firefighting lifts for other situations are optional as defined in Approved Document B5, Section 18, but will probably be required by the building insurer.

Minimum number of firefighting shafts containing lifts:

Buildings without sprinklers – 1 per 900 m^2 floor area (or part of) of the largest floor.

Buildings with sprinklers < 900 m^2 floor area = 1
900 to 2000 m^2 floor area = 2
>2000 m^2 floor area = 2 + 1 for every 1500 m^2 (or part of).

Note: Qualifying floor areas, as defined for fire service vehicle access.

Minimum distance of firefighting lift shaft to any part of a floor is 60 m. Hydrant outlets should be located in the firefighting lobby.

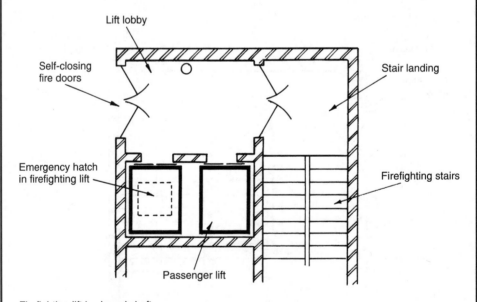

Lift lobby

Self-closing fire doors

Stair landing

Emergency hatch in firefighting lift

Firefighting stairs

Passenger lift

Firefighting lift in shared shaft

Refs: Building Regulations, Approved Document B5, Section 18: Access to buildings for fire-fighting personnel.

BS 5588-5: Code of practice for firefighting stairs and lifts.

Builder's work – machine room:

- Door and window openings sealed against the weather.
- Lockable and safe access for lift engineers and building facilities manager.
- Provide and secure a trapdoor to raise and lower machinery.
- Secure all non-structural floors, decking and temporary scaffolding in position.
- Temporary guards and ladders to be secured in position.
- Dimensions to the requirements of BS 5655 or lift manufacturer's specification.
- Provide reinforced concrete floor and plinths to include at least nine rope holes.
- Treat floor to prevent dust.
- Provide lifting beam(s) and pad stone support in adjacent walls.
- Heating and ventilation to ensure a controlled temperature between 4°C and 40°C.

Electrical work:

- Reduced voltage temporary lighting and power supplies for portable tools during construction.
- Main switch fuse for each lift at the supply company's intake.
- Run power mains from intake to the motor room and terminate with isolating switches.
- Lighting and 13 amp power supply in the machine room.
- Independent light supply from the intake to the lift car with control switchgear in the machine room or half way down the well.
- Lighting to the pit with a switch control in the lowest floor entrance.
- Permanent lighting in the well to consist of one lamp situated 500 mm maximum from the highest and lowest points with intermediate lamps at 7 m maximum spacing.

Builder's work – lift well:

- Calculations with regard to the architect's plans and structural loadings.

- Form a plumb lift well and pit according to the architect's drawings and to tolerances acceptable to the lift manufacturer (known as Nominal Minimum Plumb – the basic figures in which the lift equipment can be accommodated).

- Minimum thickness of enclosing walls – 230 mm brickwork or 130 mm reinforced concrete.

- Applying waterproofing or tanking to the pit and well as required.

- Paint surfaces to provide a dust-free finish.

- Provide dividing beams for multiple wells and inter-well pit screens. In a common well, a rigid screen extending at least 2·5 m above the lowest landing served and a full depth of the well between adjacent lifts.

- Secure lift manufacturer's car guides to lift well walls.

- Make door opening surrounds as specified and secure one above the other.

- Build or cast in inserts to secure lift manufacturer's door sills.

- Perform all necessary cutting away and making good for landing call buttons, door and gate locks, etc.

- Provide smoke vents of at least 0·1 m^2 free area per lift at the top of the shaft.

- Apply finishing coat of paintwork, to all exposed steelwork.

- Provide temporary guards for openings in the well.

- Supply and install temporary scaffolding and ladders to lift manufacturer's requirements.

- Offload and store materials, accessories, tools and clothing in a secure, dry and illuminated place protected from damage and theft.

- Provide mess rooms, sanitary accommodation and other welfare facilities in accordance with the Construction (Health, Safety and Welfare) Regulations.

- Provide access, trucking and cranage for equipment deliveries.

Escalators are moving stairs used to convey people between floor
levels. They are usually arranged in pairs for opposing directional
travel to transport up to 12 000 persons per hour between them.

The maximum carrying capacity depends on the step width and
conveyor speed. Standard steps widths are 600, 800 and 1000 mm,
with speeds of 0·5 and 0·65 m/s. Control gear is less complex than
that required for lifts as the motor runs continuously with less load
variations. In high rise buildings space for an escalator is unjustified
for the full height and the high speed of modern lifts provides for a
better service.

To prevent the exposed openings facilitating fire spread, a water
sprinkler installation (see Chapter 12) can be used to automatically
produce a curtain of water over the well. An alternative is a
fireproof shutter actuated from a smoke detector or fusible links.

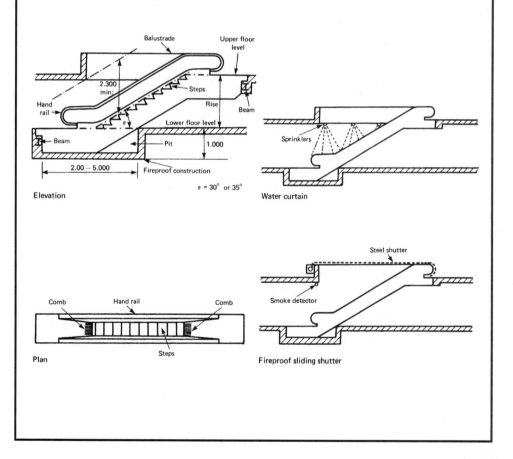

Elevation

Water curtain

Plan

Fireproof sliding shutter

Escalator configurations vary depending on the required level of service. The one-directional single bank avoids interruption of traffic, but occupies more floor space than other arrangements.

A criss-cross or cross-over arrangement is used for moving traffic in both directions.

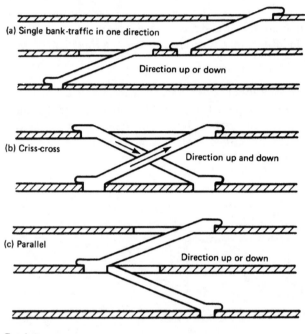

(a) Single bank-traffic in one direction

Direction up or down

(b) Criss-cross

Direction up and down

(c) Parallel

Direction up or down

Escalator arrangements

Escalator capacity formula to estimate the number of persons (N) moved per hour:

$$N = \frac{3600 \times P \times V \times \text{cosine } \theta}{L}$$

where: P = number of persons per step
V = speed of travel (m/s)
θ = angle of incline
L = length of each step (m).

E.g. an escalator inclined at 35°, operating with one person per 400 mm step at 0·65 m/s.

$$N = \frac{3600 \times 1 \times 0·65 \times 0·8192}{0·4} = 4792 \text{ persons per hour}$$

Travelators – also known as autowalks, passenger conveyors and moving pavements. They provide horizontal conveyance for people, prams, luggage trolleys, wheelchairs and small vehicles for distances up to about 300 metres. Slight inclines of up to 12° are also possible, with some as great as 18°, but these steeper pitches are not recommended for use with wheeled transport.

Applications range from retail, commercial and store environments to exhibition centres, railway and airport terminals. Speeds range between 0·6 and 1·3 m/s, any faster would prove difficult for entry and exit. When added to walking pace, the overall speed is about 2·5 m/s.

There have been a number of experiments with different materials for the conveyor surface. These have ranged from elastics, rubbers, composites, interlaced steel plates and trellised steel. The latter two have been the most successful in deviating from a straight line, but research continues, particularly into possibilities for variable speed lanes of up to 5 m/s. However, there could be a danger if bunching were to occur at the exit point.

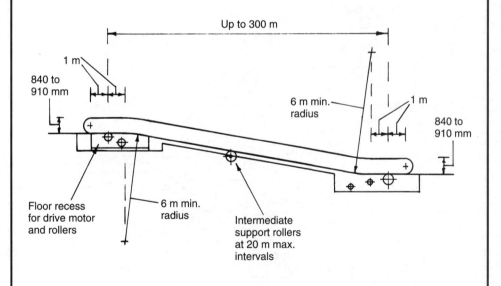

Capacity 6500 to 10 800 persons per hour

Typical inclined travelator

411

Stair Lifts

Stair lifts have been used in hospitals, homes for the elderly and convalescent homes for some time. In more recent years, manufacturers have recognised the domestic need and have produced simple applications which run on a standard steel joist bracketed to the adjacent wall. Development of Part M to the Building Regulations, 'Access and facilities for disabled people', is likely to ensure that staircases in all future dwellings are designed with the facility to accommodate and support a stair lift or a wheelchair lift. This will allow people to enjoy the home of their choice, without being forced to seek alternative accommodation.

Standard 230 volt single-phase a.c. domestic electrical supply is adequate to power a stairlift at a speed of about 0·15 m/s. A 24 volt d.c. transformed low-voltage supply is used for push button controls. Features include overspeed brake, safety belt, optional swivel seat, folding seat and armrests and a manual lowering device. The angle of support rail inclination is usually within the range of 22°–50° within a maximum travel distance of about 20 m.

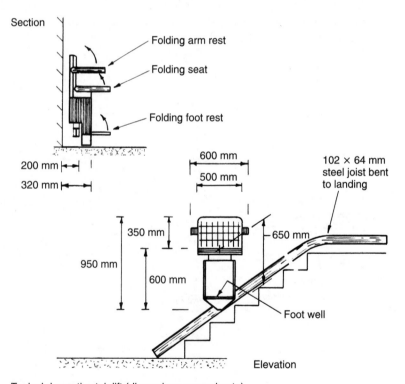

Typical domestic stair lift (dimensions approximate)

Ref: BS 5776: Specification for powered stair lifts.

12 FIRE PREVENTION AND CONTROL SERVICES

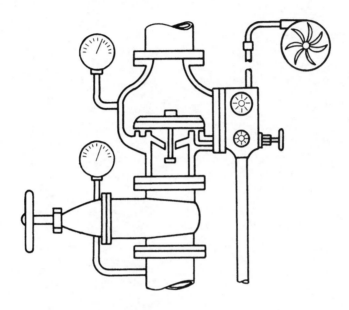

SPRINKLERS

DRENCHERS

HOSE REELS

HYDRANTS

FOAM INSTALLATIONS

GAS EXTINGUISHERS

FIRE ALARMS

SMOKE, FIRE AND HEAT DETECTORS

ELECTRICAL ALARM CIRCUITS

FIRE DAMPERS IN DUCTWORK

PRESSURISATION OF ESCAPE ROUTES

SMOKE EXTRACTION, VENTILATION AND CONTROL

PORTABLE FIRE EXTINGUISHERS

Water sprinklers provide an automatic spray dedicated to the area of fire outbreak. Sprinkler heads have temperature sensitive elements that respond immediately to heat, discharging the contents of the water main to which they are attached. In addition to a rapid response which reduces and isolates fire damage, sprinklers use less water to control a fire than the firefighting service, therefore preventing further damage from excess water.

Sprinkler systems were initially credited to an American, Henry Parmalee, following his research during the late 1800s. The idea was developed further by another American, Frederick Grinnell, and the name 'Grinnell' is still associated with the glass-type fusible element sprinkler head.

Domestic pipework – solvent cement bonded, post-chlorinated polyvinyl chloride (CPVC).

Industrial and commercial pipework – threaded galvanised mild steel.

The simplest application is to attach and suspend sprinkler heads from a water main fixed at ceiling level. However, some means of regulation and control is needed and this is shown in the domestic application indicated below.

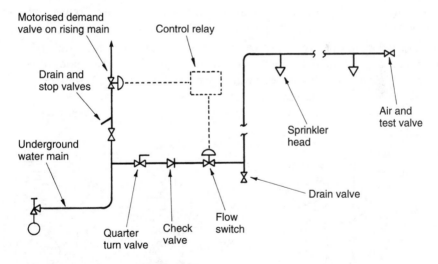

Motorised demand valve on rising main

Control relay

Drain and stop valves

Air and test valve

Sprinkler head

Underground water main

Drain valve

Flow switch

Quarter turn valve

Check valve

Typical domestic sprinkler installation
Note: When flow switch is activated, demand valve closes.

Ref: BS 5306: Fire extinguishing installations and equipment on premises, BS 5306-2: Specification for sprinkler systems. BS EN 12259: Fixed fire-fighting systems. Components for sprinkler and water supply systems.

Types of Sprinkler Head

Quartzoid bulb – a glass tube is used to retain a water valve on its seating. The bulb or tube contains a coloured volatile fluid, which when heated to a specific temperature expands to shatter the glass and open the valve. Water flows on to a deflector, dispersing as a spray over the source of fire. Operating temperatures vary with a colour coded liquid:

Orange – 57°C
Red – 68°C
Yellow – 79°C
Green – 93°C
Blue – 141°C
Mauve – 182°C
Black – 204 or 260°C

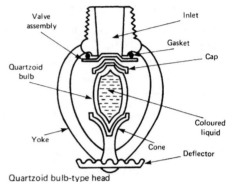

Quartzoid bulb-type head

Fusible strut – has two metal struts soldered together to retain a water valve in place. A range of solder melting temperatures are available to suit various applications. Under heat, the struts part to allow the valve to discharge water on the fire.

Duraspeed solder type – contains a heat collector which has a soldered cap attached. When heat melts the solder, the cap falls away to displace a strut allowing the head to open. Produced in a range of operating temperatures.

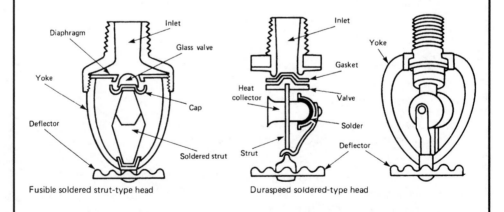

Fusible soldered strut-type head   Duraspeed soldered-type head

The type of sprinkler system, the number of sprinkler heads and their location will depend on the building insurer's requirements. These are likely to be formulated from the Building Regulations, Approved Document B: Fire safety, BS 5306-2 and the Loss Prevention Council's Technical Bulletins. Buildings are classified by fire risk or hazard according to their function and purpose, i.e.

Light – colleges and schools, hotels, hospitals, institutions, museums, offices, etc.
Ordinary – breweries, clothing factories, engineering workshops, sound and film studios, restaurants, etc.
High – fireworks factory and warehouse, paints and plastics production, volatile chemical and fluid processing, etc

See page 422 for spacing of sprinkler heads relative to hazard capacity.

Sprinkler systems:

- Wet (see page 418)
- Dry (see page 419)
- Alternative wet and dry (see page 419)
- Tail end
- Pre-action
- Recycling

Tail end – this system is mainly wet, i.e. charged with water, with the exception of one section of pipework which is fitted with an air valve to maintain that section with compressed air. It can be used where part of a building, such as a warehouse, is unheated.

Pre-action – used where there is a possibility that sprinkler heads may be accidentally damaged by tall equipment or plant, e.g. a fork-lift truck. To avoid unnecessary water damage, the system is dry. If a sprinkler head is damaged, compressed air discharges to effect an initial alarm. Water supply to the sprinkler is dependent on a fire detector which will operate a motorised valve on the water supply to effect another alarm.

Recycling – a damage limiting installation developed from the pre-action system. After sprinklers have subdued a fire, ceiling mounted heat detectors set at a slightly lower response temperature than the sprinkler heads sense the reduced temperature to effect closure of the water supply after a 5-minute time delay.

The wet system is used in heated buildings where there is no risk of the water in the pipework freezing. All pipework is permanently pressure charged with water and the sprinkler heads usually attach to the underside of the range pipes. Where water is mains supplied, it should be fed from both ends. If the main is under repair on one side, the stop valve and branch pipe can be closed and the sprinkler system supplied from the other branch pipe.

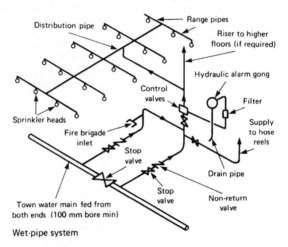

Wet-pipe system

When a sprinkler head is fractured water is immediately dispersed. Water will also flow through an annular groove in the alarm valve seating to a pipe connected to an alarm gong and turbine. A jet of water propels the turbine blades causing the alarm gong to operate. Pipeline flow switches will alert the local fire service in addition to operating an internal alarm system. Except under supervised maintenance, the main stop valve is padlocked in the open position.

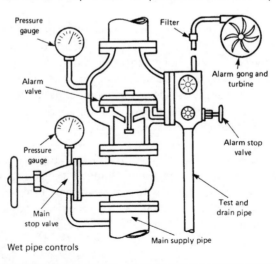

Wet pipe controls

Dry or an alternate wet-and-dry sprinkler system may be used in buildings that are unheated.

Dry system – installation pipework above the differential valve is permanently charged with compressed air. When a fire fractures a sprinkler head, the compressed air escapes to allow the retained water to displace the differential valve and flow to the broken sprinkler.

Alternate wet-and-dry system – a wet system for most of the year, but during the winter months it functions as a dry system.

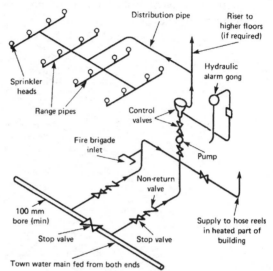

Dry pipe or alternate wet-and-dry pipe system

The dry part of the system above the diaphragm or differential valve is charged with compressed air at about 200 kPa. Any loss of pressure is automatically replenished by a small compressor, but this will not interfere with water flow if the system is activated. When a sprinkler is fractured, an automatic booster pump can be used to rapidly exhaust the air and improve water flow. Sprinkler heads are fitted above the range pipes which are slightly inclined to allow the system to be fully drained.

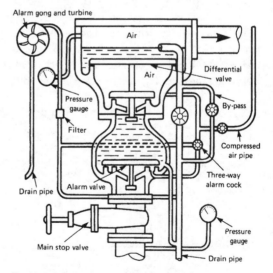

Dry pipe or alternate wet-and-dry pipe controls

Deluge system – used for specifically high fire hazards such as plastic foam manufacture, fireworks factories, aircraft hangars, etc., where there is a risk of intensive fire with a very fast rate of propagation. The pipework is in two parts, compressed air with quartzoid bulbs attached and a dry pipe with open ended spray projectors. When a fire occurs, the quartzoid bulbs shatter and compressed air in the pipeline is released allowing a diaphragm inside the deluge control valve to open and discharge water through the open pipe to the projectors.

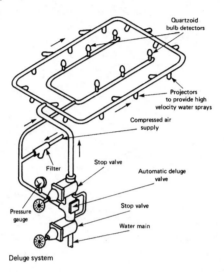

Deluge system

Multiple control system – a heat sensitive sealed valve controls the flow of water to a small group of open sprayers attached to a dry pipe. When a fire occurs, the valve quartzoid bulb shatters allowing the previously retained water to displace the valve stem and flow to the sprayers. An alternative to a heat sensitive valve is a motorised valve activated by a smoke or fire detector.

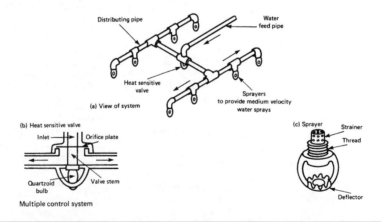

Multiple control system

There are various sources of water supply that may be used for sprinkler applications.

Elevated private reservoir – minimum volume varies between 9 m^3 and 875 m^3 depending on the size of installation served.

Suction tank – supplied from a water main. Minimum tank volume is between 2·5 m^3 and 585 m^3. A better standard of service may be achieved by combining the suction tank with a pressure tank, a gravity tank or an elevated private reservoir. A pressure tank must have a minimum volume of water between 7 m^3 and 23 m^3. A pressure switch or flow switch automatically engages the pump when the sprinklers open.

Gravity tank – usually located on a tower to provide sufficient head or water pressure above the sprinkler installation.

River or canal – strainers must be fitted on the lowest part of the suction pipes corresponding with the lowest water level. Duplicate pumps and pipes are required, one diesel and the other electrically powered.

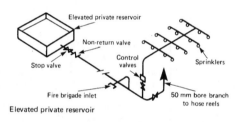

Elevated private reservoir

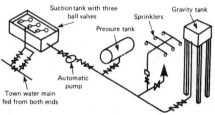

Town main suction tank automatic pump with pressure tank or gravity tank (if required)

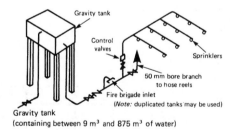

Gravity tank
(containing between 9 m^3 and 875 m^3 of water)

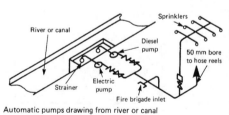

Automatic pumps drawing from river or canal

Note: Water source capacities, pressures, delivery rates, etc. vary with application. See tables in BS 5306-2 for specific situations.

Pipework Distribution to Sprinklers

The arrangement of pipework will depend on the building shape and layout, the position of the riser pipe and the number of sprinkler heads required. To provide a reasonably balanced distribution, it is preferable to have a centre feed pipe. In practice this is not always possible and end feed arrangements are used. The maximum spacing of sprinkler heads (S) on range pipes depends on the fire hazard classification of the building.

Hazard category	Max. spacing (S) of sprinkler heads (m)	Max. floor area covered by one sprinkler head (m^2)
Light	4·6	21
Ordinary	4·0 (standard)	12
	4·6 (staggered)*	12
High	3·7	9

* See next page

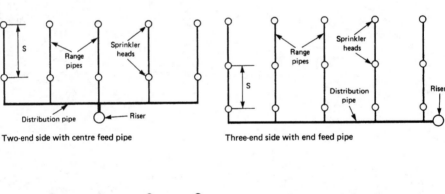

Two-end side with centre feed pipe

Three-end side with end feed pipe

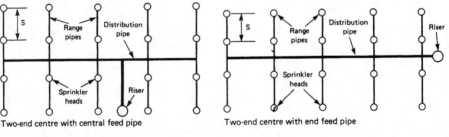

Two-end centre with central feed pipe

Two-end centre with end feed pipe

Staggered arrangement of sprinkler heads on an ordinary hazard installation:

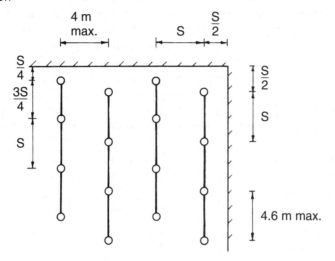

Calculating the number of sprinkler heads: e.g. an ordinary fire hazard category for a factory having a floor area 20 m × 10 m.

$$20 \times 10 = 200 \text{ m}^2.$$

Ordinary hazard requires a maximum served floor area of 12 m² per sprinkler head.

Therefore: $200 \div 12 = 16 \cdot 67$, i.e. at least 17 sprinkler heads.

For practical purposes, 18 could be installed as shown:

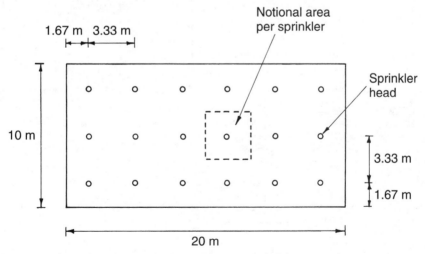

The maximum area served by each sprinkler head = $3 \cdot 33$ m × $3 \cdot 33$ m = $11 \cdot 1$ m².

This is satisfactory, being less than 12 m².

423

Drenchers

A drencher fire control system provides a discharge of water over roofs, walls and windows to prevent fire spreading from or to adjacent buildings. Automatic drenchers are similar in operating principle to individual quartzoid bulb sprinkler heads. A manually operated stop valve can also be used with dry pipes and open spray nozzles. This stop valve must be located in a prominent position with unimpeded access. Installation pipework should fall to a drain valve positioned at the lowest point above the stop valve. The number of drencher nozzles per pipe is similar to the arrangements for conventional sprinkler installations as indicated in BS 5306-2. For guidance, two drenchers can normally be supplied by a 25 mm i.d. pipe. A 50 mm i.d. pipe can supply ten drenchers, a 75 mm i.d. pipe 36 drenchers and a 150 mm i.d. pipe over 100 drenchers. An example of application is in theatres, where the drenchers may be fitted above the proscenium arch at the stage side to protect the safety curtain.

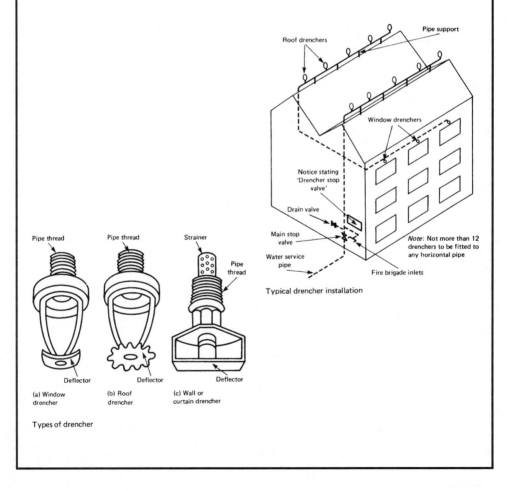

Typical drencher installation

Types of drencher

Hose reels are firefighting equipment for use as a first-aid measure by building occupants. They should be located where users are least likely to be endangered by the fire, i.e. the staircase landing. The hose most distant from the source of water should be capable of discharging 0·4 l/s at a 6 m distance from the nozzle, when the two most remote hose reels are operating simultaneously. A pressure of 200 kPa is required at the highest reel. If the water main cannot provide this, a break/suction tank and booster pumps should be installed. The tank must have a minimum volume of water of 1·6 m³. A 50 mm i.d. supply pipe is adequate for buildings up to 15 m height and a 65 mm i.d. pipe will be sufficient for buildings greater than this. Fixed or swinging hose reels are located in wall recesses at a height of about 1 m above floor level. They are supplied by a 25 mm i.d. pipe to 20 or 25 mm i.d. reinforced non-kink rubber hose in lengths up to 45 m to cover 800 m² of floor area per installation.

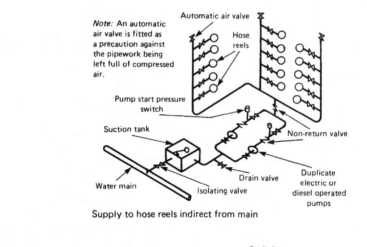

Note: An automatic air valve is fitted as a precaution against the pipework being left full of compressed air.

Automatic air valve

Hose reels

Pump start pressure switch

Suction tank

Non-return valve

Water main

Drain valve

Isolating valve

Duplicate electric or diesel operated pumps

Supply to hose reels indirect from main

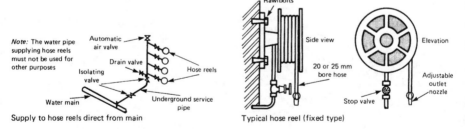

Note: The water pipe supplying hose reels must not be used for other purposes

Automatic air valve

Drain valve

Isolating valve

Hose reels

Water main

Underground service pipe

Supply to hose reels direct from main

Rawlbolts

Side view

20 or 25 mm bore hose

Typical hose reel (fixed type)

Elevation

Adjustable outlet nozzle

Stop valve

Ref: BS 5306: Fire extinguishing installations and equipment on premises, BS 5306-1: Hydrant systems, hose reels and foam inlets.

Dry Riser

A dry riser is in effect an empty vertical pipe which becomes a fire-fighter's hose extension to supply hydrants at each floor level. Risers should be disposed so that no part of the floor is more than 60 m from a landing valve. This distance is measured along a route suitable for a firefighting hose line, to include any dimension up or down a stairway. Buildings with floors up to 45 m above fire service vehicle access level require one 65 mm landing valve on each floor from a 100 mm i.d. riser. Buildings between 45 m and 60 m with one or two landing valves per floor require a 150 mm i.d. riser. For buildings above 60 m a wet riser must be installed. Two 65 mm i.d. inlet hose couplings are required for a 100 mm riser and four 65 mm i.d. inlets are required for a 150 mm riser. The riser must be electrically bonded to earth.

Note: A dry riser is installed either in unheated buildings or where the water main will not provide sufficient pressure at the highest landing valve. A hard standing for the Fire Service Vehicle is required at the base of the riser. One landing valve is required for every 900 m² of floor area

Automatic air release valve

65 mm bore landing valve

100 mm bore minimum dry riser

Fire brigade inlets

1.000 (approx)

25 mm bore drain valve

Typical arrangement of a dry riser

65 mm instantaneous coupling

600 mm

400 mm

DRY RISER INLETS

Drain holes

Wired glass

Note: Door fitted with spring lock which opens when the glass is broken

(a) Front view of Fire Brigade inlets

(b) Front view of Fire Brigade inlet box

Details of dry riser inlet

A wet riser is suitable in any building where hydrant installations are specified. It is essential in buildings where floor levels are higher than that served by a dry riser, i.e. greater than 60 m above fire service vehicle access level. A wet riser is constantly charged with water at a minimum running pressure of 400 kPa with up to three most remote landing valves operating simultaneously. A flow rate of 25 l/s is also required. The maximum pressure with one outlet open is 500 kPa to protect firefighting hoses from rupturing. Orifice plates may be fitted to the lower landing valves to restrict pressure. Alternatively, a pressure relief valve may be incorporated in the outlet of the landing valve. The discharge from this is conveyed in a 100 mm i.d. drain pipe.

To maintain water at the required pressure and delivery rate, it is usually necessary to install pumping equipment. Direct pumping from the main is unacceptable. A suction or break tank with a minimum water volume of 45 m^3 is used with duplicate power source service pumps. One 65 mm landing valve should be provided for every 900 m^2 floor area.

Note: In addition to the supply through the float valves the suction tank should also be supplied with a 150 mm Fire service inlet.

Automatic air valve

Landing valve on roof (if required)

Landing valve

Wet riser (bore, 100 mm minimum)

Drain pipe

The bore of a wet riser is the same as that given for a dry riser and the riser must be electrically earthed

50 mm bore pressure relief branch pipe

Drain pipe to discharge over the suction tank

65 mm diameter hose coupling

Suction tank

Float valves

Pump start pressure switch

Duplicate electric or diesel operated pumps

Towns main

Drain valve

Typical arrangement of a wet riser

Flange for connection to wet riser

Chain

Connection to firefighters hose

Detail of a landing valve

427

Fixed Foam Installations

A pump operated mechanical foam installation consists of a foam concentrate tank located outside of the area to be protected. The tank has a water supply pipe inlet and foam pipe outlet. A venturi is fitted in the pipeline to draw the foam out of the tank. When the water pump is switched on, the venturi effect causes a reduction in pressure at the foam pipe connection, resulting in a mixture of foam concentrate and water discharging through the outlet pipe.

A pre-mixed foam installation consists of a storage tank containing foam solution. When a fire occurs in the protected area, a fusible link is broken to release a weight which falls to open a valve on the carbon dioxide cylinder. Foam solution is forced out of the tank at a pressure of about 1000 kPa to discharge over the protected equipment, e.g. an oil tank.

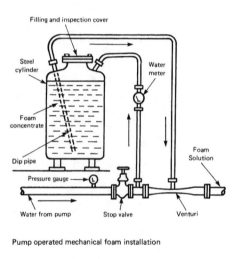

Pump operated mechanical foam installation

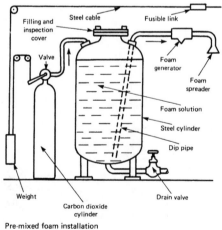

Pre-mixed foam installation

A foam installation is used for application from remote points on to flammable liquid fire risks. This type of installation is often used with oil-fired boilers and oil storage tanks. A foam access box is built into the wall at an easily accessible place for fire-fighters to attach hoses from their foam generating and mixing equipment. The box is usually located about 600 mm above adjacent ground and should be clear of any openings through which heat, smoke or flames can pass. The glass fronted box can be broken and the lock released from inside. Two 65 mm diameter inlets may be used. A 65 or 75 mm i.d. galvanised steel pipe is normally used for the distribution. A maximum pipework length of 18 m is recommended and this must slope slightly towards the spreaders. Vertical drop pipes are acceptable but vertically inclined pipes must not be used. Spreader terminals are positioned about 1 m above oil burners and about 150 mm above oil spill level of stored fuel.

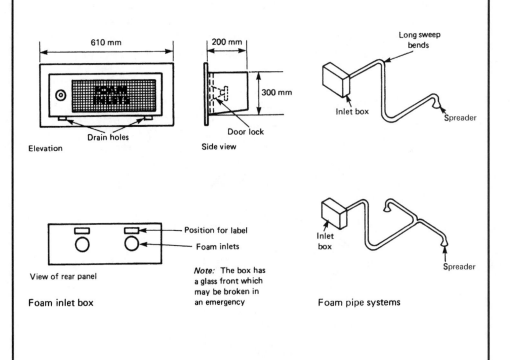

Ref: BS 5306: Fire extinguishing installations and equipment on premises, BS 5306-6: Foam systems.

The majority of gas extinguishing systems have been either halon 1301 or carbon dioxide (see next page). Halons are electrically non-conductive and safe to use where personnel remain in an area of gas discharge. They are also more effective than carbon dioxide, being five times the density of air, whilst carbon dioxide is only one-and-a-half times. Unfortunately halon or bromochlorodifluoromethane (BCF) gases are a hazard to the environment, by contributing significantly to the depleting effect of the ozone layer. In 1987 a meeting of major countries at a Montreal convention agreed to phase out the use of these gases by 2002. Therefore, except for systems installed in less co-operative countries, new installations will contain halon substitutes. These include inergen and argonite, both mixtures of nitrogen and argon, the former containing a small amount of carbon dioxide.

In principle, the systems are suitable where there is a high density of equipment, e.g. tape libraries and computer suites where an alternative wet system would be considered too damaging. Gas is stored in spherical steel containers which can be secured in a ceiling or floor void or against a wall. When activated by smoke or heat, detectors immediately open valves on the extinguishers to totally flood the protected area with a colourless and odourless gas.

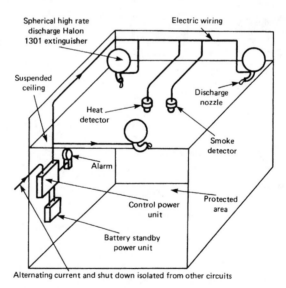

Halon 1301 installation

Ref: BS 5306: Fire extinguishing installations and equipment on premises, BS 5306-5: Halon systems.

Carbon dioxide is an alternative to halon as a dry gas extinguisher. It has been used as an extinguishing agent for a considerable time, particularly in portable extinguishers. As the gas is dry and non-conductive it is ideal for containing fires from electrical equipment, in addition to textiles, machinery, petroleum and oil fires. Carbon dioxide is heavier than air and can flow around obstacles to effectively reduce the oxygen content of air from its normal 21% to about 15%. This considerably reduces an important component of the combustion process (see page 446). Integrated high and low pressure gas systems may be used, with the former operating at up to 5800 kPa. Systems can be either electrical, pneumatic or mechanical with a manual override facility. Carbon dioxide is potentially hazardous to personnel, therefore it is essential that the system is automatically locked off when the protected area is occupied. In these circumstances it can be switched to manual control. Air tightness of a protected room is essential for the success of this system as total flooding relies on gas containment by peripheral means.

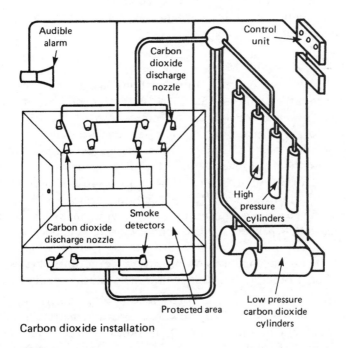

Carbon dioxide installation

Ref: BS 5306: Fire extinguishing installations and equipment on premises, BS 5306-4: Specification for carbon dioxide systems.

Fire detection and alarm systems may contain:
- system control unit
- primary (mains) electrical supply
- secondary (battery or capacitor stand-by) power supply. An emergency generator could also be used
- alarm activation devices – manual or automatic
- alarm indication devices – audible and/or visual
- remote indication on a building monitoring system
- control relay via a building management system to effect fire extinguishers and ventilation smoke control actuators.

System control unit – an alarm panel which monitors the state of all parts (zones) of the installation. It identifies the point of origin of an alarm, displays this on the panel and communicates this to remote control locations.

Zones:
- Max. 2000 m^2 floor area in one storey.
- No detachment of compartment areas within one floor area zone.
- Max. 30 m search distance into a zone.
- Single occupancy of a zone where several separate business functions occur in one building.

Requirements for dwellings

Automatic fire detection and alarm systems are to be provided to the recommendations of BS 5839: Fire detection and alarm systems in buildings. They may comply with Part 1 or 6 of the BS, i.e. Code of practice for system design, installation and servicing, or Code of practice for the design and installation of fire detection and alarm systems in dwellings, respectively. Alternatively, a smoke alarm system is acceptable if it complies with BS 5446-1: Components of automatic fire alarm systems for residential premises – Specification for self-contained smoke alarms and point-type smoke detectors. These should have primary and secondary power supplies.

Point detectors – individual heat or smoke detection units which respond to an irregular situation in the immediate vicinity.

Line detectors – a continuous type of detection comprising a pair of conducting cables separated by low temperature melting insulation to permit a short circuit alarm when the cables contact. Suitable in tunnels and service shafts.

Provision in large houses:

Floor area	Storeys (inc. basement)	System
> 200 m^2	> 3	BS 5839-1, type L2
> 200 m^2	≤ 3	BS 5839-6, type LD3

Note: prefixes used in the BS types indicates that L is a specific application to protection of life, whereas P indicates that for property.

Application:
- Optical type (photo-electric) detectors in circulation spaces, i.e. hallways, corridors and landings.
- Ionisation type detectors in living and dining areas.

Preferred location of detectors:
- Over 300 mm from light fittings.
- Min. one per storey.
- Loft conversions, with alarm linked to operate others and be operated by others in the dwelling.
- Circulation spaces between bedrooms.
- Circulation spaces < 7.5 m from doors to habitable rooms.
- Kitchens (with regard to heat/smoke producing appliances).
- Living rooms.

Requirements for buildings other than dwellings

This is less easy to define due to the variation in building types and patterns of occupancy. BS 5839 requirements may suit some buildings, but could cause panic in others, e.g. shopping centres, where people may be unfamiliar with the layout. In these situations, trained staff may be the preferred system of building evacuation. At building design stage, consultation between the local building control authority, the fire authority and the building's insurer is paramount, as alterations post-construction are always extremely expensive.

Ref. Building Regulations, Approved Document B: Fire safety.
    Section B1: Means of warning and escape.

Ionisation smoke detector – positive and negative charged plate electrodes attract opposingly charged ions. An ion is an atom or a group of atoms which have lost or gained one or more electrons, to carry a predominantly positive or negative charge. The movement of ions between the plates reduces the resistance of air, such that a small electric current is produced. If smoke enters the unit, particles attach to the ions slowing their movement. This reduction in current flow actuates an electronic relay circuit to operate an alarm.

Light scattering smoke detector – a light beam projects onto a light trap into which it is absorbed. When smoke enters the detector, some of the light beam is deflected upwards onto a photo-electric cell. This light energises the cell to produce an electric current which activates the alarm relay.

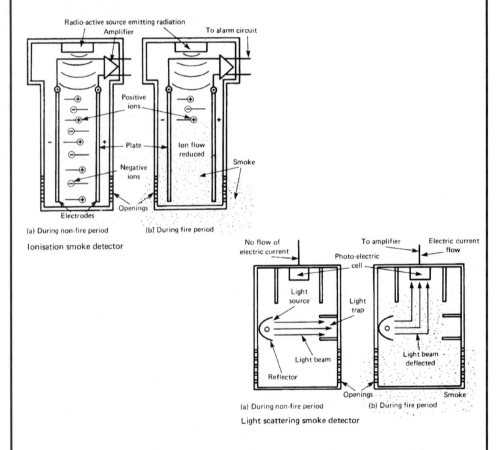

Ionisation smoke detector

Light scattering smoke detector

Refs: BS 5446: Fire detection and fire alarm devices for dwellings.
BS 5446-1: Specification for smoke alarms.

Heat detectors are used where smoking is permitted and in other situations where a smoke detector could be inadvertently actuated by process work in the building, e.g. a factory. Detectors are designed to identify a fire in its more advanced stage, so their response time is longer than smoke detectors.

Fusible type – has an alloy sensor with a thin walled casing fitted with heat collecting fins at its lower end. An electrical conductor passes through the centre. The casing has a fusible alloy lining and this functions as a second conductor. Heat melts the lining at a predetermined temperature causing it to contact the central conductor and complete an alarm relay electrical circuit.

Bi-metallic coil type – heat passes through the cover to the bi-metal coils. Initially the lower coil receives greater heat than the upper coil. The lower coil responds by making contact with the upper coil to complete an electrical alarm circuit.

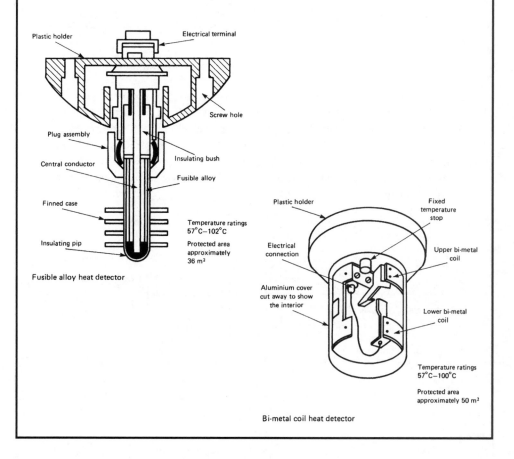

Plastic holder

Electrical terminal

Screw hole

Plug assembly

Insulating bush

Central conductor

Fusible alloy

Finned case

Temperature ratings 57°C–102°C

Protected area approximately 36 m²

Insulating pip

Fusible alloy heat detector

Plastic holder

Fixed temperature stop

Electrical connection

Upper bi-metal coil

Aluminium cover cut away to show the interior

Lower bi-metal coil

Temperature ratings 57°C–100°C

Protected area approximately 50 m²

Bi-metal coil heat detector

Light Obscuring and Laser Beam Detectors

Light obscuring – a beam of light is projected across the protected area close to the ceiling. The light falls onto a photo-electric cell which produces a small electrical current for amplification and application to an alarm circuit. Smoke rising from a fire passes through the light beam to obscure and interrupt the amount of light falling on the photo-electric cell. The flow of electric current from the cell reduces sufficiently to activate an alarm relay.

A variation is the light-scatter type. In normal use the light is widely dispersed and no light reaches the photo-electric cell receptor. In the presence of smoke, particulates deflect light on to the receptor to energise the cell.

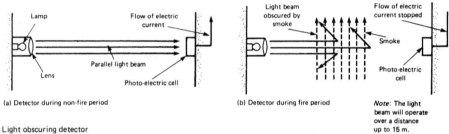

(a) Detector during non-fire period   (b) Detector during fire period   *Note*: The light beam will operate over a distance up to 15 m.

Light obscuring detector

Laser beam – a band of light which can be visible or infra-red projected onto a photo-electric cell. It does not fan out or diffuse as it travels through an uninterrupted atmosphere. The beam can operate effectively at distances up to 100 m. If a fire occurs, smoke and heat rises and the pulsating beam is deflected away from the cell or reduced in intensity. As the cell is de-energised, this effects on alarm relay.

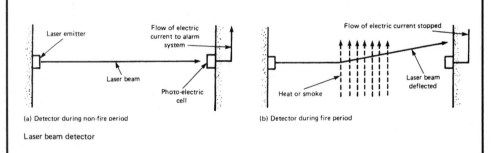

(a) Detector during non-fire period   (b) Detector during fire period

Laser beam detector

In addition to producing hot gases, fire also releases radiant energy in the form of visible light, infra-red and ultra-violet radiation. Radiant energy travels in waves from the fire.

Infra-red detector – detectors have a selective filter and lens to allow only infra-red radiation to fall on a photo-electric cell. Flames have a distinctive flicker, normally in the range of 4 to 15 Hz. The filter is used to exclude signals outside of this range. The amplifier is used to increase the current from the photo-electric cell. To reduce false alarms, a timing device operates the alarm a few seconds after the outbreak of fire.

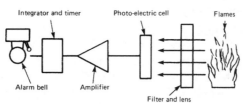

Components of an infra-red detector

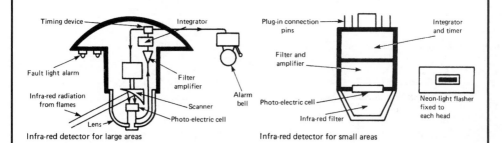

Infra-red detector for large areas

Infra-red detector for small areas

Ultra-violet detector – these detectors have a gas-filled bulb which reacts with ultra-violet radiation. When the bulb receives radiant energy, the gas is ionised to produce an electric current. When this current exceeds the set point of the amplifier the alarm circuit closes to operate the alarm system.

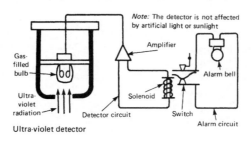

Ultra-violet detector

Fire alarm electrical circuits may be of the 'open' or 'closed' types. In addition to, or as an alternative to, automatic smoke or fire sensing switches, manual break-glass alarm switches can be wall mounted at about 1·5 m above floor level in lobbies, corridors and other common access locations. No person should have to travel more than 30 m to use an alarm. In large managed buildings, a sub-circuit will connect to the facilities manager's office or in more sophisticated situations the alarm can relay through telecommunications cables to a central controller and the fire service.

Open circuit – call points or detectors are connected to open switches, which prevent current flowing through the circuit when it is on standby. Closing a switch on the detector circuit actuates a solenoid (electromagnet) to complete the alarm circuit. As there is no current flow whilst on stand-by there is no electrical power consumption. The disadvantage of this system is that if part of the detector circuit is inadvertently damaged, some of the switches will not operate.

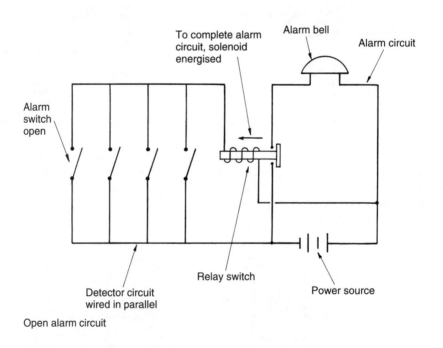

Open alarm circuit

Electrical Power to 'open' or 'closed' fire alarm circuits should be separate from any other electrical installation. To isolate it completely from any interruption to mains supply, it is usually transformed to 24–60 volts d.c. and provided with a battery back-up system in the event of the fire damaging the mains source of power.

Closed circuit – call points or detectors may be regarded as closed switches allowing current to flow in the detector circuit. This permanent current flow energises a solenoid switch which retains a break in the alarm circuit. When a detector circuit switch is operated, i.e. opened, the solenoid is de-energised allowing a spring mechanism to connect it across the alarm circuit terminals and effect the alarm.

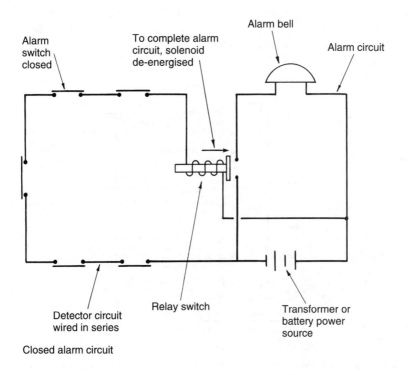

Alarm switch closed

To complete alarm circuit, solenoid de-energised

Alarm bell

Alarm circuit

Detector circuit wired in series

Relay switch

Transformer or battery power source

Closed alarm circuit

Ref: BS EN 54: Fire detection and fire alarm systems.

Ventilation of services enclosures is required to dilute flammable, toxic or corrosive gases. This can be taken to include smoke and hot gases that will occur as a result of fire, particularly where the void contains combustible PVC cable sheathing and uPVC pipes. To provide a safe level of ventilation and to prevent overheating in a restricted enclosure, permanent natural ventilation should be at least 0·05 m² and 1/150 of the cross-sectional area for enclosure areas of less than 7·5 m² and greater than 7·5 m² respectively.

Openings and access panels into services enclosures should be minimal. The enclosure itself should be gas tight and there must be no access from a stairway. Where access panels or doors are provided they should be rated at not less than half the fire resistance of the structure, and have an integrity rating of at least 30 minutes (see BS 476-22). Fire doors should be fitted with self closers.

Where ventilation ducts pass from one compartment to another or into a services enclosure, the void made in the fire resisting construction must be made good with a suitable fire stopping material. Automatic fire dampers are also required in this situation to prevent fire spreading between compartments.

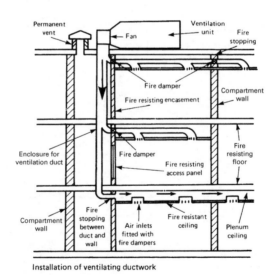

Installation of ventilating ductwork

Refs: BS 8313: Code of practice for accommodation of building services in ducts.
BS 5588-9: Code of practice for ventilation and air conditioning ductwork.
Building Regulations, Approved Document B3: Section 11, Protection of openings and fire-stopping.

Fire dampers are required in ventilation and air conditioning systems to prevent smoke and fire spreading through the ductwork to other parts of the building. Dampers should be positioned to maintain continuity of compartmentation by structural division. They can operate automatically by fusible link melting at a predetermined temperature of about 70°C, to release a steel shutter. An electro-magnet may also be used to retain the shutter in the open position. The electromagnet is deactivated to release the shutter by a relay circuit from a fire or smoke detector. The latter is preferable, as a considerable amount of smoke damage can occur before sufficient heat penetrates the ductwork to activate a heat detector or a fusible link.

An intumescent-coated honeycomb damper is an alternative. In the presence of heat, the coating expands to about a hundred times its original volume to form sufficient mass to impair the movement of fire through the duct. This type of damper has limited fire resistance and is only likely to be specified in low velocity systems.

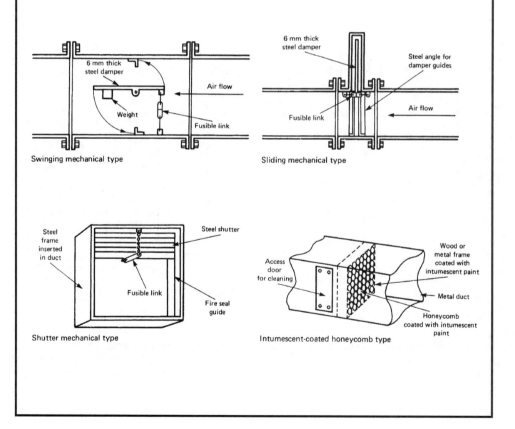

Swinging mechanical type

Sliding mechanical type

Shutter mechanical type

Intumescent-coated honeycomb type

Pressurisation of Escape Routes

In multi-storey buildings, stairways and lobbies may be air pressurised to clear smoke and provide an unimpeded escape route. The air pressurisation is usually between 25 and 50 Pa depending on the building height and degree of exposure. This pressure is insignificant for movement of personnel. A number of pressurisation methods may be used:

- Pressurisation plant is disengaged, but it is automatically switched on by a smoke or fire detector.
- Pressurisation plant runs fully during hours of occupancy as part of the building ventilation system.
- Pressurisation plant runs continuously at a reduced capacity and output during the hours of building occupancy, but fire detection automatically brings it up to full output.

It is important to provide openings so that smoke is displaced from the escape routes to the outside air. This can be through purpose-made grilles or window vents. Pressurisation will help to limit entry of rain and draughts at external openings.

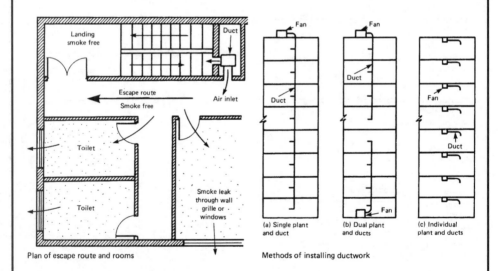

Plan of escape route and rooms

Methods of installing ductwork

(a) Single plant and duct

(b) Dual plant and ducts

(c) Individual plant and ducts

Ref: BS 5588-4: Code of practice for smoke control using pressure differentials.

Automatic fire ventilation is designed to remove heat, smoke and toxic gases from single-storey buildings. In large factories and shopping malls, the additional volume of air entering the building by fire venting is insignificant relative to the benefits of creating clear visibility. Parts of the roof can be divided into sections by using fireproof screens which may be permanent or may fall in response to smoke detection. Fire vents are fitted at the highest part of each roof section as is practical. Heat and smoke rise within the roof section above the fire outbreak. At a predetermined temperature, usually 70°C, a fusible link breaks and opens the ventilator above the fire. Heat and smoke escape to reduce the amount of smoke logging within the building. This will aid people in their escape and assist the fire service to see and promptly tackle the source of fire. The heat removed prevents risk of an explosion, flash-over and distortion to the structural steel frame.

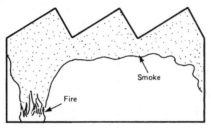

Fire in unvented building showing unrestricted spread of smoke

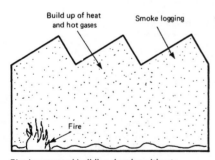

Fire in unvented building showing ultimate smoke logging

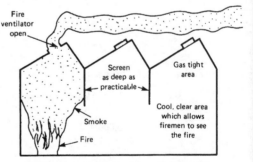

Fire in vented building showing restricted spread of smoke. The fire ventilator may also be used for normal ventilation.

Smoke and Fire Ventilators

Automatic smoke and fire ventilator:

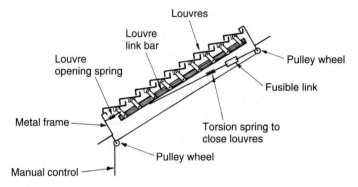

Number and area of ventilators – estimates are based on providing a smoke-free layer about 3 m above floor level.

E.g.

Floor to centre of vent height (m)	Ventilation factor (m)
4·5	0·61
7·5	0·37
10·5	0·27
13·5	0·23

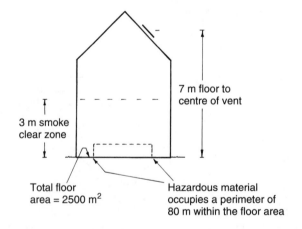

By interpolation, ventilation factor for 7 m approximates to 0·41 m.

Ventilator area can be taken as the perimeter occupied by hazardous material, multiplied by the ventilation factor, i.e. 80 m × 0·41 m. This approximates to 33 m² or (33/2500 × 100/1) = 1·3% of the floor area.

Most enclosed shopping centres have a mall with a parade of shops. The mall is the general circulation area and the obvious escape route from a fire. In these situations, a fire can generate a rapid spread of smoke and hot gases. It is therefore essential that some form of smoke control is adopted. If the central area has a normal (68°C) sprinkler system, the water may cool the smoke and hot gases to reduce their buoyancy and create an unwanted fogging effect at floor level. Therefore, consideration should be given to reducing the number of sprinkler heads and specifying a higher operating temperature. Smoke can be controlled by:

- Providing smoke reservoirs into which the smoke is retained before being extracted by mechanical or natural means.

- Allowing replacement cool air to enter the central area through low level vents to displace the smoke flowing out at higher level.

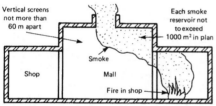

Smoke reservoir by adopting a greater ceiling height in the mall than in the shops

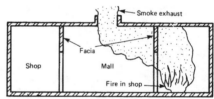

Smoke reservoir formed by facias above open fronted shops

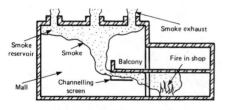

Two-storey mall showing behaviour of smoke through channelling screens

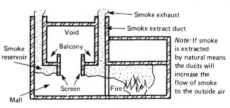

Use of smoke extract ducts through roof of mall

A portable fire extinguisher must contain the type of fire extinguishing agent suitable for the fire it is required to extinguish. It must also be clearly identifiable by colour coding for its intended purpose.

Fires can be grouped:

- Solid fuels, e.g. wood, paper, cloth, etc.
- Flammable liquids, e.g. petrol, oil, paints, fats, etc.
- Flammable gases, e.g. methane, propane, acetylene, etc.
- Flammable metals, e.g. zinc, aluminium, uranium, etc.
- Electrical.

Extinguishing agent	Extinguisher colour	Application
Water	Red	Carbonaceous fires, paper, wood, etc.
Foam	Red with cream band	Ditto and flammable liquids, oils, fats, etc.
Carbon dioxide	Red with black band	Electrical fires and flammable liquids.
Dry chemicals	Red with blue band	All fires.

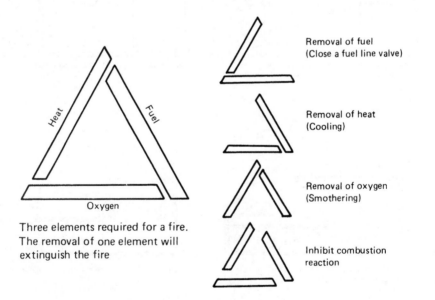

Heat

Fuel

Oxygen

Three elements required for a fire. The removal of one element will extinguish the fire

Removal of fuel
(Close a fuel line valve)

Removal of heat
(Cooling)

Removal of oxygen
(Smothering)

Inhibit combustion reaction

Ref: BS EN 3: Portable fire extinguishers.

Sand and water buckets are no longer acceptable as a first-aid fire treatment facility. Purpose provided extinguishers are now commonplace in public and commercial buildings. Under the obligations of the Health and Safety at Work, etc. Act, employees are required to undertake a briefing on the use and selection of fire extinguishers. Water in pressurised cylinders may be used for carbonaceous fires and these are commonly deployed in offices, schools, hotels, etc. The portable soda–acid extinguisher has a small glass container of sulphuric acid. This is released into the water cylinder when a knob is struck. The acid mixes with the water which contains carbonate of soda to create a chemical reaction producing carbon dioxide gas. The gas pressurises the cylinder to displace water from the nozzle. The inversion type of extinguisher operates on the same chemical principle.

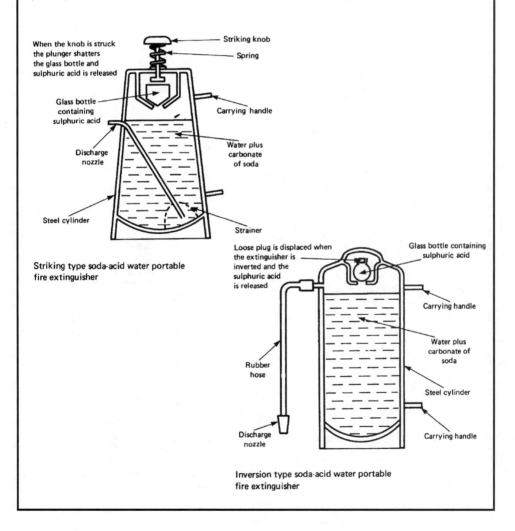

When the knob is struck the plunger shatters the glass bottle and sulphuric acid is released

Striking knob

Spring

Glass bottle containing sulphuric acid

Carrying handle

Discharge nozzle

Water plus carbonate of soda

Steel cylinder

Strainer

Striking type soda-acid water portable fire extinguisher

Loose plug is displaced when the extinguisher is inverted and the sulphuric acid is released

Glass bottle containing sulphuric acid

Carrying handle

Rubber hose

Water plus carbonate of soda

Steel cylinder

Discharge nozzle

Carrying handle

Inversion type soda-acid water portable fire extinguisher

Although water is a very good cooling agent, it is inappropriate for some types of fire. It is immiscible with oils and is a conductor of electricity. Therefore, the alternative approach of breaking the triangle of fire by depleting the oxygen supply can be achieved by smothering a fire with foam. Foam is suitable for gas or liquid fires.

Chemical foam type of extinguisher – foam is formed by chemical reaction between sodium bicarbonate and aluminium sulphate dissolved in water in the presence of a foaming agent. When the extinguisher is inverted the chemicals are mixed to create foam under pressure which is forced out of the nozzle.

Carbon dioxide extinguisher – carbon dioxide is pressurised as a liquid inside a cylinder. Striking a knob at the top of the cylinder pierces a disc to release the carbon dioxide which converts to a gas as it depressurises through the extinguisher nozzle.

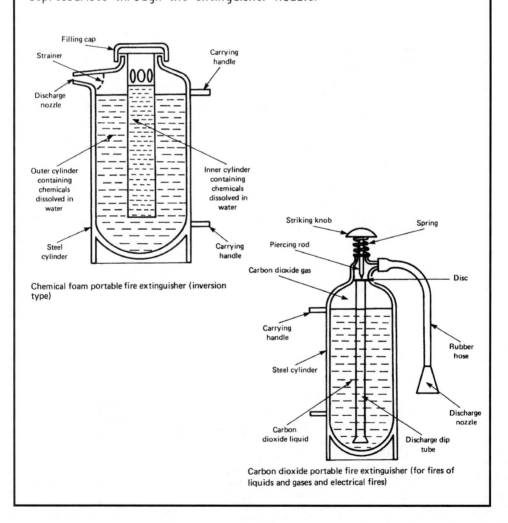

Chemical foam portable fire extinguisher (inversion type)

Carbon dioxide portable fire extinguisher (for fires of liquids and gases and electrical fires)

13 SECURITY INSTALLATIONS

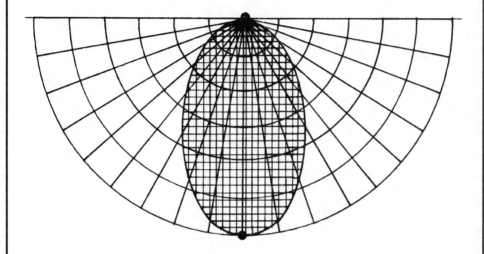

INTRUDER ALARMS

MICRO-SWITCH AND MAGNETIC REED

RADIO SENSOR, PRESSURE MAT AND TAUT WIRING

ACOUSTIC, VIBRATION AND INERTIA DETECTORS

ULTRASONIC AND MICROWAVE DETECTORS

ACTIVE INFRA-RED DETECTOR

PASSIVE INFRA-RED DETECTOR

LIGHTNING PROTECTION SYSTEMS

Intruder alarms have developed from a very limited specialist element of electrical installation work in high security buildings to the much wider market of schools, shops, offices, housing, etc. This is largely a result of the economics of sophisticated technology surpassing the efficiency of manual security. It is also a response to the increase in burglaries at a domestic level. Alarm components are an alarm bell or siren activated through a programmer from switches or activators. Power is from mains electricity with a battery back-up. Extended links can also be established with the local police, a security company and the facility manager's central control by telecommunication connection.

Selection of switches to effect the alarm will depend on the building purpose, the extent of security specified, the building location and the construction features. Popular applications include:

- Micro-switch
- Magnetic reed
- Radio sensor
- Pressure mat
- Taut wiring
- Window strip
- Acoustic detector
- Vibration, impact or inertia detector

The alternative, which may also be integrated with switch systems, is space protection. This category of detectors includes:

- Ultrasonic
- Microwave
- Active infra-red
- Passive infra-red

Circuit wiring may be 'open' or 'closed' as shown in principle for fire alarms – see pages 438 and 439. The disadvantage of an open circuit is that if an intruder knows the whereabouts of cables, the detector circuit can be cut to render the system inoperative. Cutting a closed circuit will effect the alarm.

The following references provide detailed specifications:
British Standards 4737, 6707, 6799, 7042 and 7150.

Micro-switch and Magnetic Reed

Micro-switch – a small component which is easily located in door or window openings. It is the same concept and application as the automatic light switch used in a vehicle door recess, but it activates an alarm siren. A spring loaded plunger functions in a similar manner to a bell push button in making or breaking an electrical alarm detector circuit. The disadvantage is the constant movement and associated wear, exposure to damage and possible interference.

Magnetic reed – can be used in the same situations as a micro-switch but it has the advantage of no moving parts. It is also less exposed to damage or tampering. There are, however, two parts to install. One is a plastic case with two overlapping metal strips of dissimilar polarity, fitted into a small recess in the door or window frame. The other is a magnetic plate attached opposingly to the door or window. When the magnet is close to the overlapping strips, a magnetic field creates electrical continuity between them to maintain circuit integrity. Opening the door or window demagnetises the metal strips, breaking the continuity of the closed detector circuit.

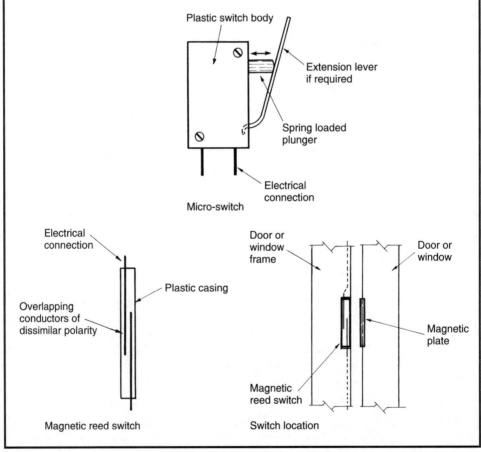

Radio sensor – these are surface mounted to windows and doors. They transmit a radio signal from an integral battery power source. This signal is picked up by a central control unit or receiver, which activates the alarm circuit. As these sensors are 'free wired' they can be moved, which is ideal for temporary premises or in buildings undergoing changes. A pocket or portable radio panic button transmitter is an option. The range without an aerial is about 60 m, therefore they can be used in outbuildings to a hard wired system from a main building.

Pressure mat – these are a 'sandwich' with metal foil outer layers as part of a detector circuit. The inner core is a soft perforated foam. Pressure on the outer upper layer connects to the lower layer through the perforations in the core to complete the circuit and activate the alarm. Location is near entrances and under windows, normally below a carpet where a small area of underlay can be removed. Sensitivity varies for different applications, such as premises where household pets occupy the building.

Taut wiring – also available as a window strip. A continuous plastic coated copper wire is embedded in floors, walls or ceilings, or possibly applied around safes and other secure compartments. As a window strip, silvered wire can be embedded between two bonded laminates of glass. Alternatively, a continuous self-adhesive lead or aluminium tape can be applied directly to the surface. In principle, it is similar to a car rear heated window. When the wire or tape is broken the closed circuit is interrupted which activates the alarm circuit.

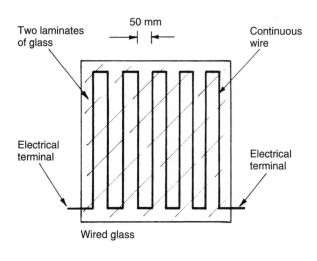

Wired glass

Acoustic – also known as sonic detectors. They are used mainly for protection against intruders in commercial and industrial premises. A sound receiver comprises a microphone, amplifier and an output relay. Also included is a filter circuit which can be tuned to respond to specific sound frequencies such as that produced by breaking glass.

Vibration – a slender leaf of steel is suspended between two electrical contacts. Hammering or structural impact produces vibration in the pendulum, sufficient for the contacts to meet and complete a detector circuit. Adjustment allows for a variety of applications, e.g. where a road or railway is adjacent and intermittent vibration would occur.

Inertia – these respond to more sensitive movements than vibrations, so would be unsuitable near roads, railways, etc. They are ideal to detect the levering or bending of structural components such as window sashes and bars. A pivotal device is part of a closed circuit, where displacement of its weight breaks the circuit continuity.

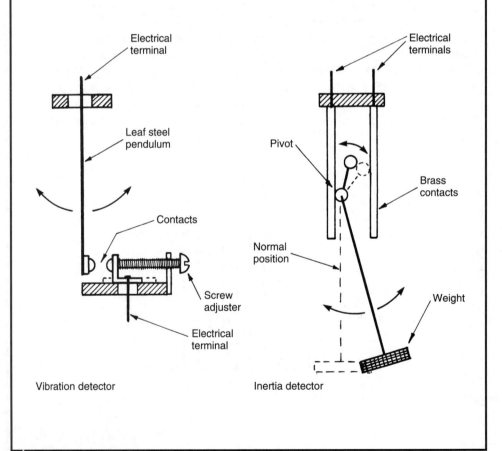

Vibration detector

Inertia detector

Ultrasonic – the equipment is simply a sound emitter and a receiver containing a microphone and sound processor. The sounds are at a very high frequency of between 20 and 40 kHz (normal hearing limit is about 15 kHz). Direct and indirect (reflected) sound distribution from the emitter to the receiver adopts a pattern which can be plotted as a polar curve. If an intruder encroaches the curve the sound frequency will be disturbed. The receiver then absorbs the original frequency, the frequency reflected off the intruder and a mixture of the two. The latter is known as the 'beat note' and it is this irregularity which effects the detector circuit. Greatest detection potential is in the depth of the lobe, therefore this should be projected towards an entry point or a window.

Microwave – operates on the same principle as ultrasonic detection, except that extremely high radio waves are emitted at a standard 10·7 GHz. Emitter and receiver occupy the same unit which is mounted at high level to extend waves over the volume of a room, warehouse, office or similar internal area. An intruder penetrating the microwaves disturbs the frequency which effects the detector circuit. Unlike ultrasonic detectors, microwave detectors are not disturbed by air currents, draughts and ultrasonic sounds from electrical equipment such as computers. They are therefore less prone to false alarms.

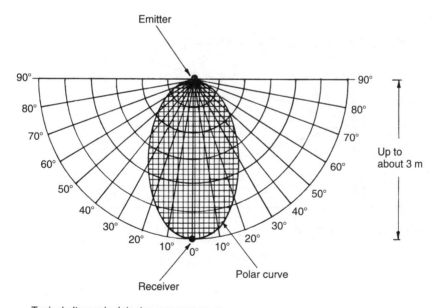

Typical ultrasonic detector response zone

Active Infra-red Detector

Otherwise known as an optical system, it uses a light beam from the infra-red part of the electromagnetic spectrum. This is imperceptible to the human eye. The system is based on a transmitter and receiver. The transmitter projects an invisible light beam at distances up to 300 m on to a photo-electric cell receiver. An intruder crossing the beam will prevent the light from activating the cell. The loss of energy source for the cell effects an alarm relay. Even though the beam has extensive range, this system is not suitable for external use.

Atmospheric changes such as fog or birds flying through the beam can affect the transmission. Mirrors may be used to reflect the beam across a room or around corners, but each reflection will reduce the beam effectiveness by about 25%. Infra-red beams will penetrate glass partitions and windows, each pane of glass reducing the beam effectiveness by about 16%. The smarter intruder may be able to fool the system by shining a portable light source at the receiver. This can be overcome by pulsing the transmission, usually at about 200 pulses per second.

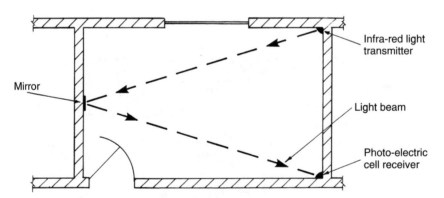

Infra-red light beam application

These detectors use highly sensitive ceramic infra-red receivers to recognise radiation from a moving body. Wall mounted detector units focus the radiation through a lens which contains curved facets to concentrate the radiation on to two sensors. Image variation between the sensors generates a small electrical differential to effect an alarm relay. These systems have enjoyed widespread application, not least the domestic market. Units of lower sensitivity can be used where pets occupy a home. A battery back-up energy source covers for periods of mains power isolation. PIR detectors can be used with other devices in the same system, e.g. radio pocket panic buttons, pressure mats, magnetic reeds, etc. PIR beam patterns vary in form and range to suit a variety of applications, both externally and internally.

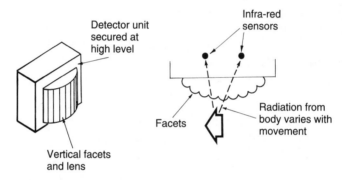

PIR detector unit, typically 75 × 50 mm

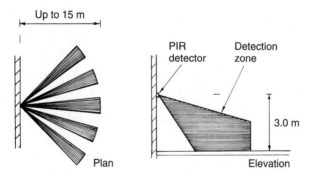

Typical pattern displacement for wall mounted detector

Lightning occurs as a result of electrostatic discharge between clouds or between a cloud and the ground. The potential is up to 100 MV with the current peaking at about 200 kA. The average current is about 20 kA. The number of days that thunderstorms occur in the UK varies between 5 and 20 per year, depending on location. Consequently, some degree of protection to buildings and their occupants is necessary.

As the risk of lightning striking a particular building is low, not all buildings are protected. Houses have least priority and are rarely protected, but other purpose groups will be assessed by their owners and their insurers. This will be on the basis of height, contents, function, type of construction (extent of metal work, e.g. lead roofing), likelihood of thunderstorms in locality, extent of isolation and the general topography. Even where a lightning protection system is provided it is unlikely to prevent some lightning damage to the building and its contents.

Function of a lightning protection system − to attract a lightning discharge which might otherwise damage exposed and vulnerable parts of a building. To provide a path of low impedance to an earth safety terminal.

Zone of protection − the volume or space around a conductor which is protected against a lightning strike. It can be measured at 45° to the horizontal, descending from the apex of the conductor. For buildings less than 20 m in height the zone around a vertical conductor is conical. For buildings exceeding 20 m, the zone can be determined graphically by applying a 60 m radius sphere to the side of a building. The volume contained between the sphere and building indicates the zone. See next page for illustrations.

Zones of protection:

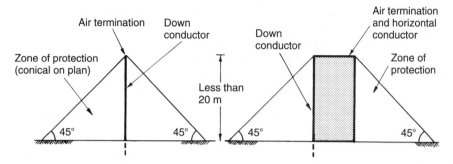

Protection zones for buildings <20 m height

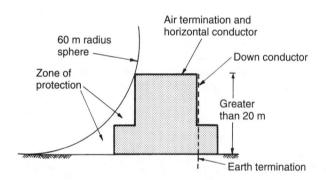

Protection zones for buildings >20 m height

Air terminations – these are provided to intercept a lightning strike. No part of a roof should exceed 5 m from part of a termination conductor, unless it is a lower level projection which falls within the zone of protection. Metallic components such as aerials, spires, cooling towers, etc., should be connected to a terminal. Apart from specific apexes such as spires, air terminations are horizontal conductors running along the ridge of a pitched roof or around the periphery of a flat roof. If the roof is of sufficient size, a 20 m × 10 m grid or lattice of parallel terminations should be provided.

Down conductors – these provide a low impedance route from the air terminations to the earth terminal. They should be direct, i.e. vertical without bends and re-entrant loops. Spacing for buildings up to 20 m in height is 1 per 20 m of periphery starting at the corners and at equal distance apart. Building in excess of 20 m height require 1 per 10 m, at corners and equi-spaced. All structural steelwork and metal pipes should be bonded to the down conductor to participate in the lightning discharge to earth.

Fixing centres for all conductors:

Horizontal and vertical – 1 m max.
Horizontal and vertical over 20 m long – 750 mm max.
        25 m long – 500 mm max.

Minimum dimensions of conductors: 20 mm × 4 mm (80 mm^2) or
         10 mm diameter (80 mm^2).

Conductor materials – aluminium, copper and alloys, phosphor-bronze,
        galvanised steel or stainless steel.

Earth termination – this is required to give the lightning discharge current a low resistance path to earth. The maximum test resistance is 10 ohms for a single terminal and where several terminals are used, the combined resistance should not exceed 10 ohms. Depth of terminal in the ground will depend on subsoil type. Vertical earthing rods of 10 or 12 mm diameter hard drawn copper are preferred, but stronger phosphor-bronze or even copper-coated steel can be used if the ground is difficult to penetrate. Alternatively, a continuous horizontal strip electrode may be placed around the building at a depth of about one metre. Another possibility is to use the reinforcement in the building's foundation. To succeed there must be continuity between the structural metalwork and the steel reinforcement in the concrete piled foundation.

Ref: BS 6651: Code of practice for protection of structures against lightning.

14 ACCOMMODATION FOR BUILDING SERVICES

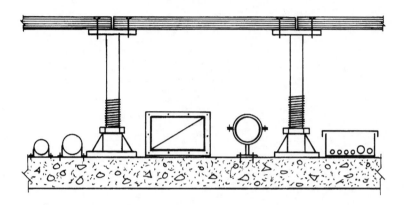

DUCTS FOR ENGINEERING SERVICES

FLOOR AND SKIRTING DUCTS

MEDIUM AND LARGE VERTICAL DUCTS

MEDIUM AND LARGE HORIZONTAL DUCTS

SUBWAYS OR WALKWAYS

PENETRATION OF FIRE STRUCTURE BY PIPES

RAISED ACCESS FLOORS

SUSPENDED AND FALSE CEILINGS

Before installing ducts for the entry of services into a building, it is essential to ascertain the location of pipes and cables provided by the public utilities companies. Thereafter, the shortest, most practicable and most economic route can be planned. For flexible pipes and cables, a purpose-made plastic pipe duct and bend may be used. For rigid pipes or large cables, a straight pipe duct to a pit will be required. Pipe ducts must be sealed at the ends with a plastic filling and mastic sealant, otherwise subsoil and other materials will encroach into the duct. If this occurs, it will reduce the effectiveness of the void around the pipe or cable to absorb differential settlement between the building and incoming service. To accommodate horizontal services, a skirting or floor duct may be used. These may be purpose made by the site joiner or be standard manufactured items. Vertical services may be housed in either a surface-type duct or a chase. The latter may only be used if the depth of chase does not affect the structural strength of the wall. The reduction in the wall's thermal and sound insulation properties may also be a consideration.

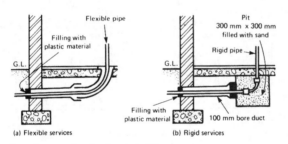

(a) Flexible services   (b) Rigid services

Ducts for entry of services into the building

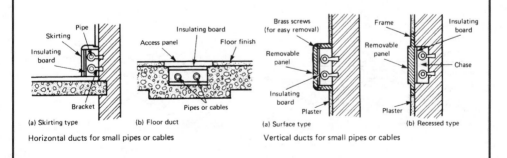

(a) Skirting type   (b) Floor duct

Horizontal ducts for small pipes or cables

(a) Surface type   (b) Recessed type

Vertical ducts for small pipes or cables

Floor and Skirting Ducts

A grid distribution of floor ducting is appropriate in open plan offices and shops where there is an absence of internal walls for power and telecommunications sockets. It is also useful in offices designed with demountable partitioning where room layout is subject to changes. Sockets are surface mounted in the floor with a hinged cover plate to protect them when not in use. The disruption to the structure is minimal as the ducts can be set in the screed, eliminating the need for long lengths of trailing cables to remote workstations. For partitioned rooms, a branching duct layout may be preferred. The branches can terminate at sockets near to the wall or extend into wall sockets. Where power supplies run parallel with telecommunications cables in shared ducts, the services must be segregated and clearly defined. For some buildings, proprietary metal, plastic or laminated plywood skirting ducts may be used. These usually have socket outlets at fixed intervals.

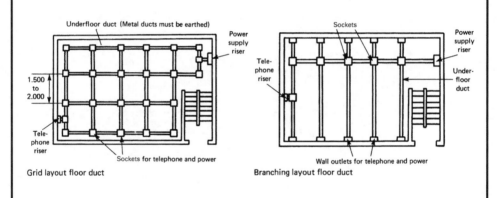

Grid layout floor duct

Branching layout floor duct

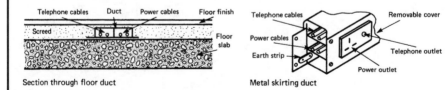

Section through floor duct

Metal skirting duct

The purpose of a service duct is to conceal the services without restricting access for inspection, repair and alterations. A duct also helps to reduce noise and protect the services from damage. When designing a service duct, the transmission of noise, possible build-up of heat in the enclosure and accessibility to the services must be considered. The number of ducts required will depend on the variation in services, the need for segregation and location of equipment served. Vertical ducts usually extend the full height of a building which is an important factor when considering the potential for spread of fire. The duct must be constructed as a protected shaft and form a complete barrier to fire between the different compartments it passes. This will require construction of at least 60 minutes' fire resistance with access doors at least half the structural fire resistance.

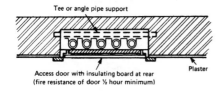

Recessed for medium-sized pipes and cables

Partially recessed for medium-sized pipes and cables

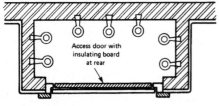

Built-out for large pipes

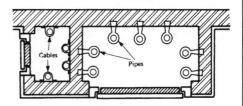

Built-out for large pipes and cables

Refs.: BS 8313: Code of practice for accommodation of building services in ducts.
Building Regulations, Approved Document B3: Internal fire spread (structure).

Floor trenches are usually fitted with continuous covers. Crawl-ways generally have access covers of minimum 600 mm dimension, provided at convenient intervals. A crawl-way should be wide enough to allow a clear working space of at least 700 mm and have a minimum headroom of at least 1 m. Continuous trench covers may be of timber, stone, reinforced concrete, metal or a metal tray filled to match the floor finish. The covers should be light enough to be raised by one person, or, at most, two. Sockets for lifting handles should be incorporated in the covers. In external situations, the cover slabs (usually of stone or concrete) can be bedded and joined together with a weak cement mortar. If timber or similar covers are used to match a floor finish, they should be fixed with brass cups and countersunk brass screws. A trench has an internal depth of less than 1 m. In internal situations where ducts cross the line of fire compartment walls, a fire barrier must be provided within the void and the services suitably fire stopped (see pages 283 and 468).

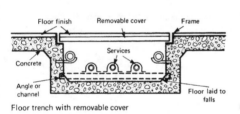

Floor trench with removable cover

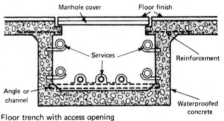

Floor trench with access opening

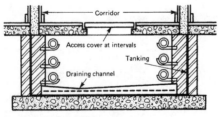

Crawl-way inside a building

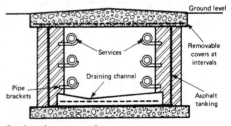

Crawl-way in open ground

Access to a subway will normally be from a plant room, control room or a basement. Additional access from the surface should also be provided at convenient junctions and direction changes. See page 202 for provision of wall step irons. The design and construction of these ducts should adequately withstand the imposed loads and pressures that will occur under extreme working conditions. They should be watertight and where used internally have adequate resistance to fire. Ducts housing boiler or control room services must be provided with a self closing fire door at the entry. Ventilation to atmosphere is essential and a shallow drainage channel should convey ground water leakage and pipe drainage residue to a pumped sump or a gully connection to a drain.

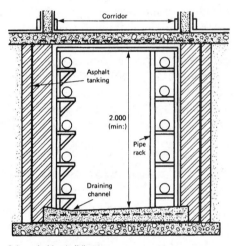

Subway inside a building

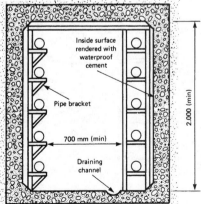

Note Lighting may be provided operated at 110 V

Subway in open ground

467

Penetration of Fire Structure by Pipes

The effect of fire spreading through the voids associated with internal pipework penetrating fire resistant walls and floors can be considered in four areas:

1. Addition of fuel to the total fire load.
2. Production of toxic gases and smoke.
3. Risk of fire spread along the pipework.
4. Reduction in fire resistance of the building elements penetrated.

Guidance in Approved Document B3 to the Building Regulations is mostly applied to sanitation pipework penetrating the structure, but could affect other services, particularly in large buildings. Acceptable sleeving and sealing methods for uPVC discharge pipes are shown on page 283. Non-combustible pipe materials up to 160 mm nominal i.d. (excluding lead, aluminium, aluminium alloys, uPVC and fibre cement) may have the structural opening around the pipe fire stopped with cement mortar, gypsum plaster or other acceptable non-combustible material. Where the pipe material is one of those listed in parentheses, and it penetrates a wall separating dwellings or a compartment wall or floor between flats, the discharge stack is limited to 160 mm nominal i.d. and branch pipes limited to 110 mm nominal i.d., provided the system they are part of is enclosed as shown.

* Any other materials, e.g. polypropylene, have a maximum nominal i.d. of 40 mm.

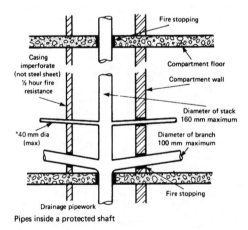

Pipes inside a protected shaft

Ref: Building Regulations, Approved Document B3: Internal fire spread (structure).

Raised flooring provides discrete housing for the huge volumes of data and telecommunications cabling, electrical power cables, pipes, ventilation ducts and other services associated with modern buildings. Proprietary raised floors use standard 600 mm square interchangeable decking panels, suspended from each corner on adjustable pedestals. These are produced in a variety of heights to suit individual applications, but most range between 100 mm and 600 mm. Panels are generally produced from wood particle board and have a galvanised steel casing or overwrap to enhance strength and provide fire resistance. Applied finishes vary to suit application, e.g. carpet, wood veneer, vinyl, etc. Pedestals are screw-threaded steel or polypropylene legs, connected to a panel support plate and a base plate. The void between structural floor and raised panels will require fire stopping at specific intervals to retain the integrity of compartmentation.

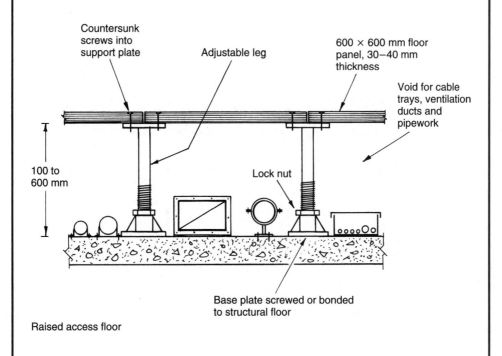

Raised access floor

Ref: BS EN 12825: Raised access floors.

Suspended and False Ceilings

A suspended ceiling contributes to the fire resistance of a structural floor. The extent of contribution can be determined by reference to Appendix A in Approved Document B of the Building Regulations. An additional purpose for a suspended ceiling is to accommodate and conceal building services, which is primarily the function of a false ceiling.

False ceiling systems may be constructed in situ from timber or metal framing. A grid or lattice support system is produced to accommodate loose fit ceiling tiles of plasterboard, particle board or composites. Proprietary systems have also become established. These are a specialised product, usually provided by the manufacturer on a design and installation basis. Most comprise a simple metal framing with interconnecting panel trays. As with raised flooring, the possibility of fire spreading through the void must be prevented. Fire stopping is necessary at appropriate intervals as determined in Approved Document B3 to the Building Regulations.

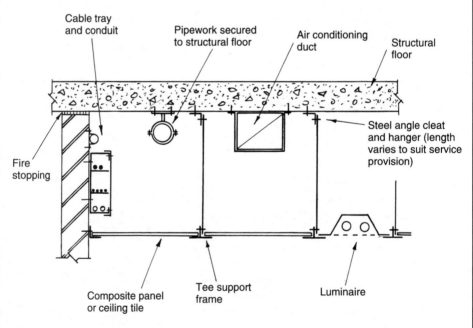

Simply suspended ceiling

Ref: BS 8290: Suspended ceilings.
Building Regulations, Approved Document B: Fire safety.

470

15 ALTERNATIVE AND RENEWABLE ENERGY

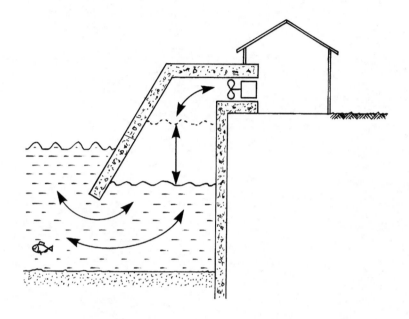

ALTERNATIVE ENERGY

WIND POWER

FUEL CELLS

WATER POWER

GEOTHERMAL POWER

SOLAR POWER

BIOMASS OR BIOFUEL

Power stations that burn conventional fossil fuels such as coal and oil, and to a lesser extent natural gas, are major contributors to global warming, production of greenhouse gases (including CO_2) and acid rain. Note: Acid rain occurs when the gaseous products of combustion from power stations and large industrial plant combine with rainfall to produce airborne acids. These can travel hundreds of miles before having a devastating effect on forests, lakes and other natural environments. Current efforts to limit the amount of combustion gases in the atmosphere include:

- CHP and district heating systems (pages 84–87).
- Condensing boilers (page 42).
- Higher standards of thermal insulation of buildings (page 99 and Building Regulations, Approved Document L – Conservation of fuel and power).
- Energy management systems (pages 96 and 97).
- Recycling of waste products for renewable energy.

Renewable energy is effectively free fuel, but remarkably few of these installations exist in the UK. Other European states, particularly the Netherlands, Germany and Scandinavian countries, have waste segregation plant and selective burners as standard equipment at many power stations. City domestic rubbish and farmers' soiled straw can be successfully blended with conventional fuels to power electricity generators and provide hot water for distribution in district heating mains. Small-scale waste-fired units from 60 kW up to 8000 kW are standard installations in many continental domestic and commercial premises, but are something of a rarity in this country.

Renewable and other alternative 'green' energy sources are also becoming viable. These include:

- Wind power.
- Wind power and hydrogen-powered fuel cells.
- Wave power.
- Geothermal power.
- Solar power.
- Biomass or biofuels.

The UK government have established the following objectives for power generation from 'green' sources:

2002 – 3%, 2010 – 10% and 2020 – 20%.

Atmospheric emissions of CO_2 should decline by 20% by 2010.

The development of wind power as an alternative energy source is well advanced. However, it is dependent on the fickle nature of the weather and can only be regarded as a supplementary energy source unless the surplus power produced is stored – see page 476.

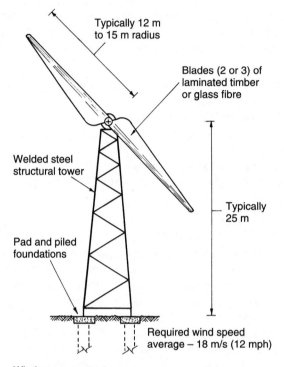

Typically 12 m
to 15 m radius

Blades (2 or 3) of
laminated timber
or glass fibre

Welded steel
structural tower

Typically
25 m

Pad and piled
foundations

Required wind speed
average – 18 m/s (12 mph)

Wind power generator

The principle is simple enough. Wind drives a propeller, which rotates a shaft through a gearbox to drive an electricity generator. The generator produces direct current, similar in concept to a much smaller bicycle dynamo. Designs include two- and three-blade variants, elevated to between 25 and 45 metres from ground level to central axis. Blades are usually made from laminated timber or glass fibre and manufactured to tip diameters of between 6 and 60 metres (25 to 30 m is typical). Electricity output is difficult to define, but claims are made of 300 kW in a 25 mph wind from one generator. This is enough electricity for about 250 houses. A wind farm of say 20 generators in an exposed location could produce 20 GW of electricity an hour averaged over a year.

Environmental issues – no release of carbon, sulphur or nitrogen oxides, methane and other atmospheric pollutants. Conservation of finite fossil fuels. Aesthetically undesirable and noisy.

Costs – produces electricity for a minimal amount. Foundation costs are very high to anchor the units against lateral wind forces and dynamic forces during rotation. The capital cost of generators and their installation costs must be calculated against the long-term savings and environmental benefits. The purchase costs of wind turbines commence at about £1200 per kW of output, with a life expectancy of about 30 years. The smallest of units may take about a week to install.

Savings – estimates vary from speculative projections to realistic comparisons. A small generator such as that used at Wansbeck General Hospital, Northumberland, can produce up to 450 kW daily. On a greater scale, it is anticipated that by the year 2025, up to 20% of the UK's electrical energy requirements could be wind generated.

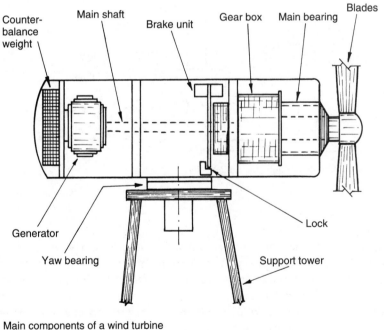

Main components of a wind turbine

Wind Power and Fuel Cells

Wind is limited as a source of electrical power because of the unreliable nature of the weather. To use the potential of the wind effectively, it is necessary to store the energy generated when the wind blows and release it in response to demand.

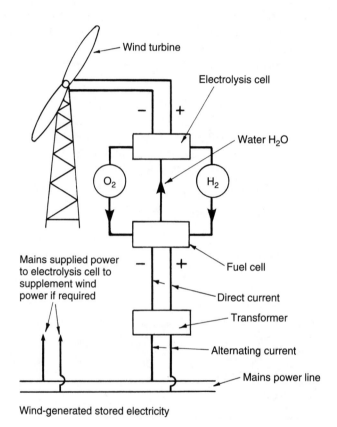

Wind-generated stored electricity

Instead of using the wind-generated electricity directly, it is used to electrolytically decompose water. This means separation of the hydrogen and the oxygen in water into different storage vessels. The stored hydrogen and oxygen are supplied to a fuel cell or battery in regulated amounts to produce a direct current. As the two gases combine they give water, which is returned to the electrolysis cell for reprocessing. Direct current is transformed to alternating current for compatibility with electricity distribution power lines.

The energy potential in differing water levels has been exploited for centuries through water mills and subsequently hydro-electric power. Another application is to build tidal barrages across major estuaries such as the Severn or Mersey. As the tide rises the water would be impounded, to be released back as the tide recedes, using the head or water level differential as a power source. This has been used to good effect since the 1960s at La Rance near St Malo in France.

Another application uses a series of floats moored in the sea to generate an electrical potential as each float moves with the waves. Attempts have also been made to use the floats to rotate a crankshaft. There are limitations with this, not least the obstruction it creates in the sea.

Power potential from waves can also be harnessed by using their movement to compress air in shoreline chambers. Air pressure built up by the wave oscillations is used to propel an air turbine/ electricity generator.

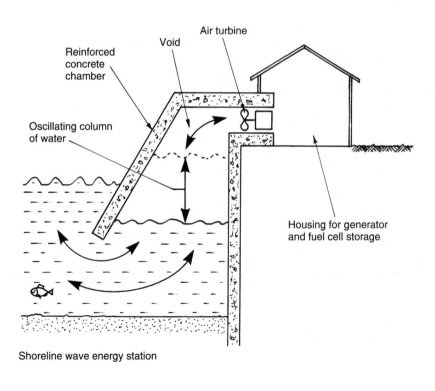

Shoreline wave energy station

Geothermal Power

This is otherwise known as 'hot-dry-rock' technology, a name which gives some indication of the energy source. Heat energy is produced by boring two or more holes into the granite fissures found at depths up to 4·5 miles (7·2 km) below the earth's surface. Cold water pumped down one borehole and into the fissures converts into hot water or steam which is extracted from the other borehole(s). The hot water can then be used directly for heating or it can be reprocessed into steam to drive turbines and electricity generators on the surface.

Enormous quantities of heat are believed to exist in underground rock formations throughout the world. New Zealand and Iceland are well known for having hot volcanic springs and established use of naturally occurring hot water from geysers. In the UK there are a few isolated examples of spas, but the greatest potential lies below the impermeable granite sub-strata in the south-west corner of England. This concentrates in Cornwall and ranges up to Dartmoor and the Scilly Isles. Geological surveys suggest that the heat energy potential here is twice that elsewhere in the UK. Since the 1970s the centre of research has been at Rosemanowes Quarry, near Falmouth. Indications from this and other lesser sites in the locality are that there may be enough geothermal energy in the west country to provide up to 20% of the UK's electricity needs. Exploration by boreholes into aquifers in other parts of the country have met with some success. In Marchwood, Southampton, water at over 70°C has been found at depths of less than 2 km. However, this resource was found to be limited and not cost effective for long-term energy needs.

Exploitation of hot water from naturally occurring springs is not new. All over the world there are examples of spas which are known to have been enjoyed since Roman times. More recently in the early 1900s, a natural source of steam was used to generate electricity in Italy. Now it is very much a political and economic decision as to whether it is cost effective to expend millions of pounds exploiting this possibly limited source of heat energy.

The potential of solar energy as an alternative fuel is underrated in the UK. It is generally perceived as dependent solely on hot sunny weather to be effective. In fact it can be successfully used on cloudy days, as it is the solar radiation which is effective. The average amount of solar radiation falling on a south facing inclined roof is shown to vary between about 900 and 1300 kW/m^2 per year depending on the location in the UK.

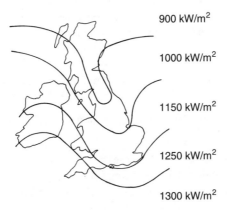

900 kW/m^2

1000 kW/m^2

1150 kW/m^2

1250 kW/m^2

1300 kW/m^2

Solar radiation averaged over a year
for a 30° pitched roof facing south

The reluctance to accept solar panels in this country is understandable. The capital outlay is quite high and even though it is possible to achieve up to 40% of the average household's hot water requirements from solar energy, the payback period may be in excess of 10 years. It could also be argued that the panels are visually unattractive. The typical installation is shown on page 52. It has a flat plate 'black radiator' solar panel to absorb solar energy in water, which is transferred for storage in an insulated cylinder. From here it supplements hot water from a conventional boiler source. This application is also suitable for heating swimming pools.

An improvement uses collectors inside clear glass vacuum cylinders. These 'evacuated tube collectors' are capable of absorbing more heat at low levels of light. Other types of solar panel which can be used to power batteries or fuel cells include the photovoltaic system. This uses expensive crystalline silicon as a power generator. A less expensive alternative is amorphous silicon. Although less efficient, it is still capable of providing a trickle feed to batteries.

Biomass or Biofuel

Biomass is current terminology for the combustion of traditional fuels such as wood, straw and cow dung. The difference is that today we have the facility to process and clean the waste products. Gas scrubbers and electrostatic precipitators can be installed in the flues to minimise atmospheric pollution. Intensive farming methods produce large quantities of potentially harmful residues, including straw and chicken droppings. The latter combines with wood shavings and straw from the coops. Instead of burning these as waste, they can be reprocessed. A pioneer scheme at Eye in Suffolk burns the waste in a 10 MW steam turbine electricity generator and sells the ash as an environmentally friendly fertiliser. This has the additional benefits of:

- Eliminating the traditional unregulated burning of farm waste which contaminates the atmosphere with carbon dioxide.

- Destroying the harmful nitrates which could otherwise be released into the soil.

- Destroying the potential for methane generation from decomposition. When this is released into the atmosphere it is far more active than carbon dioxide as a greenhouse gas.

Farm wastes can also be used to produce methane gas for commercial uses. The waste is processed in a controlled environment in large tanks called digesters. The gas is siphoned off and used for fuel, whilst the remains are bagged for fertiliser.

The potential for forest farming wood as a fuel for power generation is also gaining interest. Trees naturally clean the atmosphere by absorbing carbon dioxide. However, when they die, they rot, releasing as much carbon dioxide as absorbed during growth and a significant amount of methane. By controlled burning the carbon dioxide is emitted, but the gains are destruction of the methane and an economic, sustainable fuel supply.

16 APPENDICES

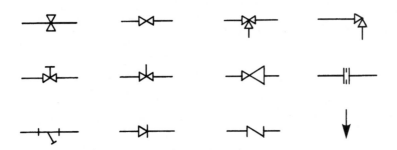

BBA – British Board of Agrément. The function of the BBA is to assess, test and establish the quality of new products and innovations not represented by existing British (BSI) or European (CEN) Standards.

BRE – Building Research Establishment. Critically examines products and materials applicable to construction and issues certificates of conformity. Publishes research digests, good practice guides and information papers.

BS – British Standard. Publications issued by the British Standards Institution as support documents and recommendations for minimum practice and product manufacturing standards. Materials and components which comply are kitemarked:

BS EN – A British Standard which is harmonised with the European Standards body, CEN.

$C\!\!\in$ – Communauté Européenne (European Community). This is a product mark which indicates presumption of conformity with the minimum legal requirements of the Construction Product Regulations 1991. Compliance is manufacture to a British Standard, a harmonised European Standard or a European Technical Approval (ETA).

CEN – Comité Européen de Normalisation. European standardisation body recognised by the European Commission (EC) for harmonising standards of product manufacturers in support of the CPD. Membership of CEN is composed of the standardisation bodies of the participating members of the European Union (EU) and the European Free Trade Association (EFTA). The standardisation body representing the UK is the BSI.

CIRIA – Construction Industry Research and Information Association. An independent research organisation which addresses all key aspects of construction business practice. Its operating principles are on a 'not-for-profit' basis for the benefit of industry and public good.

CPD – Construction Products Directive. Determines that construction products satisfy all or some of (depending on the application) the following essential requirements:

- Mechanical resistance and stability
- Hygiene, health and the environment
- Protection against noise
- Safety in case of fire
- Safety in use
- Energy economy and heat retention

EC – European Commission. The executive organisation of the European Union (EU).

EEA – European Economic Area. Austria, Belgium, Denmark, Finland, France, Germany, Greece, Iceland, Ireland, Italy, Luxemburg, Liechtenstein, Netherlands, Norway, Portugal, Spain, Sweden and the United Kingdom.

EOTA – European Organisation for Technical Approvals. Operates over the same area as CEN, complementing the work of this body by producing guidelines for new and innovative products.

ETA – European Technical Approval. A technical assessment of products which indicate suitability and fitness for use for the CPD. Authorised bodies working with ETA include the BBA and WIMLAS Ltd (now part of BRE Certification). These bodies also produce technical specifications against which product compliance can be measured for approval.

EU – European Union. A unification of states, composed of 15 countries: Austria, Belgium, Denmark, Finland, France, Germany, Greece, Ireland, Italy, Luxemburg, Netherlands, Portugal, Spain, Sweden and the United Kingdom.

ISO – International Organisation for Standardisation. This authority issues standards which are appropriate throughout the world. Products are identified with a number following the prefix ISO. Some of these may be adopted by the CPD, e.g. BS ISO 6341: Water quality and BS EN ISO 10960: Rubber and plastic hoses.

UKAS – United Kingdom Accreditation Service. An independent certification body that may be used by manufacturers to test and assess the suitablity of their material products. UKAS issue certificates to show that materials conform to the criteria required of a recognised document, appropriate for the intended product use and application.

WRC – Water Research Council. A specialist testing agency with its own established brand of approval.

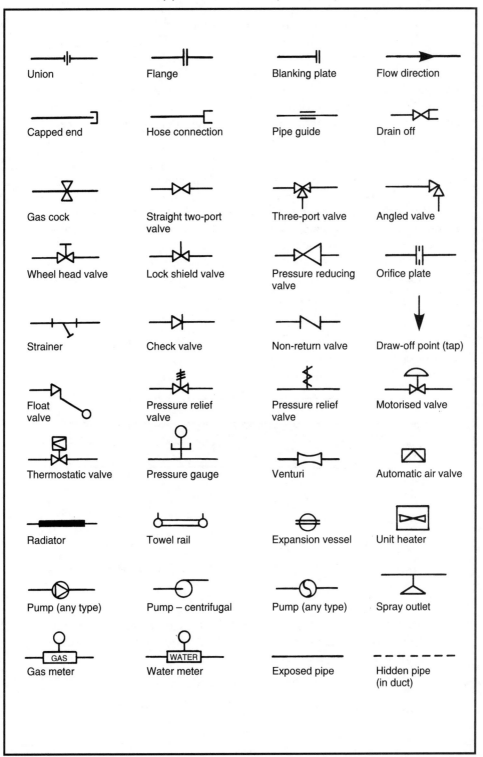

Union	Flange	Blanking plate	Flow direction
Capped end	Hose connection	Pipe guide	Drain off
Gas cock	Straight two-port valve	Three-port valve	Angled valve
Wheel head valve	Lock shield valve	Pressure reducing valve	Orifice plate
Strainer	Check valve	Non-return valve	Draw-off point (tap)
Float valve	Pressure relief valve	Pressure relief valve	Motorised valve
Thermostatic valve	Pressure gauge	Venturi	Automatic air valve
Radiator	Towel rail	Expansion vessel	Unit heater
Pump (any type)	Pump – centrifugal	Pump (any type)	Spray outlet
Gas meter	Water meter	Exposed pipe	Hidden pipe (in duct)

Where a large quantity of piped services are deployed in boiler rooms, process plant service areas, etc., identification of specific services, e.g. compressed air, chilled water, etc., can be very difficult and time consuming. The situation is not helped when installation drawings are lost or may not even have existed. Also, modifications could have occurred since original installation. This is made more difficult where a common pipe material such as galvanised steel is used for a variety of services.

The recommendations of BS 1710 have improved the situation considerably by providing a uniformly acceptable colour coding. This has also been endorsed by the Health & Safety (Safety Signs & Signals) Regulations which require visible markings on all pipework containing or transporting dangerous substances. Direction of flow arrows should also complement coloured markings. Colours can be applied by paint to BS 4800 schedules or with proprietory self-adhesive tape.

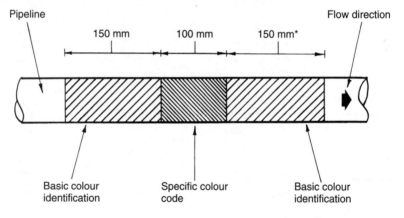

Refs. BS 1710: Specification for identification of pipelines and services.
        BS 4800: Schedule of paint colours for building purposes.
        Health & Safety (Safety Signs & Signals) Regulations 1996.

Contents	Basic i.d. colour	Specific colour	Basic i.d. colour
Water:			
Drinking	Green	Auxiliary blue	Green
Cooling (primary)	Green	White	Green
Boiler feed	Green	Crimson.White.Crimson	Green
Condensate	Green	Crimson.Emerald green. Crimson	Green
Chilled	Green	White.Emerald green. White	Green
Heating <100°C	Green	Blue.Crimson.Blue	Green
Heating >100°C	Green	Crimson.Blue.Crimson	Green
Cold down service	Green	White.Blue.White	Green
Hot water supply	Green	White.Crimson.White	Green
Hydraulic power	Green	Salmon pink	Green
Untreated water	Green	Green	Green
Fire extinguishing	Green	Red	Green
Oils:			
Diesel fuel	Brown	White	Brown
Furnace fuel	Brown	Brown	Brown
Lubricating	Brown	Emerald green	Brown
Hydraulic power	Brown	Salmon pink	Brown
Transformer	Brown	Crimson	Brown
Refrigeration:			
Refrigerant 12	Yellow ochre	Blue	Yellow ochre
Refrigerant 22	Yellow ochre	Green	Yellow ochre
Refrigerant 502	Yellow ochre	Brown	Yellow ochre
Ammonia	Yellow ochre	Violet	Yellow ochre
Others	Yellow ochre	Emerald green	Yellow ochre
Other pipelines:			
Natural gas	Yellow ochre	Yellow	Yellow ochre
Compressed air	Light blue	Light blue	Light blue
Vacuum	Light blue	White	Light blue
Steam	Silver grey	Silver grey	Silver grey
Drainage	Black	Black	Black
Conduit/ducts	Orange	Orange	Orange
Acids/alkalis	Violet	Violet	Violet

Appendix 4 – Graphical Symbols for Electrical Installation Work

Switches (rows 1 and 2)

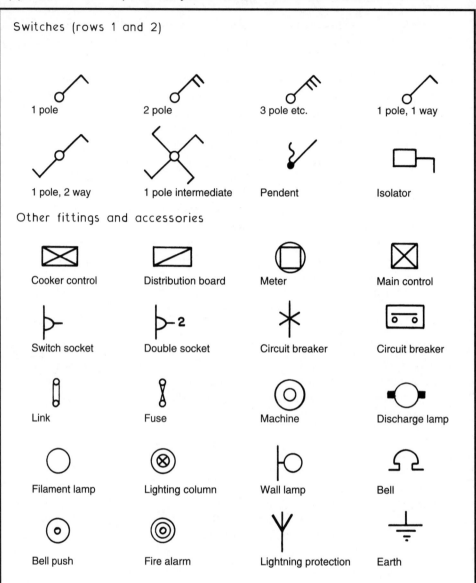

1 pole 2 pole 3 pole etc. 1 pole, 1 way

1 pole, 2 way 1 pole intermediate Pendent Isolator

Other fittings and accessories

Cooker control Distribution board Meter Main control

Switch socket Double socket Circuit breaker Circuit breaker

Link Fuse Machine Discharge lamp

Filament lamp Lighting column Wall lamp Bell

Bell push Fire alarm Lightning protection Earth

Note: In addition to established office practice, the following standards provide recommendations for drawing representations: BS 1192-3: 'Recommendations for symbols and other graphic conventions', and its international successors – BS EN ISO 3766, 7518 and 11091.

Metric measurements have been officially established in the UK since the Council of Ministers of the European Community met in 1971 to commit member countries to an International System of Units (SI). This has been endorsed by the International Organisation for Standardisation (ISO).

Basic or primary units:

Quantity	Unit	Symbol
Length	metre	m
Mass	kilogram	kg
Time	second	s
Electric current	ampere	A
Temperature	Kelvin	K
Luminous intensity	candela	cd

Some commonly used supplementary and derived units:

Quantity	Unit	Symbol
Area	square metre	m^2
Volume	cubic metre	m^3
Velocity	metres per second	m/s
Acceleration	metres per second squared	m/s^2
Frequency	hertz (cycles per second)	Hz
Density	kilogram per cubic metre	kg/m^3
Force	newton	N
Moment of force	newton metre	N m
Pressure	newton per square metre	N/m^2 (pascal – Pa)
Work, energy and heat	joule	J
Power, heat flow rate	watt	W (J/s)
Temperature – customary unit	degree Celsius	°C
Temperature – interval	degree Kelvin	K

Note: degree Celsius and Kelvin have the same temperature interval. Kelvin is absolute temperature with a zero factor equivalent to –273·15°C, i.e. 0°C = 273·15 K.

Further derived units:

Quantity	Unit	Symbol
Density of heat flow	watt per square metre	W/m^2
Thermal conductivity	watt per metre degree	W/m K
Heat transfer (U value)	watt per square metre degree	W/m^2 K
Heat capacity	joule per degree	J/K
Specific heat capacity	joule per kilogram degree	J/kg K
Entropy	joule per degree	J/K
Specific entropy	joule per kilogram degree	J/kg K
Specific energy	joule per kilogram	J/kg

Derived units for electrical applications:

Quantity	Unit	Symbol
Electric charge	coulomb	C (As)
Potential difference	volt	V (W/A)
Electromotive force	volt	V (W/A)
Electric field strength	volt per metre	V/m
Electric resistance	ohm	Ω (V/A)
Electric capacitance	farad	F (As/V)
Magnetic flux	weber	Wb (Vs)
Magnetic field strength	ampere per metre	A/m
Inductance	henry	H (Vs/A)
Luminous flux	lumen	lm
Luminance	candela per square metre	cd/m^2
Illuminance	lux (lumens per square metre)	lx (lm/m^2)

Multiples and submultiples:

Factor	Unit	Name	Symbol
One billion	10^{12}	tera	T
One million million	10^{12}	tera	T
One thousand million	10^9	giga	G
One million	10^6	mega	M
One thousand	10^3	kilo	k
One hundred	10^2	hecto	h
Ten	10^1	deca	da
One tenth	10^{-1}	deci	d
One hundreth	10^{-2}	centi	c
One thousandth	10^{-3}	milli	m
One millionth	10^{-6}	micro	μ
One thousand millionth	10^{-9}	nano	n
One million millionth	10^{-12}	pico	p
One billionth	10^{-12}	pico	p
One thousand billionth	10^{-15}	femto	f
One trillionth	10^{-18}	atto	a

Common units for general use:

Quantity	Unit	Symbol
Time	minute	min
	hour	h
	day	d
Capacity	litre	l (1 l = 1 dm^3)
		(1000 l = 1 m^3)
Mass	tonne or kilogram	t (1 t = 1000 kg)
Area	hectare	ha (100 m × 100 m)
		(10 000 m^2)
Pressure	atmospheric	atm
		(1 atm = 101.3 kN/m^2)
Pressure	bar	b
		(1 bar = 100 kN/m^2)

Appendix 6 – Conversion of Common Imperial Units to Metric (1)

Length	1 mile = 1·609 km 1 yd = 0·914 m 1 ft = 0·305 m (305 mm)
Area	1 sq. mile = 2·589 km^2 or 258·9 ha 1 acre = 4046·86 m^2 or 0·404 ha 1 yd^2 (square yard) = 0·836 m^2 1 ft^2 (square foot) = 0·093 m^2 1 in^2 (square inch) = 645·16 mm^2
Volume	1 yd^3 (cubic yard) = 0·765 m^3 1 ft^3 (cubic foot) = 0·028 m^3 1 in^3 (cubic inch) = 16387 mm^3 (16·387 cm^3)
Capacity	1 gal = 4·546 l 1 qt = 1·137 l 1 pt = 0·568 l
Mass	1 ton = 1·016 tonne (1016 kg) 1 cwt = 50·8 kg 1 lb = 0·453 kg 1 oz = 28·35 g
Mass per unit area	1 lb/ft^2 = 4·882 kg/m^2 1 lb/in^2 = 703 kg/m^2
Mass flow rate	1 lb/s = 0·453 kg/s
Volume flow rate	1 ft^3/s = 0·028 m^3/s 1 gal/s = 4·546 l/s
Pressure	1 lb/in^2 = 6895 N/m^2 (68·95 mb) 1 in (water) = 249 N/m^2 (2·49 mb) 1 in (mercury) = 3386 N/m^2 (33·86 mb)

Energy	1 therm = 105·5 MJ 1 kWh = 3·6 MJ 1 Btu (British thermal unit) = 1·055 kJ
Energy flow	1 Btu/h = 0·293 W (J/s) (see note below)
Thermal conductance	1 Btu/ft^{2}h °F = 5.678 W/m^2 K ('U' values)
Thermal conductivity	1 Btu ft/ft^{2}h °F = 1.730 W/m K
Illumination	1 lm/ft^2 = 10·764 lx (lm/m^2) 1 foot candle = 10·764 lx
Luminance	1 cd/ft^2 = 10·764 cd/m^2 1 cd/in^2 = 1550 cd/m^2
Temperature	32°F = 0°C 212°F = 100°C
Temperature conversion	Fahrenheit to Celsius (°F − 32) × 5/9 e.g. 61°F to °C (61 − 32) × 5/9 = 16·1°C
Temperature conversion	Fahrenheit to Kelvin (°F + 459·67) × 5/9 e.g. 61°F to K (61 + 459·67) × 5/9 = 289·26 K, i.e. 289·26 − 273.15 = 16·1°C

Note regarding energy flow:
Useful for converting boiler ratings in Btu/h to kW,
e.g. a boiler rated at 65 000 Btu/h equates to:
65 000 × 0·293 = 19 045 W, i.e. approx. 19 kW.

Index

501